Signal Detection and Estimation

For a complete listing of the *Artech House Radar Library*,
turn to the back of this book . . .

Signal Detection and Estimation

Mourad Barkat
State University of New York,
Stony Brook

Artech House
Boston • London

Library of Congress Cataloging-in-Publication Data

Barkat, Mourad.
  Signal detection and estimation / Mourad Barkat.
    p.   cm.
  Includes bibliographical references and index.
  ISBN 0-89006-454-7
  1. Signal detection. 2. Stochastic processes. 3. Estimation
theory. I. Title.
  TK5102.5.B338   1991   91-13451
  621.382'2--dc20   CIP

British Library Cataloguing in Publication Data

Barkat, Mourad
  Signal detection and estimation.
  1. Statistical mathematics
  I. Title
  519.50246213

  ISBN 0-89006-454-7

© 1991 Artech House, Inc.
685 Canton Street
Norwood, MA 02062

International Standard Book Number: 0-89006-454-7
Library of Congress Catalog Card Number: 91-13451

10   9   8   7   6   5   4   3   2   1

Contents

Preface

This book provides an overview and introduction to signal detection and estimation. It is intended as a textbook for a one-semester graduate course. The material is covered in such a way that one need not use other references to understand the present one. In addition, the book contains *numerous* examples solved in *detail.* Since some of the material on signal detection and estimation could be really complex and requires a lot of background in engineering math, a chapter to cover such background is included, so that one can easily understand the rest of the intended material. Probability theory and stochastic processes are prerequisites to the fundamentals of signal detection and parameter estimation. Consequently, Chapters 1 and 2 carefully cover these topics.

In a one-semester course on signal detection and parameter estimation, the material to cover should be:

> *Chapter 3* *Signal Detection*
> *Chapter 4* *Parameter Estimation*
> *Chapter 6* *Representation of Signals*
> *Chapter 7* *The General Gaussian Problem*
> *Chapter 8* *Detection and Parameter Estimation*

and perhaps part of Chapter 5 on *Filtering.* However, one may choose to cover all of Chapter 5 for a relatively fast course. The book can also be used as a two-semester course. The first semester should cover the first four chapters, which will be an introductory course on "Random

Signals: Probability Theory, Stochastic Processes and Decision and Estimation". The second semester, which can be a continuation (part 2) of the first one, should cover Chapters 5 to 8.

This book may also be used as a textbook for an advanced senior-level. In this case, one may cover Chapters 1 to 4, and possibly part of Chapter 5.

Since this material is essential in many applications of communications and signal processing, this book can be used as a reference by practicing engineers and physicists. The detailed examples and the problems presented at the end of each chapter make this book suitable for self-study and facilitate teaching a class.

Acknowledgements

This book would have not been completed without the will and help of God, and then, the help of many people. I am grateful to my teachers, professor D. Weiner, professor H. Schwarzlander, and professor P. Varshney, who taught me about probability theory, stochastic processes, and signal detection and estimation. Their lectures were an excellent guide for the preparation of this manuscript. I am thankful to my friend Mohd Rodzi Ismail for typing and retyping this manuscript. My thanks also go to my students, especially, Shen-de Lin and Yung-Dar Huang. I acknowledge the support of all my friends and colleagues who contributed in one way or another toward this book. The efforts of Lois Koh in preparing the figures are appreciated.

I express my special thanks to the team at Artech House for their cooperation and encouragement during the course of this work. The reviewer's constructive and encouraging comments are also well acknowledged.

I am extremely grateful to my entire family to whom I dedicate this book: my parents, my brothers and sisters, my brother-in-law General Khaled Nezzar and my in-laws; with a special mention to my mother. Most of all, I thank my wife and my children for their love and patience.

Chapter 1

Probability Concepts

1.1 INTRODUCTION

This book is primarily designed for the study of statistical signal detection and parameter estimation. We start this chapter with the set theory because it provides the most fundamental concepts in the theory of probability. Probability theory is a prerequisite for Chapter 2 in which we cover stochastic processes. Similarly, the fundamentals of stochastic processes will be essential for proper understanding of the subsequent topics, which cover the fundamentals of signal detection and parameter estimation.

In this chapter, we derive some basic results, to which we shall refer throughout the book, and establish the notation to be used.

1.2 SETS AND PROBABILITY

1.2.1 Basic Definitions

A set may be defined as a collection of objects. The individual objects forming the set are the "elements" of the set, or "members" of the set. In general, sets are denoted by capital letters as A, B, C, and elements or particular members of the set by lower case letters as a, b, c.

If an element a "belongs" to or is a "member" of A, we write

$$a \epsilon A \tag{1.1}$$

Otherwise, we say that a is not a member of A or does not belong to A and write

$$a \notin A \tag{1.2}$$

A set can be described in three possible ways. The first is listing all the members of the set. For example, $A = \{1, 2, 3, 4, 5, 6\}$. It can also be described in words. For example, say that A consists of integers between 1 and 6 inclusive. Another method would be to describe the set in the form shown below.

$$A = \{a | a \text{ integer and } 1 \leq a \leq 6\}.$$

The symbol | is read as "such that", and the above expression is read in words as "the set of all elements a such that a is an integer between 1 and 6 inclusive".

A set is said to be *countable* if its elements can be put in a one-to-one correspondence with the integers $1, 2, 3$, etc. Otherwise, it is called *uncountable*.

A finite set has a number of elements equal to zero or some specified positive integer. If the number of elements is greater than any conceivable positive integer, then it is considered *infinite*.

The set of all elements under consideration is called the *universal* set and is denoted by U. The set containing no elements is called the *empty* set or *null* set and is denoted by $\emptyset$.

Given two sets A and B, if every element in B is also an element of A, then B is a *subset* of A. This is denoted as

$$B \subseteq A \tag{1.3}$$

and is read as "B is a subset of A". If at least one element in A is not in B, then B is a *proper subset* of A, denoted by

$$B \subset A \qquad (1.4)$$

On the other hand, if every element in B is in A, and every element in A is in B, so that $B \subseteq A$ and $A \subseteq B$, then

$$A = B \qquad (1.5)$$

If the sets A and B have no common elements, they are called *disjoint* or *mutually exclusive*.

Example 1.1

In this example, we apply the definitions that we have just discussed above. Consider the sets $A, B, C, D,$ and E as shown below.
$A = \{$numbers that show in the upper face of a rolling die$\}$,
$B = \{x | x$ odd integer and $1 < x < 6\}$,
$C = \{x | x$ real and $x \geq 1\}$,
$D = \{2, 4, 6, 8, 10\}$,
$E = \{1, 3, 5\}$,
$F = \{1, 2, 3, 4, ...\}$,
$G = \{0\}$.

Solution.
Note that the sets A, B can be written as $A = \{1, 2, 3, 4, 5, 6\}$ and $B = \{1, 3, 5\}$. A, B, D, E and G are countable and finite. C is uncountable and infinite. F is countable but infinite. Since the elements in A are the numbers that show in the upper face of a rolling die and if the problem under consideration (game of chance) are the numbers on the upper face of the rolling die, then the set A is actually the universal set U.
$A \subset F$, $B \subset F$, $D \subset F$, and $E \subset F$. $B \subset A$ and $E \subset A$. $B \subseteq E$ and $E \subseteq B$, then $E = B$. D and E are mutually exclusive. Note that G is not the empty set but a set with element zero. The empty set is a subset of all sets. If the universal set has n elements, then there are 2^n subsets. In the case of the rolling die we have $2^6 = 64$ subsets.

4

1.2.2 Venn Diagrams and Some Laws

In order to provide a geometric intuition and a visual relationship between sets, sets are represented by Venn diagrams. The universal set, U, is represented by a rectangle, while the other sets are represented by circles or some geometrical figures.

Union: Set of all elements that are members of A or B or both and is denoted as $A \cup B$. This is shown in Figure 1.1.

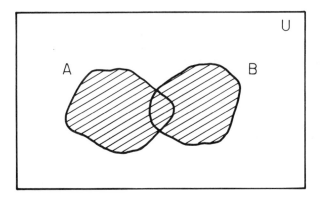

Figure 1.1 Union.

Intersection: Set of all elements which belong to both A and B and is denoted as $A \cap B$. This is shown in Figure 1.2.

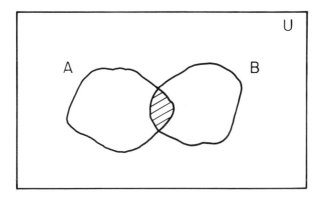

Figure 1.2 Intersection.

Difference: Set consisting of all elements in A which are not in B and is denoted as $A - B$. This is shown in Figure 1.3.

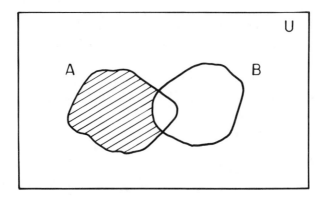

Figure 1.3 A-B.

6

Complement: The set composed of all members in U not in A is the complement of A and is denoted as $\overline{A}$. This is shown in Figure 1.4.

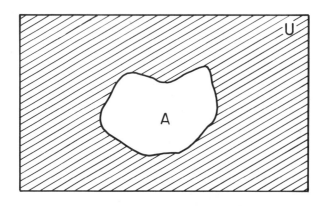

Figure 1.4 Complement of A.

Some Laws and Theorems.

(i)   If A and B are sets, then $A \cup B$ and $A \cap B$ are sets.

(ii)   There is only one set $\emptyset$ and one universal set U, such that $A \cup \emptyset = A$ and $A \cap U = A$ for any A.

(iii)   Commutative laws: $A \cup B = B \cup A$ and $A \cap B = B \cap A$.

(iv)   Associative laws: $(A \cup B) \cup C = A \cup (B \cup C)$ and $(A \cap B) \cap C = A \cap (B \cap C)$.

(v)   Distributive laws: $A \cup (B \cap C) = (A \cup B) \cap (A \cup C)$, and $A \cap (B \cup C) = (A \cap B) \cup (A \cap C)$.

(vi) $A \cup \overline{A} = U$ and $A \cap \overline{A} = \emptyset$.

(vii)   De Morgan's laws: $\overline{(A \cup B)} = \overline{A} \cap \overline{B}$ and $\overline{(A \cap B)} = \overline{A} \cup \overline{B}$.

(viii)   If $A = B$ then $\overline{A} = \overline{B}$. If $A = B$ and $C = D$ then $A \cup C = B \cup D$ and $A \cap C = B \cap D$.

(ix) $\overline{\overline{A}} = A$

1.2.3 Basic Notions of Probability

Originally, the theory of probability was developed to serve as a model of games of chances such as rolling a die, spinning a roulette wheel, or dealing from a deck of cards. Later, this theory developed to model scientific physical experiments.

In building the relationship between the set theory and the notion of probability, we call the set of all possible distinct outcomes of interest in a particular experiment as the *sample space S*. An event is a particular outcome or a combination of outcomes. According to the set theory, the notion of an event is a subset of the sample space.

Example 1.2

Consider the experiment of two six-sided dice and that each die has its sides marked 1 through 6. The sample space, S, in this case is

$$S = \left\{ \begin{array}{cccccc} (1,1) & (1,2) & (1,3) & (1,4) & (1,5) & (1,6) \\ (2,1) & (2,2) & (2,3) & (2,4) & (2,5) & (2,6) \\ (3,1) & (3,2) & (3,3) & (3,4) & (3,5) & (3,6) \\ (4,1) & (4,2) & (4,3) & (4,4) & (4,5) & (4,6) \\ (5,1) & (5,2) & (5,3) & (5,4) & (5,5) & (5,6) \\ (6,1) & (6,2) & (6,3) & (6,4) & (6,5) & (6,6) \end{array} \right\}$$

Let the event A be that "the sum is 7", the event B is that "one die shows an even number and the other an odd number". The events A and B are
$A = \{(1,6),(2,5),(3,4),(4,3),(5,2),(6,1)\}$

$$B = \left\{ \begin{array}{ccc} (2,1) & (4,1) & (6,1) \\ (1,2) & (3,2) & (5,2) \\ (2,3) & (4,3) & (6,3) \\ (1,4) & (3,4) & (5,4) \\ (2,5) & (4,5) & (6,5) \\ (1,6) & (3,6) & (5,6) \end{array} \right\}$$

We now formalize the concept of obtaining an outcome lying in a specified subset A of the sample space S into a definition of probability.

Definition: Given the sample space S and an event A, a probability function, $P(\cdot)$, associates to the event A a real number such that
(i) $P(A) \geq 0$ for every event A
(ii) $P(S) = 1$
(iii) If there exists some countable events $A_1, A_2, \ldots, A_n$, mutually exclusive $(A_i \cap A_j = \emptyset, i \neq j)$, then
$P(A_1 \cup A_2 \cup \cdots \cup A_n) = P(A_1) + P(A_2) + \cdots + P(A_n)$.

Example 1.3

From Example 1.2, we can obtain the probabilities of events A, B, $A \cap B$ and $\overline{A}$ to be $P(A) = 6/36$, $P(B) = 1/2$, $P(A \cap B) = P(A) = 1/6$, and $P(\overline{A}) = 1 - P(A) = 5/6$.

1.2.4 Properties; Conditional Probability; Bayes Rule

Now that we have defined probability, we can state some useful properties.

Properties

(i) For every event A, its probability is between 0 and 1.

$$0 \leq P(A) \leq 1 \tag{1.6}$$

(ii) The probability of the impossible event is zero.

$$P(\emptyset) = 0 \tag{1.7}$$

(iii) If $\overline{A}$ is the complement of A, then,

$$P(\overline{A}) = 1 - P(A) \tag{1.8}$$

(iv) If A and B are two events, then,

$$P(A \cup B) = P(A) + P(B) - P(A \cap B) \qquad (1.9)$$

(v) If the sample space consists of n mutually exclusive events such that $S = A_1 \cup A_2 \cup \cdots \cup A_n$, then,

$$P(S) = P(A_1) + P(A_2) + \cdots + P(A_n) = 1 \qquad (1.10)$$

Conditional Probability and Independent Events

Let A and B be two events such that $P(B) \geq 0$. The probability of event B *given that* event A has occurred is

$$P(A|B) = \frac{P(A \cap B)}{P(B)} \qquad (1.11)$$

$P(A|B)$ is the probability that A will occur given that B has occurred and is called the conditional probability of A given B. However, if the occurrence of event B has no effect on A, we say that A and B are *independent* events. In this case:

$$P(A|B) = P(A) \qquad (1.12)$$

which is equivalent to

$$P(A \cap B) = P(A)P(B) \qquad (1.13)$$

For any three events A_1, A_2, A_3, we have

$$P(A_1 \cap A_2 \cap A_3) = P(A_1)P(A_2|A_1)P(A_3|A_1 \cap A_2) \qquad (1.14)$$

If the three events are independent, then they must be pairwise independent, i.e.,

$$P(A_i \cap A_j) = P(A_i)P(A_j) \quad i \neq j \text{ and } i, j = 1, 2, 3. \qquad (1.15)$$

and also

$$P(A_1 \cap A_2 \cap A_3) = P(A_1)P(A_2)P(A_3) \tag{1.16}$$

Note that both conditions (1.15) and (1.16) must be satisfied for A_1, A_2, A_3, to be independent.

Bayes Rule

If we have n mutually exclusive events $A_1, A_2, \ldots, A_n$, the union of which is the sample space S, $S = A_1 \cup A_2 \cup \cdots \cup A_n$, then, for every event A, Bayes rule says that

$$P(A_k|A) = \frac{P(A_k \cap A)}{P(A)} \tag{1.17}$$

where

$$P(A_k \cap A) = P(A_k)P(A|A_k) \, , \, k = 1, 2, \cdots, n$$

and

$$P(A) = P(A|A_1)P(A_1) + P(A|A_2)P(A_2) + \cdots + P(A|A_n)P(A_n)$$

Example 1.4

A ball is drawn at random from a box containing 7 white balls, 3 red balls and 6 green balls.
(a) Determine the probability that the ball drawn is
    (i) white, (ii) red, (iii) green, (iv) not red, (v) red or white
(b) If 3 balls are drawn successively from the box unstead of one. Find the probability that they are drawn in the order red, white, and green if each ball is
    (i) replaced, (ii) not replaced.

Solution.

Let W, R and G denote the events of drawing a white ball, a red ball, and a green ball. The total number of balls in the sample space is $7 + 3 + 6 = 16$.

(a) (i) $P(W) = 7/16$,

(ii) $P(R) = 3/16$,

(iii) $P(G) = 6/16 = 3/8$,

(iv) $P(\overline{R}) = 1 - P(R) = 1 - 7/16 = 9/16$

(v) $P(\text{red or white}) = P(R \cup W) = P(R) + P(W) - P(R \cap W)$.

Since the events R and W are mutually exclusive, then $P(R \cap W) = 0$ and $P(R \cup W) = P(R) + P(W) = (7 + 3)/16 = 5/8$.

(b) In this case the order becomes a factor. Let the events R_1, W_2, and G_3 represent "red on first draw", "white on second draw", and "green on third draw"; respectively.

(i) Since each ball is replaced before the next draw, the events are independent and from (1.16) we can write

$P(R_1 \cap W_2 \cap G_3) = P(R_1)P(W_2|R_1)P(G_3|R_1 \cap W_2)$

$= P(R_1)P(W_2)P(G_3) = (3/16) \cdot (7/16) \cdot (3/8) = 0.0308$

(ii) When the ball is not replaced before the next draw, the events are dependent. Then,

$P(R_1 \cap W_2 \cap G_3) = P(R_1)P(W_2|R_1)P(G_3|R_1 \cap W_2)$

but $P(W_2|R_1) = 7/(6 + 3 + 6) = 7/15$ and

$P(G_3|R_1 \cap W_2) = 6/(6 + 2 + 6) = 1/2, \Longrightarrow P(R_1 \cap W_2 \cap G_3) = 0.04375$

Example 1.5

A digital communication source transmits symbols of 0s and 1s independently with probability 0.6 and 0.4, respectively, through some noisy channel. At the receiver we obtain symbols of 0s and 1s but with the chance that any particular symbol was garbled at the channel is 0.2. What is the probability of receiving a zero.

Solution.

Let the probability to transmit a zero be $P(0) = 0.6$ and the probability to transmit a one be $P(1) = 0.4$. The probability that a particular symbol is garbled is 0.2; i.e., the probability to receive a zero when a one is transmitted and the probability to receive a one when a zero is transmitted are $P(\text{receive } 0|1 \text{ transmitted}) = P(\text{receive } 1|0 \text{ transmitted}) = 0.2$. Hence, the probability to receive a zero is

$$
\begin{aligned}
P(\text{receive a zero}) &= P(0|1)P(1) + P(0|0)P(0) \\
&= (0.2)(0.4) + (0.8)(0.6) = 0.56
\end{aligned}
$$

Example 1.6

A receiver receives a string of 0s and 1s transmitted from a certain source. The receiver uses a majority decision rule. In other words, if the receiver acquires 3 symbols and out of these 3 symbols 2 or 3 are zeros, it will decide that these symbols represent that a 0 was transmitted. The receiver is correct only 80% of the time. What is $P(C)$ the probability of a correct decision if the probabilities of receiving 0s and 1s are equally likely.

Solution.
This case is known as the Bernoulli trials. From combinatorial analysis, the possibilities that an event A occurs k times out of n trials can be represented as

$$
\binom{n}{k} = \frac{n!}{k!(n-k)!} \tag{1.18}
$$

The quantity $\binom{n}{k}$ is called the binomial coefficients. The probability that the event A occurs exactly k times is

$$
P(A \text{ occurs exactly } k \text{ times}) = \binom{n}{k} p^k (1-p)^{n-k} \tag{1.19}
$$

where p represents $P(A)$ the probability that event A occurs in a given trial. Define D as the event decide 0 or 1. $P(D) = 0.8$. The number of symbols received is $n = 3$. From (1.19), we have

$$
P(0 \text{ correct decisions}) = \binom{3}{0} (0.8)^0 (1-0.8)^3 = 0.008
$$

$$P(1 \text{ correct decision}) = \binom{3}{1} (0.8)^1 (1 - 0.8)^2 = .096$$

$$P(2 \text{ correct decisions}) = \binom{3}{2} (0.8)^2 (1 - 0.8)^1 = .384$$

$$P(3 \text{ correct decisions}) = \binom{3}{3} (0.8)^3 (1 - 0.8)^0 = .512$$

Therefore, the probability of a correct decision is

$$P(C) = P(D = 2) + P(D = 3) = 0.896$$

1.3 RANDOM VARIABLES AND PROBABILITY DISTRIBUTIONS

We define a *random variable* as a *real function* (point function) that maps the elements of the sample space S into points of the real axis. The random variable is represented by a capital letter $(X, Y, Z, \ldots)$ and any particular real value of the random variable is denoted by a lower-case letter $(x, y, z, \ldots)$.

1.3.1 Discrete Random Variables

If a random variable X can assume only a particular finite or countably infinite set of values, $x_1, \ldots, x_n$, then X is said to be a *discrete* random variable. If we associate with each outcome x_i a number $P(x_i) = P(X = x_i)$ called the probability of x_i. The number $P(x_i)$, sometimes denoted P_i for simplicity, $i = 1, 2, \ldots$, must satisfy the following conditions:

$$P(x_i) \geq 0 \text{ for all } i \tag{1.20}$$

14

and

$$\sum_{i=1}^{\infty} P(x_i) = 1 \tag{1.21}$$

That is, the probability of each value that X can assume must be non-negative and the sum of the probabilities over all of the different values must be one. If X is a random variable, its *distribution function* or *cumulative distribution function* (cdf) is defined as

cumulative probability distr. fn.

$$F_X(x) = P(X \le x) \text{ for all } x \tag{1.22}$$

The *probability density function* (pdf) of a discrete random variable that assumes values $x_1, x_2, \ldots$, is $P(x_1), P(x_2), \ldots$, where $P(x_i) = P(X = x_i)$, $i = 1, 2, \ldots$. If there is more than one random variable, we denote the pdf of a particular random variable X by a subscript X on P as $P_X(x)$.

Example 1.7

Consider the experiment of rolling two dice. Let X represent the total number that shows up on the upper faces of the two dice. What is the probability that X is between 4 and 6 inclusive? Determine $P(X \ge 5)$. Sketch the pdf and the cdf of X.

Solution.

how it can be.

lesteu due? then
the sum #?

Since the possible events are mutually exclusive,

$$P(4 \le X \le 6) = P(X = 4) + P(X = 5) + P(X = 6)$$

where

$$P(X = 4) = \frac{1}{12}, \ P(X = 5) = \frac{1}{9}, \ P(X = 6) = \frac{5}{36}$$

$$\Rightarrow P(4 \le X \le 6) = \frac{11}{36}$$

$$P(X \ge 5) = 1 - P(X < 4) = 1 - [P(X = 2) + P(X = 3) + P(X = 4)] = \frac{5}{6}$$

The density function and the distribution function of X are shown in Figures 1.5(a) and 1.5(b), respectively.

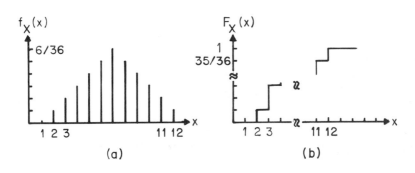

Figure 1.5 (a) Density function of X; (b) Distribution function of X.

1.3.2 Continuous Random Variable $F_X(x)$

X is called a continuous random variable if its distribution function
$F_X(x)$ may be represented as

$$F_X(x) = P(X \leq x) = \int_{-\infty}^{x} f_X(u)du \qquad (1.23)$$

where $f_X(x)$ is a probability density function. By definition, $f_X(x)$
must satisfy

$$f_X(x) \geq 0 \text{ for all } x \qquad (1.24)$$

and

$$\int_{-\infty}^{\infty} f_X(x)dx = 1 \qquad (1.25)$$

$f_X(x)$ is often called the *density function*.

Example 1.8

(a) Find the constant c such that the function

$$f_X(x) = \begin{cases} cx & ,0 < x < 3 \\ 0 & ,\text{otherwise} \end{cases}$$

is a density function.

(b) Compute $P(1 < X < 2)$.

(c) Find the distribution function $F_X(x)$.

Solution.

(a) $f_X(x)$ is a nonnegative function for the given range of x. For $f_X(x)$ to be a pdf, we need to find c such that $\int_{-\infty}^{\infty} cx\,dx = 1$. Solving the integral, we obtain $c = 2/9$ and thus $f_X(x) = (2/9)x$ for $0 < x < 3$.

(b) $P(1 < X < 2) = \int_1^2 (2/9)x\,dx = 1/3$

(c) $F_X(x) = \int_0^x f_X(u)du = (1/9)x^2$ for $0 \le x < 3$ and $F_X(x) = 1$ for $x \ge 3$. Thus,

$$F_X(x) = \begin{cases} 0 & ,x < 0 \\ (1/9)x^2 & ,0 \le x < 3 \\ 1 & ,x \ge 3 \end{cases}$$

The density function can be obtained directly from the distribution function by simply taking the derivative; i.e.,

$$f_X(x) = \frac{d}{dx}F_X(x) \tag{1.26}$$

where $F_X(x) = \int_{-\infty}^x f_X(u)du$. This is a special case of Leibnitz's rule for differentiation of an integral which is

$$\frac{d}{dx}\int_{a(x)}^{b(x)} F(u,x)du = \int_{a(x)}^{b(x)} \frac{\partial F}{\partial x}du + F[b(x),x]\frac{db(x)}{dx} - F[a(x),x]\frac{da(x)}{dx}$$

$$\tag{1.27}$$

1.4 MOMENTS

1.4.1 Expectations

An important concept in the theory of probability and statistics is the *mathematical expectations*, or *expected value*, or *mean value*, or *statistical average* of a random variable X. The expected value of a random variable is denoted by $E[X]$ or $\overline{X}$. If X is a discrete random variable having possible values $x_1, x_2, \ldots, x_n$, the expected value of X is defined to be

$$E[X] = \sum_x x P(X = x) = \sum_x x P(x) \qquad (1.28)$$

where the sum is taken over all the appropriate values that X can assume. Similarly, for a continuous random variable X with density function $f_X(x)$, the expectation of X is defined to be

$$E[X] = \int_{-\infty}^{\infty} x f_X(x) dx \qquad (1.29)$$

Example 1.9

Find the expected value of the total points in tossing a fair die.

Solution.
    In tossing a fair die, each face shows up with probability $1/6$. Let X be the points showing on the top face of the die. Then,

$$E[X] = 1(\frac{1}{6}) + 2(\frac{1}{6}) + 3(\frac{1}{6}) + 4(\frac{1}{6}) + 5(\frac{1}{6}) + 6(\frac{1}{6}) = 3.5$$

Example 1.10

Consider the random variable X with the distribution shown in Figure 1.6. Find $E[X]$.

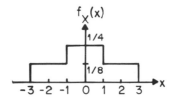

Figure 1.6 Density funcion of X.

Solution.

$$E[X] = \int_{-3}^{-1} x\frac{1}{8}dx + \int_{-1}^{1} x\frac{1}{4}dx + \int_{1}^{3} x\frac{1}{8}dx = 0$$

Let X be a random variable. Then, the function $g(X)$ is also a random variable and its expected value, $E[g(X)]$, is

$$E[g(X)] = \int_{-\infty}^{\infty} g(x)f_X(x)dx \tag{1.30}$$

Properties:

(i) If c is any constant, then,

$$E[cX] = cE[X] \tag{1.31}$$

(ii) If the function $g(X) = X^n$, $n = 0, 1, 2, \ldots$, then,

$$E[g(X)] = E[X^n] = \int_{-\infty}^{\infty} x^n f_X(x)dx \tag{1.32}$$

is called the nth moment of the random variable X about the origin. For $n = 2$, we obtain the second moment of X. Because of its importance, the second moment of X defined as

$$E[X^2] = \int_{-\infty}^{\infty} x^2 f_X(x)dx \qquad (1.33)$$

is called the *mean square value*.

Another quantity of importance is the central moment about the mean. It is called the variance, it is denoted by σ_X^2, and it is defined as

$$\sigma_X^2 = E[(X - E[X])^2] = E[X^2] - (E[X])^2 \qquad (1.34)$$

The quantity σ_X is called the standard deviation.

Example 1.11

Find the variance of the random variable given in Example 1.10.

Solution.
The mean was found previously to be zero. From Equation (1.33) the mean square value is $E[X^2] = 2[\int_0^1 x^2(1/4)dx + \int_1^3 x^2(1/8)dx] = 7/3$. Since the mean is zero, the mean square value is the variance $\sigma_X^2 = 7/3$.

1.4.2 Moment Generating Function and Characteristic Function

The moment generating function $M_X(t)$ of a random variable X is defined by

$$M_X(t) = E[e^{tX}] \qquad (1.35)$$

If X is a discrete random variable with probability distribution $P(x_i) = P(X = x_i)$, $i = 1, 2, \ldots$, then,

$$M_X(t) = \sum_x e^{tx} P_X(x) \qquad (1.36)$$

If X is a continuous random variable with density function $f_X(x)$, then its moment generating function is

$$M_X(t) = \int_{-\infty}^{\infty} e^{tx} f_X(x)dx \qquad (1.37)$$

A nice advantage of the moment generating function is its ability to give the moments. Recall that the McLaurin series of the function e^x is

$$e^x = 1 + x + \frac{x^2}{2!} + \frac{x^3}{3!} + \cdots + \frac{x^n}{n!} + \cdots \qquad (1.38)$$

This is a convergent series. Thus, e^{tx} can be expressed in the series as

$$e^{tx} = 1 + tx + \frac{(tx)^2}{2!} + \cdots + \frac{(tx)^n}{n!} + \cdots \qquad (1.39)$$

By using the fact that the expected value of the sum equals the sum of the expected values, we can write the moment generating function (mgf) as

$$\begin{aligned}
M_X(t) = E[e^{tX}] &= E[1 + tX + \frac{(tX)^2}{2!} + \cdots + \frac{(tX)^n}{n!} + \cdots] \\
&= 1 + tE[X] + \frac{t}{2!}E[x^2] + \cdots + \frac{t^n}{n!}E[X^n] + \cdots (1.40)
\end{aligned}$$

Since t is a constant, taking the derivative of $M_X(t)$ with respect to t, we obtain

$$\frac{dM_X(t)}{dt} = M'_X(t) = E[X] + \frac{t}{2!}E[X^2] + \cdots + \frac{t^{n-1}}{(n-1)!}E[X^n] + \cdots$$
$$(1.41)$$

Setting $t = 0$, all terms become zero except $E[X]$. We have

$$M'_X(0) = E[X] \qquad (1.42)$$

Similarly, taking the second derivative of $M_X(t)$ with respect to t and setting it equal to zero, we obtain

$$M''_X(0) = E[X^2] \qquad (1.43)$$

Continuing in this manner, we obtain all moments to be

$$M_X^{(n)}(0) = E[X^n] \quad n = 1, 2, \ldots \tag{1.44}$$

where $M_X^{(n)}(t)$ denotes the nth derivative of $M_X(t)$ with respect to t.

If we let $t = j\omega$, where j is the complex imaginary unit, in the moment generating function we obtain the *characteristic function*. Hence, the characteristic function $E[e^{j\omega X}]$ and denoted $\Phi_X(\omega)$ is actually the Fourier transform of the density function $f_X(x)$. It follows that

$$\begin{aligned} \Phi_X(\omega) &= E[e^{j\omega X}] \\ &= \int_{-\infty}^{\infty} f_X(x) e^{j\omega x} dx \end{aligned} \tag{1.45}$$

As before, differentiating $\Phi_X(\omega)$ n times with respect to ω and setting $\omega = 0$ in the derivative, we obtain the nth moment of X to be

$$E[X^n] = (-j)^n \frac{d^n \Phi_X(\omega)}{d\omega^n}\Big|_\omega = 0 \tag{1.46}$$

where $\sqrt{j} = -1$. An important role of the characteristic function is to give the density function of a random variable using the theory of Fourier transform. The inverse Fourier transform of the characteristic function is

$$f_X(x) = \frac{1}{2\pi} \int_{-\infty}^{\infty} e^{-j\omega x} \Phi_X(\omega) d\omega \tag{1.47}$$

It is preferable to use the characteristic function over the moment generating function because it always exists, whereas the moment generating function may not exist.

If X is a discrete random variable, its characteristic function is defined as

$$\Phi_X(\omega) = \sum_x e^{j\omega x} P_X(x) \tag{1.48}$$

Example 1.12

Find the characteristic function of the random variable X having density function $f_X(x) = e^{-\frac{1}{2}|x|}$ for all x.

Solution.
From Equation (1.45) the characteristic function is

$$
\begin{aligned}
\Phi_X(\omega) &= \int_{-\infty}^{0} e^{j\omega x} e^{\frac{1}{2}x} dx + \int_{0}^{\infty} e^{j\omega x} e^{-\frac{1}{2}x} dx \\
&= \frac{1}{4(.5 + j\omega)} + \frac{1}{4(.5 - j\omega)} = \frac{1}{1 + 4\omega^2}
\end{aligned}
$$

1.5 SPECIAL DISTRIBUTION FUNCTIONS

In the previous sections, we have defined the concepts of probability, random variables, and statistical moments. In this section, we shall study some important distribution functions which are frequently encountered. Since these distributions have a wide range of applications we shall study them in their general form. Some of the notions defined will be applied to these special distributions which yield some standard results to be used later.

Bernoulli Distribution

The simplest distribution is one with only two possible events. For example, a coin is tossed and the events are heads or tails, which must occur with some probability. Tossing the coin n times consists of a series of independent trials, each of which yields one of the possible outcomes: heads or tails. These two possible outcomes are also referred to as "success" associated with the value 1 and "failure" associated with the value 0. The outcome 1 occurs with probability p, whereas the outcome 0 occurs with probability $1 - p$; $0 \leq p \leq 1$. These are called the Bernoulli trials.

A random variable X is said to have a Bernoulli distribution if for some p, $0 \leq p \leq 1$, its probability density function is

$$P_X(x) = \begin{cases} p^x(1-p)^{1-x} & , x = 0, 1 \\ 0 & , \text{otherwise} \end{cases} \tag{1.49}$$

$(1-p)$ is often denoted by q. The mean and variance of X are $E[X] = p$ and $\text{var}(X) = pq$.

Binomial Distribution

The probability of observing exactly k successes in n independent Bernoulli trials yields the binomial distribution. The probability of success is p and the probability of failure is $q = 1 - p$. The random variable X is said to have a binomial distribution with parameters n and p if

$$P(X = k) = \binom{n}{k} p^k q^{n-k} \text{ for } k = 0, 1, 2, \ldots \tag{1.50}$$

The mean, the variance, and the characteristic function of X are

$$E[X] = np \tag{1.51}$$

$$\text{var}(X) = npq \tag{1.52}$$

$$\Phi_X(\omega) = (p\, e^{j\omega} + q)^n \tag{1.53}$$

Poisson Distribution

In many applications we are concerned about the number of occurrences of an event in a given period of time t. Let the occurrence (or nonoccurrence) of the event in any interval be independent of its occurrence (nonoccurence) in another interval. Furthermore, let the probability of occurrence of the event in a given period be the same irrespective of the starting or ending of the period. Then we say that the distribution of X, the number of occurrences of the event in the time period t, is given by a Poisson distribution. Applications of such a random phenomenon may be the occurrence of telephone traffic, random failures of equipment, desintegration of radio active material, claims in an insurance company, or arrival of customers in a service facility.

The random variable X has a *Poisson* distribution with parameter λ, $\lambda > 0$, if

$$P(X = k) = P_X(k) = \begin{cases} e^{-\lambda}\frac{\lambda^k}{k!} & , k = 0, 1, 2, \ldots, \text{ and } \lambda > 0 \\ 0 & , \text{otherwise} \end{cases} \tag{1.54}$$

The mean and variance of X are equal and can be computed to be

$$E[X] = \sigma_X^2 = \lambda \tag{1.55}$$

while the mean square value is $E[X^2] = \lambda^2 + \lambda$. It can also be shown that the characteristic function of the random variable X is

$$\Phi_X(\omega) = e^{\lambda(e^{j\omega} - 1)} \tag{1.56}$$

Uniform Distribution

A random variable X is said to be uniformly distributed on the interval from a to b, $a < b$, as shown in Figure 1.7, if its density function is

$$f_X(x) = \begin{cases} \frac{1}{b-a} & , a \leq x \leq b \\ 0 & , \text{otherwise} \end{cases} \tag{1.57}$$

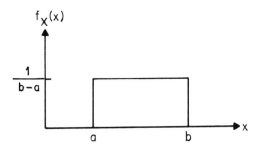

Figure 1.7 Uniform density function.

The mean, the variance, and the characteristic function are, respectively,

$$E[X] = \frac{1}{2}(a + b) \tag{1.58}$$

$$\sigma_X^2 = \frac{1}{12}(b - a)^2 \tag{1.59}$$

and

$$\Phi_X(\omega) = \frac{e^{j\omega b} - e^{j\omega a}}{j\omega(b - a)} \tag{1.60}$$

Normal Distribution

One of the most important continuous random variables of a probability distribution is the normal distribution; often called the Gaussian distribution shown in Figure 1.8. The density function for this distribution is

$$f_X(x) = \frac{1}{\sqrt{2\pi}\sigma} \, e^{-\frac{1}{2\sigma^2}(x-m)^2} \text{ for all } x \tag{1.61}$$

where m and σ are the mean and standard of deviation respectively, and satisfy the conditions $-\infty < m < \infty$ and $\sigma > 0$.

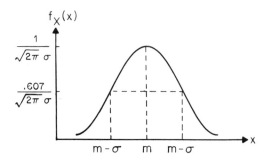

Figure 1.8 Normal density function.

We define the standard normal distribution as the normal distribution with mean $m = 0$ and variance $\sigma = 1$, denoted $N(0,1)$ and expressed as

$$f_Z(z) = \frac{1}{\sqrt{2\pi}} e^{-\frac{z^2}{2}} \tag{1.62}$$

The corresponding distribution function is determined in terms of the *error function*. The error function denoted by erf $(\cdot)$ is defined in many different ways in the literature. We define the error function as

$$\text{erf}(x) = \frac{2}{\sqrt{\pi}} \int_0^x e^{-u^2} du \tag{1.63}$$

Additional information on the error function and its tabulated values is given in Appendix A.

Other important results that we need to define are the complementary error function and the Q function, given by

$$\text{erfc}(x) = \frac{2}{\sqrt{\pi}} \int_x^\infty e^{-u^2} du \tag{1.64}$$

and

$$Q(x) = \int_x^\infty \frac{1}{\sqrt{2\pi}} e^{-\frac{u^2}{2}} du \tag{1.65}$$

where

$$Q(0) = \frac{1}{2} \tag{1.66}$$

and

$$Q(-x) = 1 - Q(x) \text{ , for } x \geq 0 \tag{1.67}$$

Also,

$$Q(x) \simeq \frac{1}{x\sqrt{2\pi}} e^{-\frac{x^2}{2}} \text{ , for } x > 4 \tag{1.68}$$

The characteristic function is known to be

$$\begin{aligned}
\Phi_X(\omega) &= E[e^{j\omega X}] = \int_{-\infty}^\infty f_X(x) e^{j\omega x} dx \\
&= e^{(jm\omega - \frac{\sigma^2\omega^2}{2})} \tag{1.69}
\end{aligned}$$

As in (1.44), the moments can be obtained from the characteristic function to be

$$\begin{aligned}
E[X^n] &= \frac{1}{j^n} \frac{d^n}{dw^n} \Phi_X(\omega)|_{\omega=0} \\
&= n! \sum_{k=0}^K \frac{m^{n-2k}\sigma^{2k}}{2^k k!(n-2k)!} \tag{1.70}
\end{aligned}$$

where

$$K = \left\{ \begin{array}{ll} \frac{n}{2} & \text{for } n \text{ even} \\ \frac{n-1}{2} & \text{for } n \text{ odd} \end{array} \right.$$

If the random variable is zero mean, the characteristic function is

$$
\begin{aligned}
\Phi_X(\omega) &= e^{-\frac{1}{2}\sigma^2\omega^2} \\
&= 1 - \frac{\sigma^2\omega^2}{2} + \frac{1}{2!}\frac{\sigma^4\omega^4}{4} - \frac{1}{3!}\frac{\sigma^6\omega^6}{8} + \cdots
\end{aligned}
\tag{1.71}
$$

Therefore, the moments are

$$
E[X^n] = \begin{cases} 0 & \text{for } n \text{ odd} \\ \dfrac{n!\sigma^n}{(\frac{n}{2})!2^{\frac{n}{2}}} & \text{for } n \text{ even} \end{cases}
\tag{1.72}
$$

Example 1.13

Suppose that X has the distribution $N(m,\sigma^2)$. We want to find the value λ such that $P(X > \lambda) = \alpha$, α, and λ constants.

Solution.

$$
P(X > \lambda) = \int_\lambda^\infty \frac{1}{\sqrt{2\pi}} e^{-\frac{1}{2\sigma^2}(x-m)^2} dx
$$

We need to make a change of variables to obtain the standard normal. Let $z = (x - m)/(\sqrt{2}\sigma)$, then $dx = \sqrt{2}\sigma dz$ and the integral becomes

$$
P(X > \lambda) = \frac{1}{2}\frac{2}{\sqrt{\pi}} \int_{\frac{\lambda-m}{\sqrt{2}\sigma}}^\infty e^{-z^2} dz
$$

thus, $\alpha = (1/2)\mathrm{erfc}\,[(\lambda - m)/(\sqrt{2}\sigma)]$.

Exponential Distribution

Although the normal distribution is the most important distribution, there are many applications in which the normal distribution would not be appropriate. A particular distribution of interest is the exponential distribution. A random variable X has an exponential distribution with parameter λ, $\lambda > 0$, and $\alpha = 1/\lambda$, as shown in Figure 1.9, if its density function is

$$f_X(x) = \begin{cases} \alpha\, e^{-\alpha x} & , x \geq 0 \\ 0 & , \text{otherwise} \end{cases} \tag{1.73}$$

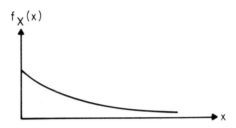

Figure 1.9 Exponential density function.

The mean and variance of X are $E[X] = \lambda$ and $\text{var}(X) = \lambda^2$. The characteristic function is

$$\Phi_X(\omega) = \left(1 + \frac{j\omega}{\alpha}\right)^{-1} = (1 + j\lambda\omega)^{-1} \tag{1.74}$$

Gamma Distribution

In this section, we first describe the gamma function before we introduce the gamma distribution. The gamma function, denoted by Γ, is defined as

$$\Gamma(x) = \int_0^\infty x^{\alpha-1} e^{-x} dx \quad \alpha > 0 \tag{1.75}$$

The above improper integral converges for $\alpha > 0$. Integrating by parts, using $u = x^{\alpha-1}$ and $dv = e^{-x} dx$, we obtain

$$\begin{aligned} \Gamma(\alpha) &= (\alpha-1) \int_0^\infty e^{-x} x^{\alpha-2} dx \\ &= (\alpha-1)\Gamma(\alpha-1) \end{aligned} \tag{1.76}$$

Continuing in this manner and letting α be some positive integer, $\alpha = n$, we obtain

$$\begin{aligned} \Gamma(n) &= (n-1)(n-2)\cdots\Gamma(1) \\ &= (n-1)! \end{aligned} \qquad (1.77)$$

where $\Gamma(1) = \int_0^\infty e^{-x}dx = 1$. Another important result about the gamma function is

$$\Gamma(\tfrac{1}{2}) = \int_0^\infty x^{-\frac{1}{2}}e^{-x}dx = \sqrt{\pi} \qquad (1.78)$$

Now, we are ready to define the gamma distribution. A random variable X is said to have a gamma distribution, or to be gamma distributed, as shown in Figure 1.10, if its density function is given by

$$f_X(x) = \begin{cases} \frac{\alpha}{\Gamma(r)}(\alpha x)^{r-1}e^{-\alpha x} & , x > 0 \\ 0 & , \text{otherwise} \end{cases} \qquad (1.79)$$

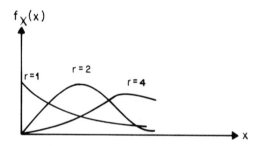

Figure 1.10 Gamma density function.

The mean and variance are respectively $E[X] = m = r/\alpha$ and $\text{var}(x) = \sigma^2 = r/\alpha^2$, while the characteristic function is

$$\Phi_X(\omega) = (1 - \frac{j\omega}{\alpha})^{-r} \qquad (1.80)$$

Chi-Square Distribution

The chi-square distribution is an important distribution function. It may be considered as a special case of the gamma distribution with $\alpha = 1/2$ and $r = n/2$, where n is a positive integer. We say that a random variable Z has a chi-square distribution with n degrees of freedom, denoted χ_n^2, if its density function is given by

$$f_Z(z) = \begin{cases} \frac{1}{2^{\frac{n}{2}}\Gamma(\frac{n}{2})} z^{(\frac{n}{2})-1} e^{-\frac{z}{2}} & , z > 0 \\ 0 & , \text{otherwise} \end{cases} \tag{1.81}$$

It should be noted that the chi-square distribution represents the distribution of the random variable χ_n^2, where $\chi_n^2 = X_1^2 + X_2^2 + \cdots + X_n^2$ and X_i, $i = 1, 2, \ldots, n$ is the standard random variable $N(0,1)$. The mean and variance of the chi-square distribution are $E[\chi_n^2] = n$ and $\text{var}(\chi_n^2) = 2n$, while the characteristic function is $\Phi_{\chi_n^2} = (1 - 2j\omega)^{-\frac{n}{2}}$.

There are many other distributions, but, in practice, they may not occur as often as the ones mentioned above. Some other distributions are mentioned in the appendix.

1.6 TWO AND HIGHER DIMENSIONAL RANDOM VARIABLES

In the previous sections we have developed the concept of random variables and some other related topics such as statistical averages, moment generating functions and characteristic functions. We have also studied in detail some distribution functions that are most frequently encountered in many applications.

Often, we are not interested in one random variable, but in the relationship between two or more random variables. We now generalize the above concepts to N random variables. We will mainly consider continuous random variables since the appropriate modifications for the discrete or mixed cases are easily made by analogy. If X and Y are two continuous random variables, we define the *joint probability density function* or simply the *joint density function* of X and Y by

$$f_{X,Y}(x,y) \geq 0 \qquad (1.82)$$

and

$$\int_{-\infty}^{\infty} \int_{-\infty}^{\infty} f_{X,Y}(x,y)\,dx\,dy = 1 \qquad (1.83)$$

Geometrically, $f_{X,Y}(x,y)$ represents a surface as shown in Figure 1.11. The total volume bounded by this surface and the xy-plane is unity as given in Equation (1.83). The probability that X lies between x_1 and x_2 and Y lies between y_1 and y_2, as shown in the shaded area of Figure 1.11 is given by

$$P(x_1 < X < x_2, y_1 < Y < y_2) = \int_{x_1}^{x_2} \int_{y_1}^{y_2} f_{X,Y}(x,y)\,dx\,dy \qquad (1.84)$$

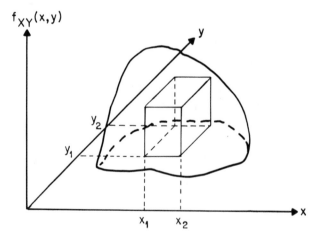

Figure 1.11 Two-dimensional density function.

The joint distribution of X and Y is the probability of the joint event $\{X \leq x, Y \leq y\}$ given by

$$
\begin{aligned}
F_{X,Y}(x,y) &= P(X \leq x, Y \leq y) \\
&= \int_{-\infty}^{x} \int_{-\infty}^{y} f_{X,Y}(u,v) du dv
\end{aligned} \tag{1.85}
$$

Consequently, the joint density function can be obtained from the distribution function by taking the derivative of $F_{X,Y}(x,y)$ with respect to x and y to be

$$
f_{X,Y}(x,y) = \frac{\partial^2}{\partial x \partial y} F_{X,Y}(x,y) \tag{1.86}
$$

The marginal distribution function of X, $F_X(x) = P(X \leq x)$, is obtained from (1.85) by integrating y over all possible values. Hence,

$$
F_X(x) = \int_{-\infty}^{x} \int_{\infty}^{\infty} f_{X,Y}(u,v) dv du \tag{1.87}
$$

Similarly, the marginal distribution of Y is

$$
F_Y(y) = \int_{-\infty}^{y} \int_{-\infty}^{\infty} f_{X,Y}(u,v) du dv \tag{1.88}
$$

1.6.1 Conditional Distributions

The marginal density functions of the random variables X and Y are obtained by taking the derivatives of the respective distribution functions $F_X(x)$ and $F_Y(y)$ given in Equations (1.87) and (1.88). Using the joint density function of X and Y, the marginal density functions $f_X(x)$ and $f_Y(y)$ are

$$
f_X(x) = \int_{-\infty}^{\infty} f_{X,Y}(x,y) dy \tag{1.89}
$$

and

$$
f_Y(y) = \int_{-\infty}^{\infty} f_{X,Y}(x,y) dx \tag{1.90}
$$

If X and Y are independent random variables then the events $\{X \leq x\}$ and $\{Y \leq y\}$ are independent events for all x and y. This yields

$$P(X \leq x, Y \leq y) = P(X \leq x)P(Y \leq y) \qquad (1.91)$$

That is,

$$F_{X,Y}(x,y) = F_X(x)F_Y(y) \qquad (1.92)$$

Equivalently,

$$f_{X,Y}(x,y) = f_X(x)f_Y(y) \qquad (1.93)$$

where $f_X(X)$ and $f_Y(y)$ are the marginal density functions of X and Y. If the joint distribution functions or the joint density functions cannot be written in a product form as given in Equations (1.92) and (1.93), then the random variables X and Y are not independent.

Once the marginal distribution functions are known it becomes simple to determine the conditional distribution functions. In many practical problems we are interested in the distribution of the random variable X given that the random variable Y assumes some specific value or that the random variable Y is between some interval from y_1 to y_2. When the random variable assumes some specific value, we say we have *point conditioning*. The other case is called *interval conditioning*. The conditional density function for $X = x$ given that $Y = y$ is defined as

$$f_{X|Y}(x|y) = \frac{f_{X,Y}(x,y)}{f_Y(y)} \qquad (1.94)$$

where $f_{X,Y}(x,y)$ is the joint density function of X and Y, and $f_Y(y)$ is the marginal density function of Y. However, the conditional density function of X given that $y_1 \leq Y \leq y_2$ is defined as

$$f_{X|Y}(x|y_1 \leq Y \leq y_2) = \frac{\int_{y_1}^{y_2} f_{X,Y}(x,y)dy}{\int_{y_1}^{y_2} f_Y(y)dy} \qquad (1.95)$$

Note that if the random variables X and Y are independent, using (1.93) in (1.94) results in $f_{X|Y}(x|y) = f_X(x)$ and $f_{Y|X}(y|x) = f_Y(y)$ as expected.

Example 1.14

Let X and Y be two random variables with the joint density function

$$f_{X,Y}(x,y) = \begin{cases} x^2 + \frac{xy}{3} & ,0 \le x \le 1 \text{ and } 0 \le y \le 1 \\ 0 & ,\text{otherwise} \end{cases}$$

(a) Check that $f_{X,Y}(x,y)$ is a density function.
(b) Find the marginal density functions $f_X(x)$ and $f_y(y)$.
(c) Compute $P(X > \frac{1}{2})$, $P(Y < X)$ and $P(Y < \frac{1}{2}|X < \frac{1}{2})$.

Solution.
(a) For $f_{X,Y}(x,y)$ to be a density function, it must satisfy Equations (1.82) and (1.83). The first one is easily verified while the second one says that the integral over all possible values of x and y must be one. That is,

$$\int_0^2 \int_0^1 (x^2 + \frac{xy}{3})dx\,dy = \int_0^2 (\frac{1}{3} + \frac{1}{6}y)dy = 1$$

(b) The marginal density functions of X and Y are direct applications of Equations (1.89) and (1.90). Thus,

$$f_X(x) = \int_0^2 (x^2 + \frac{xy}{3})dy = 2x^2 + \frac{2}{3}x \text{ for } 0 < x < 1$$

and

$$f_Y(y) = \int_0^1 (x^2 + \frac{xy}{3})dx = \frac{1}{6}y + \frac{1}{3} \text{ for } 0 < y < 2$$

(c) $\quad P(X > \frac{1}{2}) = \int_{\frac{1}{2}}^1 f_X(x)dx = \int_{\frac{1}{2}}^1 (2x^2 + \frac{2}{3}x)dx = \frac{5}{6}$

$$P(Y < X) = \int_0^1 \int_0^x (x^2 + \frac{xy}{3})dy\,dx = \frac{7}{24}$$

$$P(Y < \frac{1}{2}|X < \frac{1}{2}) = \frac{P(Y < \frac{1}{2}, X < \frac{1}{2})}{P(X < \frac{1}{2})}$$

We have already found $P(X > 1/2)$ to be 5/6. Hence, $P(X < 1/2) = 1 - P(X > 1/2) = 1/6$. We now need only find $P(Y < 1/2, X < 1/2)$ which is

$$P(Y < \frac{1}{2}, X < \frac{1}{2}) = \int_0^{\frac{1}{2}} \int_0^{\frac{1}{2}} (x^2 + \frac{xy}{3}) dx dy = \frac{5}{192}$$

Hence,

$$P(Y < \frac{1}{2} | X < \frac{1}{2}) = \frac{5/192}{1/6} = \frac{5}{32}$$

1.6.2 Expectations and Correlations

We have seen in Section 1.4 that if X is a continuous random variable having density function $f_X(x)$, then the expected value of $g(X)$, a function of the random variable X, is

$$E[g(X)] = \int_{-\infty}^{\infty} g(x) f_X(x) dx \qquad (1.96)$$

This concept is easily generalized to functions of two random variables. In fact, if X and Y are two random variables with joint density function $f_{X,Y}(x, y)$, then,

$$E[g(X, Y)] = \int_{-\infty}^{\infty} \int_{-\infty}^{\infty} g(x, y) f_{X,Y}(x, y) dx dy \qquad (1.97)$$

Also, the expected value of the sum of random variables is just the sum of expected values. Specifically, if X and Y are two random variables then,

$$E[X + Y] = E[X] + E[Y] \qquad (1.98)$$

If, in addition X and Y are independent, the expected value of the product of X and Y can be expressed as the product of expected values. That is,

$$E[XY] = E[X]E[Y] \qquad (1.99)$$

$E[XY]$ is known as the correlation, R_{XY}, between X and Y. The variance of X and Y is the sum of the variances of X and Y; that is,

$$\text{var}(X + Y) = \text{var}(X) + \text{var}(Y) \tag{1.100}$$

When X and Y are not independent, we often try to determine the "degree of relation" between X and Y by some meaningful parameter. This parameter is the *correlation coefficient* defined as

$$\rho_{XY} = \frac{E\{[X - m_X][Y - m_Y]\}}{\sigma_X \sigma_Y} \tag{1.101}$$

where ρ_{XY} is the correlation coefficient between X and Y , m_X is the mean of X, m_Y is the mean of Y, and σ_X, σ_Y are the standard deviations of X and Y, respectively. The degree of correlation, which is the value of the coefficient ρ, is between -1 and $+1$ inclusive:

$$-1 \leq \rho \leq 1 \tag{1.102}$$

Observe that when X and Y are independent ($R_{XY} = E[X]E[Y]$), $\rho = 0$. The numerator of Equation (1.101), known as the covariance of X and Y, and given by

$$C_{XY} = E[(X - m_X)(Y - m_Y)] \tag{1.103}$$

becomes equal to zero. If X and Y are jointly Gaussian and uncorrelated, then they are independent. In general, uncorrelated random variables are not independent.

When the random variables X and Y are not independent, we can define the conditional expectation of one random variable in terms of its conditional density function. The conditional expectation of X given that $Y = y$ is defined as

$$E[X|y] = \int_{-\infty}^{\infty} x f_{X|Y}(x|y) dx \tag{1.104}$$

It can also be easily shown that

$$E[E\{X|Y\}] = E[X] \tag{1.105}$$

and

$$E[E\{Y|X\}] = E[Y] \tag{1.106}$$

where

$$E[X|Y] = \int_{-\infty}^{\infty} E[X|y]f_Y(y)dy \tag{1.107}$$

Note that if X and Y are independent, $E[X|Y] = E[X]$ and $E[Y|X] = E[Y]$. Another important result is that if $g(X,Y)$ is a function of two random variables X and Y, then,

$$E[E\{g(X,Y)|X\}] = E[g(X,Y)] \tag{1.108}$$

If $g(X)$ is a function of X and $h(Y)$ is a function of Y, then $g(X)$ and $h(Y)$ are independent, provided that X and Y are independent. Consequently, the characteristic function of $(X + Y)$ is the product of the individual characteristic functions of X and Y. That is,

$$
\begin{aligned}
\Phi_{X+Y}(\omega) &= E[e^{j\omega(X+Y)}] \\
&= E[e^{j\omega X}]E[e^{j\omega Y}] \\
&= \Phi_X(\omega)\Phi_Y(\omega)
\end{aligned}
\tag{1.109}
$$

Example 1.15

Consider the two dimensional random variable (X,Y) with joint density

$$f_{X,Y}(x,y) = \begin{cases} kxy & , x \le y \text{ and } 0 \le y \le 1 \\ 0 & , \text{otherwise} \end{cases}$$

Find
(a) the constant k
(b) $f_{X|Y}(x|y)$, and
(c) $E[X|Y = y]$

Solution.

(a) To find the constant k we solve the integral in (1.83). From Figure 1.12, we see that the integral we have to solve is

$$\int_0^1 \int_0^y kxy\,dx\,dy = 1 \ , \ \Longrightarrow k = 8.$$

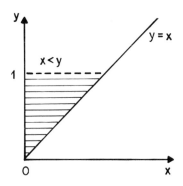

Figure 1.12 Boundaries of $f_{XY}(x,y)$.

(b) In order to use the definition of (1.94), we need to determine $f_Y(y)$:

$$f_Y(y) = \int_0^y 8xy\,dx = 4y^3 \text{ for } 0 \leq y \leq 1$$

Hence,

$$f_{X|Y}(x|y) = \begin{cases} \frac{2x}{y^2} & , \text{for } 0 \leq x \leq y \\ 0 & , \text{otherwise} \end{cases}$$

(c) $E[X|Y=y] = \int_{-\infty}^{\infty} x f_{X|Y}(x|y)\,dx = \int_0^y [(2x^2)/(y^2)]\,dx = (2/3)y.$

1.7 TRANSFORMATION OF RANDOM VARIABLES

In this section, we shall learn techniques for obtaining the density functions for functions of one and two random variables. We shall also mention the generalized approach to n random variables.

1.7.1 Functions of One Random Variable

Consider the problem of determining the density function of a random variable Y, where Y is a function of X, $Y = g(X)$, and the density function of X, $f_X(x)$, is known. We assume that the function $y = g(x)$ is monotonically increasing and differentiable as shown in Figure 1.13. The distribution function of Y in terms of X is

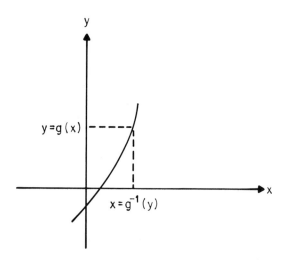

Figure 1.13 Monotone function of x.

$$F_Y(y) = P(Y \leq y) = P[X \leq g^{-1}(y)] \qquad (1.110)$$

where $g^{-1}(\cdot)$ is the inverse transformation. Since we know the density function of X, we can then write

$$F_Y(y) = \int_{-\infty}^{g^{-1}(y)} f_X(x)dx \qquad (1.111)$$

Differentiation of both sides of Equation (1.111) yields

$$f_Y(y) = f_X[g^{-1}(y)] \cdot \frac{d}{dy}[g^{-1}(y)] \qquad (1.112)$$

If the function g were monotonically decreasing, we would have

$$F_Y(y) = \int_{g^{-1}(y)}^{\infty} f_X(x)dx \qquad (1.113)$$

and, consequently,

$$f_Y(y) = -f_X[g^{-1}(y)] \frac{d}{dy}[g^{-1}(y)] \qquad (1.114)$$

In this case, the derivative of $d[g^{-1}(y)]/dy$ is negative. Combining both results of (1.112) and (1.114), the density function of Y is

$$f_Y(y) = |f_X[g^{-1}(y)]| \cdot \{\frac{d}{dy}[g^{-1}(y)]\} \qquad (1.115)$$

This result can be generalized to the case where the function $g(x)$ has many real roots $x_1, x_2, \ldots, x_n, \ldots$. In this case the density function of the random variable Y, $Y = g(X)$, is

$$f_Y(y) = \frac{f_X(x_1)}{|g'(x_1)|} + \cdots + \frac{f_X(x_n)}{|g'(x_n)|} + \cdots \qquad (1.116)$$

where $f_X(x)$ is the density function of X and x_i, $i = 1, 2, \ldots$, is expressed in terms of y and $g'(x)$ is the derivative of $g(x)$ with respect to x. A special case of this fundamental theorem is when $Y = aX + b$. The function $y = g(x) = ax + b$ has one root $x_1 = (y - a)/b$. The derivative of $g(x)$ is just the constant a; $g'(x) = a$. Therefore,

$$f_Y(y) = \frac{f_x(x_1)}{|g'(x_1)|} = \frac{1}{|a|} f_X \left(\frac{y-b}{a} \right) \tag{1.117}$$

Example 1.16

Determine the density function of the random variable Y where $Y = g(X) = aX^2$ given that a is positive and that the density function of X is $f_X(x)$.

Solution.

There are two ways of solving this problem. We either apply directly the fundamental theorem or use the formal derivation starting from the distribution function. We will try both methods and see if the results agree.

Method 1. As shown in Figure 1.14, we have two roots $x_1 = -\sqrt{y/a}$ and $x_2 = +\sqrt{y/a}$.

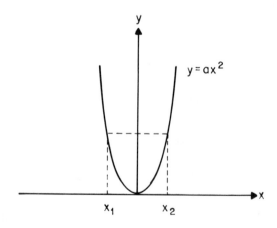

Figure 1.14 $y = g(x) = ax^2$.

$$F_Y(y) \;=\; P(Y \le y) = P\left(-\sqrt{\tfrac{y}{a}} \le X \le -\sqrt{\tfrac{y}{a}}\right)$$
$$=\; P\left(X \le \sqrt{\tfrac{y}{a}}\right) - P\left(X \le -\sqrt{\tfrac{y}{a}}\right)$$
$$=\; F_X\left(\sqrt{\tfrac{y}{a}}\right) - F_X\left(-\sqrt{\tfrac{y}{a}}\right)$$

Differentiation of both sides of the above relation yields

$$f_Y(y) = \frac{1}{2a\sqrt{\tfrac{y}{a}}}[f_X\left(\sqrt{\tfrac{y}{a}}\right) + f_X\left(-\sqrt{\tfrac{y}{a}}\right)] \quad y > 0$$

Method 2. In this case we use the fundamental theorem. We have two roots and consequently the density function of Y is

$$f_Y(y) = \frac{f_X(x_1)}{|g'(x_1)|} + \frac{f_X(x_2)}{|g'(X_2)|}$$

where $g'(x) = 2ax$, $x_1 = -\sqrt{y/a}$ and $x_2 = -\sqrt{y/a}$. Thus, $g'(x_1) = 2a(-\sqrt{y/a}) = -2a\sqrt{ay}$ and $g'(x_2) = 2a(\sqrt{y/a}) = 2\sqrt{ay}$, $\Longrightarrow$

$$f_Y(y) = \frac{f_X(-\sqrt{\tfrac{y}{a}})}{2\sqrt{ay}} + \frac{f_X(\sqrt{\tfrac{y}{a}})}{2\sqrt{ay}}$$

Both results agree.

1.7.2 Functions of Two Random Variables

We shall state some important results for some specific operations without giving any proof. The problem is to determine the density function of Z, where Z is a function of the random variables X and Y. That is,

$$Z = g(X, Y) \tag{1.118}$$

The joint density function of (X, Y), $f_{X,Y}(x, y)$, is also known. Let Z be a random variable equal to the sum of two independent random variables X and Y:

$$Z = X + Y \tag{1.119}$$

The density function of Z can be shown to be the convolution of the density functions of X and Y,

$$f_Z(z) = f_X(x) * f_Y(y) = \int_{-\infty}^{\infty} f_Y(y)f_X(z-y)dy \qquad (1.120)$$

where $*$ denotes convolution and we used the fact that $f_{X,Y}(x,y) = f_X(x)f_Y(y)$, since X and Y are independent.

Example 1.17

Find the density function of $Z = X + Y$ for X and Y independent and uniformly distributed over the intervals $0 \le x \le a$ and $0 \le y \le b$, $a < b$, respectively.

Solution.
The density functions of X and Y are shown in Figure 1.15.

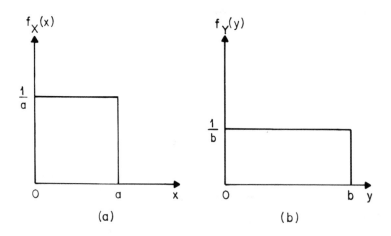

(a) (b)

Figure 1.15 Density functions of: (a) X; (b) Y.

It is much easier to solve convolutions graphically. For $z < a$, there is no overlap between the areas representing the density functions as shown in Figure 1.16(a). This yields $f_z(z) = 0$ for $z < a$.

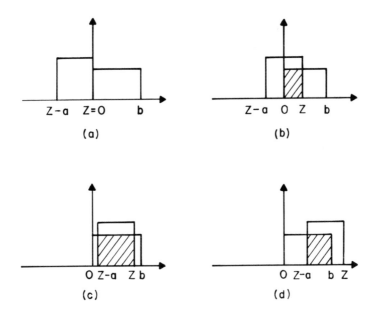

Figure 1.16 Areas of integration for convolution of X and Y.

For $0 \leq z < a$, we have an increasing area as z moves from 0 to a as shown in Figure 1.16(b). Thus,

$$f_Z(z) = \int_0^z \frac{1}{a}\frac{1}{b}dy = \frac{z}{ab}$$

For $a \leq z < b$, we have a constant area as shown in Figure 1.16(c) $\implies$

$$\int_{z-a}^z \frac{1}{a}\frac{1}{b}dy = \frac{1}{b}$$

For $b \leq z < a + b$, from Figure 1.16(d), we have

$$\int_{z-a}^b \frac{1}{a}\frac{1}{b}dy = \frac{a+b-z}{ab}$$

For $z \geq a + b$, there is no overlap between the two curves and, consequently, $f_Z(z) = 0$ for $z \geq a + b \Longrightarrow$

$$f_Z(z) = \begin{cases} z/ab & ,0 \leq z < a \\ 1/b & ,a \leq z < b \\ (a+b-z)/ab & ,b \leq z < a+b \\ 0 & ,z \geq a+b \text{ or } z < 0 \end{cases}$$

The density function of Z, $f_Z(z)$, is shown in Figure 1.17.

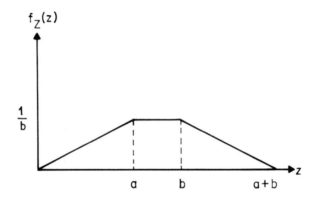

Figure 1.17 Density function of Z.

If X and Y are not independent, then,

$$f_Z(z) = \int_{-\infty}^{\infty} f_{X,Y}(z-y,y)dy = \int_{-\infty}^{\infty} f_{X,Y}(x,z-x)dx \qquad (1.121)$$

Similar results can be obtained for other operations. Let

$$Z = X - Y \qquad (1.122)$$

$$U = XY \qquad (1.123)$$

$$V = \frac{X}{Y} \qquad (1.124)$$

$$M = Max(X,Y) \qquad (1.125)$$

and

$$N = Min(X, Y) \qquad (1.126)$$

Then, the density functions of Z, U, V, M, and N are respectively, given by

$$f_Z(z) = \int_{-\infty}^{\infty} f_{X,Y}(z+y, y)dy = \int_{-\infty}^{\infty} f_{X,Y}(x, x-z)dx \qquad (1.127)$$

$$f_U(u) = \int_{-\infty}^{\infty} f_{X,Y}(x, \frac{u}{x})\frac{1}{|x|}dx = \int_{-\infty}^{\infty} f_{X,Y}(\frac{u}{y}, y)\frac{1}{|y|}dy \qquad (1.128)$$

$$f_V(v) = \int_{-\infty}^{\infty} f_{X,Y}(vy, y)|y|dy \qquad (1.129)$$

$$f_M(m) = \int_{-\infty}^{m} f_{X,Y}(m, y)dy + \int_{-\infty}^{m} f_{X,Y}(x, m)dm \qquad (1.130)$$

and

$$f_N(n) = f_X(n) + f_Y(y) - \int_{-\infty}^{n} f_{X,Y}(n, y)dy - \int_{-\infty}^{n} f_{X,Y}(x, n)dn \qquad (1.131)$$

If X and Y are independent with marginal densities $f_X(x)$ and $f_Y(y)$, then,

$$f_Z(z) = \int_{-\infty}^{\infty} f_X(z+y)f_Y(y)dy = \int_{-\infty}^{\infty} f_X(x)f_Y(x-z)dx \qquad (1.132)$$

$$f_U(u) = \int_{-\infty}^{\infty} f_X(x)f_y(\frac{u}{x})\frac{1}{|x|}dx = \int_{-\infty}^{\infty} f_X(\frac{u}{y})f_Y(y)\frac{1}{|y|}dy \qquad (1.133)$$

$$f_V(v) = \int_{-\infty}^{\infty} f_X(vy)f_Y(y)|y|dy \qquad (1.134)$$

$$f_M(m) = f_X(m)[1 - F_Y(m)] + f_Y(y)[1 - F_X(m)] \qquad (1.135)$$

and

$$f_N(n) = f_X(n)F_Y(n) + f_Y(n)F_X(n) \qquad (1.136)$$

where $F_X(x)$ and $F_Y(y)$ are the marginal distributions of X and Y, respectively.

Example 1.18

Find the density function of $U = XY$ where X and Y are independent random variables with density functions

$$f_X(x) = \begin{cases} 2x & ,0 \leq x \leq 1 \\ 0 & ,\text{otherwise} \end{cases}$$

and

$$f_Y(y) = \begin{cases} \frac{3y^2}{8} & ,0 \leq y \leq 2 \\ 0 & ,\text{otherwise} \end{cases}$$

Solution.

Using Equation (1.133) and the given boundaries of x and y, we obtain

$$f_U(u) = \int_{\frac{u}{2}}^{1} 2x \frac{3}{8} \left(\frac{u}{x}\right)^2 \frac{1}{x} dx = \frac{3}{4} u^2 (2 - u) \quad , 0 \leq u \leq 2$$

1.8 THE GAUSSIAN DENSITY FUNCTION

Because of the importance of the Gaussian distribution and its many applications, we extend the concepts developed earlier to the two-dimensional and n-dimensional Gaussian distribution.

1.8.1 The Bivariate Gaussian Density Function

Let X_1 and X_2 be two jointly Gaussian random variables with means, $E[X_1] = m_1$ and $E[X_2] = m_2$, and variances σ_1^2 and σ_2^2. The bivariate Gaussian density function is defined as

$$f_{X_1,X_2}(x_1, x_2) = \frac{1}{2\pi\sigma_1\sigma_2\sqrt{1-\rho^2}}$$

$$\cdot e^{-\frac{1}{2(1-\rho^2)}[\frac{(x_1-m_1)^2}{\sigma_1^2} + \frac{(x_2-m_2)^2}{\sigma_2^2} - 2\rho\frac{(x_1-m_1)(x_2-m_2)}{\sigma_1\sigma_2}]} \qquad (1.137)$$

where ρ is the correlation coefficient between X_1 and X_2. The two-dimensional characteristic function of X_1 and X_2 is

$$\begin{aligned}
\Phi_{X_1,X_2}(\omega_1,\omega_2) &= E[e^{j(\omega_1 X_1 + \omega_2 X_2)}] \\
&= \int_{-\infty}^{\infty}\int_{-\infty}^{\infty} f_{X_1,X_2}(x_1, x_2)e^{j(\omega_1 x_1 + \omega_2 x_2)}dx_1 dx_2 \\
&= e^{-\frac{1}{2}[\sigma_1^2\omega_1^2 + \sigma_2^2\omega_2^2 + 2\sigma_1\sigma_2\rho\omega_1\omega_2] + j[m_1\omega_1 + m_2\omega_2]} \qquad (1.138)
\end{aligned}$$

The moments are obtained from the characteristic function to be

$$E[X_1^n X_2^m] = (-j)^{n+m}\frac{\partial^n}{\partial\omega_1^n}\frac{\partial^m}{\partial\omega_2^m}\Phi_{X_1,X_2}(\omega_1,\omega_2)|_{\omega_1=\omega_2=0} \qquad (1.139)$$

Sometimes, it is easier to represent the joint density function and the characteristic function in matrix form. Especially, when the number of random variables is greater than 2.

Let $\mathbf{C} = \mathbf{C}[X_1, X_2]$ denote the covariance matrix of the two random variables X_1 and X_2,

$$\mathbf{C} = \begin{bmatrix} c_{11} & c_{12} \\ c_{21} & c_{22} \end{bmatrix} \qquad (1.140)$$

where $c_{11} = \sigma_1^2$, $c_{12} = \rho\sigma_1\sigma_2$, $c_{21} = \rho\sigma_1\sigma_2$, and $c_{22} = \sigma_2^2$. The correlation coefficient is

$$\rho = \frac{c_{12}}{\sqrt{c_{11}c_{22}}} \qquad (1.141)$$

The determinant of the covariance matrix $\mathbf{C}$ is

$$|\mathbf{C}| = \sigma_1^2\sigma_2^2(1-\rho^2) \qquad (1.142)$$

Consequently,

$$C^{-1} = \frac{1}{|C|} \begin{bmatrix} \sigma_2^2 & \rho\sigma_1\sigma_2 \\ -\rho\sigma_1\sigma_2 & \sigma_1^2 \end{bmatrix} \qquad (1.143)$$

Let $\mathbf{X} = \begin{bmatrix} x_1 \\ x_2 \end{bmatrix}$, $\boldsymbol{\omega} = \begin{bmatrix} \omega_1 \\ \omega_2 \end{bmatrix}$, and the mean vector $\mathbf{m} = \begin{bmatrix} m_1 \\ m_2 \end{bmatrix}$, then the bivariate density function is

$$f_{X_1,X_2}(x_1, x_2) = \frac{1}{2\pi\sqrt{|C|}} e^{-\frac{1}{2}[(\mathbf{X}^T - \mathbf{m}^T)\mathbf{C}^{-1}(\mathbf{X} - \mathbf{m})]} \qquad (1.144)$$

where T denotes matrix transpose. The characteristic function becomes

$$\Phi_{X_1,X_2}(\omega_1, \omega_2) = e^{-\frac{1}{2}\boldsymbol{\omega}^T \mathbf{C}\boldsymbol{\omega} + j\mathbf{m}^T\boldsymbol{\omega}} \qquad (1.145)$$

1.8.2 The Multivariate Gaussian Density Function

The results of the previous section can easily be generalized to n random variables. Let $(X_1, X_2, \cdots, X_n)$ be n jointly Gaussian random variables. We define the means as

$$E[X_k] = m_k \; ; \; k = 1, 2, \cdots, n \qquad (1.146)$$

the covariances as

$$c_{jk} = E[(X_j - m_j)(X_k - m_k)] \; ; j, k = 1, 2, \cdots, n \qquad (1.147)$$

and the correlation coefficients as

$$\rho_{jk} = \frac{c_{jk}}{\sqrt{c_{jj}c_{kk}}} \qquad (1.148)$$

We define the vectors:

$$\mathbf{X} = \begin{bmatrix} X_1 \\ X_2 \\ \vdots \\ X_n \end{bmatrix}, \quad \mathbf{x} = \begin{bmatrix} x_1 \\ x_2 \\ \vdots \\ x_n \end{bmatrix}, \quad \mathbf{m} = \begin{bmatrix} m_1 \\ m_2 \\ \vdots \\ m_n \end{bmatrix} \qquad (1.149)$$

and the covariance matrix:

$$\mathbf{C} = \begin{bmatrix} c_{11} & c_{12} & \cdots & c_{1n} \\ c_{21} & c_{22} & \cdots & c_{2n} \\ \vdots & \vdots & \vdots & \vdots \\ c_{n1} & c_{n2} & \cdots & c_{nn} \end{bmatrix} \qquad (1.150)$$

The multivariate Gaussian density function is given by

$$f_{\mathbf{X}}(\mathbf{x}) = \frac{1}{(2\pi)^{\frac{n}{2}}\sqrt{|\mathbf{C}|}} \, e^{-\frac{1}{2}\{(\mathbf{X}^T - \mathbf{m}^T)\mathbf{C}^{-1}(\mathbf{X} - \mathbf{m})\}} \qquad (1.151)$$

The characteristic function is

$$\begin{aligned} \Phi_{\mathbf{X}}(\boldsymbol{\omega}) &= e^{-\frac{1}{2}\boldsymbol{\omega}^T \mathbf{C}\boldsymbol{\omega} + j\mathbf{m}^T \boldsymbol{\omega}} \\ &= e^{-\frac{1}{2}\sum_{j=1}^{n}\sum_{k=1}^{n} c_{jk}\omega_j \omega_k + j\sum_{k=1}^{n} m_k \omega_k} \end{aligned} \qquad (1.152)$$

The Standard Ellipse

The standard ellipse of the bivariate Gaussian density function is obtained from Equation (1.137) by setting the term in the exponent equal to one:

$$\frac{(x_1 - m_1)^2}{\sigma_1^2} + \frac{(x_2 - m_2)^2}{\sigma_2^2} - 2\rho\frac{(x_1 - m_1)(x_2 - m_2)}{\sigma_1 \sigma_2} = 1 \qquad (1.153)$$

Equation (1.153) represents the equation of an ellipse centered at $x_1 = m_1$ and $x_2 = m_2$, as shown in Figure 1.18.

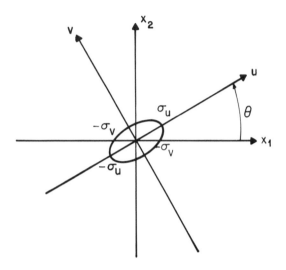

Figure 1.18 Ellipse centered at $m_1 = m_2 = 0$.

For simplicity, let $m_1 = m_2 = 0$. The ellipse is easily represented by assuming two independent random variables U, V with zero mean, and respective variances σ_u^2 and σ_v^2. The standard ellipse is given by

$$\frac{u^2}{\sigma_u^2} + \frac{v^2}{\sigma_v^2} = 1 \tag{1.154}$$

and the joint density function of U and V is

$$f_{U,V}(u, v) = \frac{1}{2\pi\sigma_u\sigma_v}\, e^{-\frac{1}{2}(\frac{u^2}{\sigma_u^2}+\frac{v^2}{\sigma_v^2})} \tag{1.155}$$

Applying a rotation by an angle θ to the uv-axes yields the coordinate system x_1, x_2 given by

$$x_1 = u\,\cos\theta - v\,\sin\theta \tag{1.156}$$

$$x_2 = u\,\sin\theta + v\,\cos\theta \tag{1.157}$$

The random variables X_1, X_2 are obtained by the transformation of (1.156) and (1.157). Specifically,

$$X_1 = U \cos\theta - V \sin\theta \tag{1.158}$$

$$X_2 = U \sin\theta + V \cos\theta \tag{1.159}$$

where

$$\sigma_X^2 = \sigma_u^2 \cos^2\theta + \sigma_v^2 \sin^2\theta \tag{1.160}$$

$$\sigma_Y^2 = \sigma_u^2 \sin^2\theta + \sigma_v^2 \cos^2\theta \tag{1.161}$$

$$E[X_1 X_2] = (\sigma_u^2 - \sigma_v^2) \sin\theta \cos\theta \tag{1.162}$$

$E[X_1 X_2]$ is the covariance between X_1 and X_2, since $E[X_1] = E[X_2] = E[U] = E[V] = 0$.

In a similar manner, the distributions of U and V are derived given the distributions of X_1 and X_2. We obtain

$$\sigma_u^2 = \frac{\sigma_X^2 \cos^2\theta - \sigma_Y^2 \sin^2\theta}{\cos^2\theta - \sin^2\theta} \tag{1.163}$$

$$\sigma_v^2 = \frac{\sigma_Y^2 \cos^2\theta - \sigma_X^2 \sin^2\theta}{\cos^2\theta - \sin^2\theta} \tag{1.164}$$

$$\rho = \frac{E[X_1 X_2]}{\sigma_X \sigma_Y} = \frac{1}{2}\left(\frac{\sigma_X}{\sigma_Y} - \frac{\sigma_Y}{\sigma_X}\right)\tan 2\theta \qquad \theta \neq \pm\frac{\pi}{4}, \pm\frac{3\pi}{4} \tag{1.165}$$

or

$$\theta = \frac{1}{2}\arctan\frac{2\rho\sigma_X \sigma_Y}{\sigma_X^2 - \sigma_Y^2} \tag{1.166}$$

PROBLEMS

1.1 Determine which sets are finite and countable or infinite and uncountable. $A = \{1, 2, 3, 4\}$, $B = \{x | x \text{ integer and } x < 9\}$, $C = \{x | x \text{ real and } 1 \leq x < 3\}$, $D = \{2, 4, 7\}$ and $E = \{4, 7, 8, 9, 10\}$

1.2 Using the sets A, B, D, and E of Problem 1.1, determine the following sets:
(a) $A \cap B$
(b) $A \cup B \cup D \cup E$
(c) $(B \cup E) \cap D$
(d) $B - E$
(e) $A \cap B \cap D \cap E$

1.3 Let the universal set be $U = \{x | x \text{ integer and } 0 \leq x \leq 12\}$. For the subsets of U given as $A = \{0, 1, 4, 6, 7, 9\}$, $B = \{x | x \text{ even}\}$, and $C = \{x | x \text{ odd}\}$, find
(a) $A \cap B$
(b) $(A \cup B) \cap C$
(c) $\overline{B \cup C}$
(d) $B - A$
(e) $(A \cup B) \cap (A \cup C)$
(f) $A \cap \overline{C}$
(g) $B - \overline{C}$
(h) $B \cap \overline{C}$

1.4 Using Venn diagrams for the four sets A, B, C, D within the universal set U, show the areas corresponding to the following sets:
(a) $A - B$
(b) $(A \cup B) \cap C$
(c) $A \cap B \cap C \cap D$
(d) $\overline{A}$
(e) $\overline{A \cap B}$

1.5 Show that if $A \subset B$ and $B \subset C$, then $A \subset C$.

1.6 Find all the mutually exclusive sets defined in Problem 1.3

1.7 A ball is drawn at random from a box containing 10 red balls, 3 white balls, and 7 blue balls. Determine the probability that it is
(a) red
(b) white
(c) blue
(d) not red
(e) red or white

1.8 Assume that three balls are drawn successively from the box of Problem 1.7. Find the probability that they are drawn in the order blue, white, and red if each ball is
(a) replaced before the next draw
(b) not replaced

1.9 If, in addition to the box of Problem 1.7, we have another box containing 2 red balls, 6 white balls, and 1 blue ball. One ball is drawn from each box. Find the probability that
(a) both are red
(b) both are white
(c) one is white and one is blue

1.10 A small box, B_1, contains 4 white balls and 2 black balls; another larger box, B_2, contains 3 white balls and 5 black balls. We first select a box and then draw a ball from the selected box. The probability of selecting the larger box is twice that of the small box. Find the probability that
(a) a black ball is drawn given box B_2
(b) a black ball is drawn given box B_1
(c) a black ball is drawn
(d) a white ball is drawn

1.11 Determine the probability of obtaining three 1s in four tosses of a fair die.

1.12 In a special training, a parachutist is expected to land in a specified zone 90% of the time. Ten of them jumped to land in the zone.
(a) Find the probability that at least 6 of them will land in the specified zone.
(b) Find the probability that none lands in the specified zone.
(c) The training is considered successful if the probability that at least 70% of them land in the prescribed zone is .93. Is the training successful?

1.13 Consider the random variable X in Example 1.10. Find
(a) the distribution function
(b) the probability that $|X| < 1$
(c) the variance σ_X^2

1.14 A random variable X is uniformly distributed over the interval $[-2, 2]$. Determine
(a) $P(X \leq x)$
(b) $P(|X| \leq 1)$
(c) the mean and variance
(d) the characteristic function

1.15 A random variable X is Gaussian with zero mean and variance unity. What is the probability that
(a) $|X| > 1$
(b) $X > 1$

1.16 A random variable X is Poisson distributed with parameter λ and $P(X = 0) = 0.2$. Calculate $P(X > 2)$.

1.17 The incoming calls to a particular station have a Poisson distribution with intensity 12 per hour. What is the probability that
(a) more than 15 calls will come in any given hour
(b) no calls will arrive in a 15 minute break

1.18 A random variable X is Poisson distributed with $P(X = 2) = \frac{2}{3}P(X = 1)$. Calculate $P(X = 0)$ and $P(X = 3)$.

1.19 The random variable X has density function:

$$f_X(x) = \begin{cases} x & ,0 < x \leq 1 \\ 2 - x & ,1 \leq x < 2 \\ 0 & ,\text{otherwise} \end{cases}$$

(a) What is the probability that $\frac{1}{2} < X < \frac{3}{2}$?
(b) Find the mean and variance of X.
(c) Obtain the moment generating function of X.
(d) Obtain the mean of X from the moment generating function and compare it with the value obtained by direct application of the definition.

1.20 The random variable X has mean $E[X] = \frac{2}{3}$ and density function:

$$f_X(x) = \begin{cases} \alpha + \beta x^2 & ,0 \leq x \leq 1 \\ 0 & ,\text{otherwise} \end{cases}$$

(a) Find α and β.
(b) Determine $E[X^2]$ and σ_X^2.

1.21 Solve Problem 1.17 assuming that X has an exponential distribution.

1.22 A random variable X has the distribution $N(0,1)$. Find the probability that $X > 3$.

1.23 A random variable X has the following exponential distribution with parameter α:

$$f_X(x) = \begin{cases} \alpha e^{-\alpha x} & ,x > 0 \\ 0 & ,\text{otherwise} \end{cases}$$

Show that X has the "lack of memory property". That is, show that

$$P(X \geq x_1 + x_2 | X > x_1) = P(X \geq x_2)$$

for x_1, x_2 positive.

1.24 The joint density function of two random variables X and Y is

$$f_{X,Y}(x,y) = \begin{cases} k(x+y) & ,0 \le x \le 2 \text{ and } 0 \le y \le 2 \\ 0 & ,\text{otherwise} \end{cases}$$

Find
(a) k
(b) The marginal density functions of X and Y
(c) $P(X < 1|Y < 1)$
(d) $E[X]$, $E[Y]$, $E[XY]$ and $\rho_{X,Y}$
(e) Are X and Y independent?

1.25 The joint density function of the two random variables X and Y is

$$f_{X,Y}(x,y) = \begin{cases} kxy & ,1 < x < 5 \text{ and } 0 < y < 4 \\ 0 & ,\text{otherwise} \end{cases}$$

Find
(a) The constant k
(b) $P(X \ge 3, Y \le 2)$ and $P(1 < X < 2, 2 < Y < 3)$
(c) $P(1 < X < 2|2 < Y < 3)$
(d) $E[X|Y = y]$

1.26 The joint density function of the two random variables X and Y is

$$f_{X,Y}(x,y) = \begin{cases} kxy & ,1 < x < 3 \text{ and } 1 < y < 2 \\ 0 & ,\text{otherwise} \end{cases}$$

(a) What is the probability that $X + Y < 3$?
(b) Are X and Y independent?

1.27 The joint density function of two random variables X and Y is

$$f_{X,Y}(x,y) = \begin{cases} 16\frac{y}{x^3} & ,x > 2 \text{ and } 0 < y < 1 \\ 0 & ,\text{otherwise} \end{cases}$$

Find the mean of X and Y.

1.28 The density function of two independent random variables X and Y are

$$f_X(x) = \begin{cases} 2e^{-2x} & ,x\geq0 \\ 0 & ,\text{otherwise} \end{cases} \qquad f_Y(y) = \begin{cases} kye^{-3y} & ,y>0 \\ 0 & ,\text{otherwise} \end{cases}$$

Find
(a) $P(X+Y>1)$
(b) $P(1<X<2, Y\geq1)$
(c) $P(1<X<2)$
(d) $P(Y\geq1)$
(e) $P(1<X<2|Y\geq1)$

1.29 Find the density function of the random variable $Y=2X$, where

$$f_X(x) = \begin{cases} 2e^{-2x} & ,x>0 \\ 0 & ,\text{otherwise} \end{cases}$$

Compute $E[Y]$ in two ways:
(a) directly using $f_X(x)$
(b) using the density function of Y

1.30 Two independent random variables X and Y have the following density functions:

$$f_X(x) = \begin{cases} e^{-x} & ,x\geq0 \\ 0 & ,\text{otherwise} \end{cases} \qquad f_Y(y) = \begin{cases} \frac{1}{4} & ,0\leq x\leq4 \\ 0 & ,\text{otherwise} \end{cases}$$

What is the probability of $Z=X+Y$?

1.31 The joint density function of X and Y is

$$f_{X,Y}(x,y) = \begin{cases} e^{-(x+y)} & ,x\geq0 \text{ and } y\geq0 \\ 0 & ,\text{otherwise} \end{cases}$$

Find the density of
(a) $Z=XY$
(b) $Z=X+Y$

1.32 The joint density function of (X, Y) is

$$f_{X,Y}(x,y) = \begin{cases} 1 & , 0 \leq x \leq 1 \text{ and } 0 \leq y \leq 1 \\ 0 & , \text{otherwise} \end{cases}$$

Find the density function of $Z = XY$.

REFERENCES

1. Brunk, H.D., *An Introduction to Mathematical Statistics,* Blaisdell, New York, 1965.

2. Dudewics, E.J., *Introduction to Statistics and Probability,* Holt, Rinehart and Winston, New York, 1976.

3. Ehrenfeld S. and S.B. Littauer, *Introduction to Statistical Method,* McGraw-Hill, New York, 1964.

4. Feller, W., *An Introduction to Probability Theory and its Application,* John Wiley and Sons, New York, 1968.

5. Freeman, H., *Introduction to Statistical Inference,* Addison-Wesley, Reading, MA, 1963.

6. Harnett, D.L., *Introduction to Statistical Method,* Addison-Wesley, Reading, MA, 1972.

7. Hayes, W.L. and R.L. Winkler, *Statistics: Probability, Inference and Decision,* Holt, Rinehart and Winston, New York, 1971.

8. Haykin, S., *Communication Systems,* John Wiley and Sons, New York, 1983.

9. Helstron, C.W., *Probability and Stochastic Processes for Engineers,* Macmillan, New York, 1984.

10. Hoel, P.G., *Introduction to Mathematical Statistics,* John Wiley and Sons, New York, 1978.

11. Hogg, R.V. and A.T. Craig, *Introduction to Mathematical Statistics,* Macmillan, New York, 1978.

12. Meyer, P.L., *Introductory Probability and Statistical Applications,* Addison-Wesley, Reading, MA, 1970.

13. Papoulis, A., *Probability, Random Variables, and Stochastic Processes,* McGraw-Hill, New York, 1986.

14. Peebles, P.Z., *Probability, Random Variables, and Random Signal Principles,* McGraw-Hill, New York, 1980.

15. Rohatgi, V.K., *An Introduction to Probability Theory and Mathematical Statistics,* John Wiley and Sons, New York, 1976.

16. Spiegel, M.R., *Schaum's Outline Series: Probability and Statistics,* McGraw-Hill, New York, 1975.

17. Spiegel, M.R., *Schaum's Outline Series: Statistics,* McGraw-Hill, New York, 1961.

18. Stark, H., and J.W. Woods, *Probability, Random Processes, and Estimation Theory for Engineers,* Prentice-Hall, Englewood Cliffs, N.J., 1986.

19. Urkowitz, H., *Signal Theory and Random Processes,* Artech House, Norwood, MA, 1983.

20. Wozencraft, J.M. and I.M. Jacobs, *Principles of Communication Engineering,* John Wiley and Sons, New York, 1965.

Chapter 2

Random Processes

2.1 INTRODUCTION

A random process may be viewed as a collection of random variables, with time t a parameter running through all real numbers. In Chapter 1, we defined a random variable as a mapping of the elements of the sample space S, into points of the real axis. For random processes, the sample space would map into a family of time functions. Formally, we say a random process $X(t)$ is a mapping of the elements of the sample space, into functions of time. Each element of the sample space is associated with a time function, as shown in Figure 2.1.

Associating a time function to each element of the sample space results in a family of time functions called *ensemble*. Hence, ensemble is the set of sample functions with associated probabilities. Observe that we are denoting the random process by $X(t)$, and not $X(t,\xi)$, where the dependence on the ξ is omitted. A sample function is denoted by $x(t)$.

64

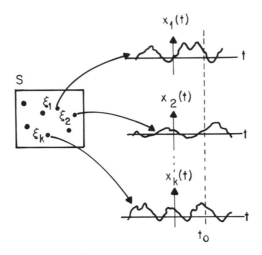

Figure 2.1 Mapping of sample space into sample functions.

Example 2.1

Consider a random process $X(t) = A \cos(\omega t + \Theta)$ where Θ is a random variable uniformly distributed between 0 and 2π as shown in Figure 2.2. That is,

$$f_\Theta(\theta) = \begin{cases} \frac{1}{2\pi} & ,0 \leq \theta \leq 2\pi \\ 0 & ,\text{otherwise} \end{cases}$$

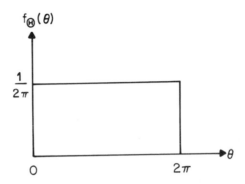

Figure 2.2 Density function of Θ.

Some sample functions of this random process are as shown in Figure 2.3.

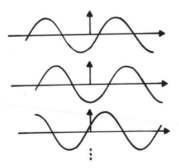

Figure 2.3 Some sample functions of $X(t)$.

This variation in the sample functions of this particular process is due to the phase only. Such a random process for which future values are predicted from knowledge of past ones is said to be predictable or deterministic. In fact, fixing the phase to some particular value, $\pi/4$, the sample function (corresponding to the particular element ξ_k of the sample space) becomes a deterministic time function, i.e.,

$$x_k(t) = A \, \cos(\omega t + \frac{\pi}{4})$$

When the parameter t is fixed to some instant t_0, the random process $X(t)$ becomes the random variable $X(t_0)$, and $x(t_0)$ would be a sample value of the random variable. In this case, the techniques we use with random variables apply. Consequently, we may characterize a random process by the first-order distribution as

$$F_X(x;t) = P[X(t_0) \le x] \tag{2.1}$$

or by the first-order density function as

$$f_X(x;t) = \frac{d}{dx} F_X(x;t) \tag{2.2}$$

for all possible values of t. The second-order distribution function is the joint distribution of the two random variables $X(t_1)$ and $X(t_2)$ for each t_1 and t_2. This results in

$$F_{X_1,X_2}(x_1, x_2; t_1, t_2) = P[X(t_1) \le x_1 \text{ and } X(t_2) \le x_2] \tag{2.3}$$

while the second-order density function is

$$f_X(x_1, x_2; t_1, t_2) = \frac{\partial^2}{\partial x_1 \partial x_2} F_{X_1,X_2}(x_1, x_2; t_1, t_2) \tag{2.4}$$

Normally, a complete probabilistic description of an arbitrary random process requires the specification of distributions of all orders given by

$$F_{X_1,X_2,\ldots,X_n}(x_1, x_2, \cdots, x_n; t_1, t_2, \cdots, t_n) =$$
$$P[X(t_1) \leq x_1, X(t_2) \leq x_2, \cdots, X(t_n) \leq x_n] \tag{2.5}$$

or given by the nth order density function:

$$f_{X_1,X_2,\ldots,X_n}(x_1, x_2, \cdots, x_n; t_1, t_2, \cdots, t_n) =$$
$$\frac{\partial^n F_{X_1,X_2,\ldots,X_n}(x_1, x_2, \cdots, x_n; t_1, t_2, \cdots, t_n)}{\partial x_1 \partial x_2 \cdots \partial x_n} \tag{2.6}$$

Fortunately, we are usually interested in processes that may possess some regularity so that they can be described more simply and knowledge of the first- and second-order density functions may be sufficient to generate higher-order density functions.

2.2 EXPECTATIONS

In many problems of interest only the first-order and the second-order statistics may be necessary to characterize a random process. Given a real random process $X(t)$, its mean value function is

$$m_X(t) = E[X(t)] = \int_{-\infty}^{\infty} x f_X(x, t) dx \tag{2.7}$$

The autocorrelation is defined to be

$$\begin{aligned} R_{XX}(t_1, t_2) &= E[X(t_1)X(t_2)] \\ &= \int_{-\infty}^{\infty} \int_{-\infty}^{\infty} x_1 x_2 f_{X1,X2}(x_1, x_2; t_1, t_2) dx_1 dx_2 \end{aligned} \tag{2.8}$$

When the autocorrelation $R_{XX}(t_1, t_2)$ of the random process $X(t)$ varies only with the time difference $|t_1 - t_2|$, and the mean m_X is constant, $X(t)$ is said to be *stationary in the wide-sense*, or *wide-sense stationary.* In this case, the autocorrelation is written as a function of one argument $\tau = t_1 - t_2$. If we let $t_2 = t$ and $t_1 = t + \tau$ the autocorrelation, in terms of τ only, is

$$R_{XX}(t + \tau, t) = R_{XX}(\tau) \tag{2.9}$$

A random process $X(t)$ is *strictly stationary* or *stationary in the strict sense* if its statistics are unchanged by a time shift in the time origin. Equivalently, its distributions of all orders are independent of the time origin. Note that a stationary process in the strict sense is stationary in the wide-sense but not the opposite. Also, the condition for wide-sense stationary is weaker than the condition for the second order stationary because, for wide-sense stationary processes, only the second-order statistics, autocorrelation, is constrained.

Example 2.2

Is the random process given in Example 2.1 wide-sense stationary ?

Solution.
For a random process to be stationary in the wide sense it must satisfy two conditions :

$$E[X(t)] = constant$$
$$R_{XX}(t + \tau, t) = R_{XX}(\tau)$$

To compute the mean of $X(t)$, we use the concept that

$$E[g(\theta)] = \int_{-\infty}^{\infty} g(\theta) f_\Theta(\theta) d\theta$$

where in this case $g(\theta) = A \cos(\omega t + \theta)$ and $f_\Theta(\theta) = 1/(2\pi)$ in the interval between 0 and 2π. Then,

$$E[X(t)] = \int_0^{2\pi} A \cos(\omega t + \theta) \frac{1}{2\pi} d\theta = 0$$

The autocorrelation is

$$
\begin{aligned}
E[X(t+\tau)X(t)] &= E[A\,\cos\{\omega(t+\tau)+\theta\}A\,\cos\{\omega t+\theta\}] \\
&= \frac{A^2}{2}E[\,\cos(\omega\tau)+\,\cos(2\omega t+\omega\tau+2\theta)]
\end{aligned}
$$

where we have used the trigonometric identity that says

$$
\cos a\,\cos b = \frac{1}{2}[\,\cos(a-b)+\,\cos(a+b)]
$$

The second term evaluates to zero. Thus, the autocorrelation is

$$
R_{XX}(t+\tau,t) = \frac{A^2}{2}\,\cos\omega\tau = R_{XX}(\tau)
$$

since the mean is constant and the autocorrelation depends on τ only, $X(t)$ is a wide-sense stationary process.

When dealing with two random processes $X(t)$ and $Y(t)$, we say that they are *jointly wide-sense stationary* if, in addition, each process is stationary in the wide sense,

$$
R_{XY}(t+\tau,t) = E[X(t+\tau)Y(t)] = R_{XY}(\tau) \qquad (2.10)
$$

$R_{XY}(t_1,t_2)$ represents the cross-correlation function of $X(t)$ and $Y(t)$. We also define the autocovariance function $C_{XX}(t_1,t_2)$ of $X(t)$ and the cross-covariance function $C_{XY}(t_1,t_2)$ between $X(t)$ and $Y(t)$ as

$$
C_{XX}(t_1,t_2) = E[\{X(t_1)-m_X(t_1)\}\{X(t_2)-m_X(t_2)\}] \qquad (2.11)
$$

and

$$
C_{XY}(t_1,t_2) = E[\{X(t_1)-m_X(t_1)\}\{Y(t_2)-m_Y(t_2)\}] \qquad (2.12)
$$

If $Z(t)$ is a complex random process such that $Z(t) = X(t)+jY(t)$, the autocorrelation and the autocovariance functions of $Z(t)$ are

$$R_{ZZ}(t_1, t_2) = E[Z(t_1)Z^*(t_2)] \qquad (2.13)$$

and

$$C_{ZZ}(t_1, t_2) = E[\{Z(t_1) - m_Z(t_1)\}\{Z(t_2) - m_Z(t_2)\}^*] \qquad (2.14)$$

where $*$ denotes complex conjugate and $m_Z(t)$ is the mean function of $Z(t)$. The cross-correlation and cross-covariance functions between the complex random process $Z(t)$ and another complex random process $W(t)$, $W(t) = U(t) + jV(t)$, is

$$R_{ZW}(t_1, t_2) = E[Z(t_1)W^*(t_2)] \qquad (2.15)$$

and

$$C_{ZW}(t_1, t_2) = E[\{Z(t_1) - m_Z(t_1)\}\{W(t_2) - m_W(t_2)\}^*] \qquad (2.16)$$

Example 2.3

Consider an experiment of tossing a coin in an infinite number of interval times. A sample function of random process $X(t)$ is defined as

$$x(t) = \begin{cases} 1 & \text{for } (n-1)T \leq t < nT \text{ if heads at } n\text{th toss} \\ -1 & \text{for } (n-1)T \leq t < nT \text{ if tails at } n\text{th toss} \end{cases}$$

where n takes all possible integer values. Is the process stationary in the wide-sense?

Solution.

For the process to be wide-sense stationary, it must be verified that it has a constant mean and an autocorrelation function which is a function of τ only.

Let $P(H) = P(head)$ and $P(T) = P(tail)$, then, from Figure 2.4,

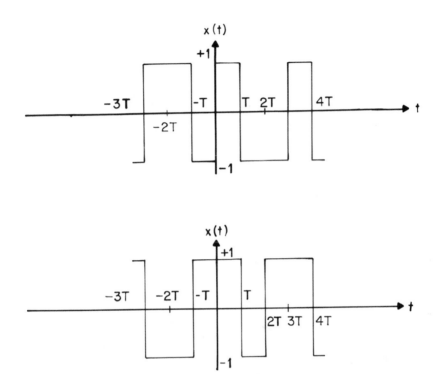

Figure 2.4 Sample functions of $X(t)$.

$$E[X(t)] = (+1)P(H) + (-1)P(T) = (1)\frac{1}{2} + (-1)\frac{1}{2} = 0$$

Since the mean is constant, the process may be wide-sense stationary. The mean square value is

$$E[X^2(t)] = (1)^2 P(H) + (-1)^2 P(T) = 1$$

We now consider the autocorrelation function :

$$R_{XX}(t_1, t_2) = E[X(t_1)X(t_2)]$$

We have two cases to consider :

Case 1 : t_1 and t_2 in the same tossing interval, that is

$$(n - 1)T \leq t_1, t_2 \leq nT$$

Hence,

$$R_{XX}(t_1, t_2) = E[X(t_1)X(t_2)] = E[X^2(t)] = 1$$

Case 2 : t_1 and t_2 in different tossing intervals, i.e.,

$$(j - 1)T \leq t_1 \leq jT \text{ and } (k - 1)T \leq t_2 \leq kT \text{ for } j \neq k$$

Since succcessive tosses are statistically independent, $X(t_1)$ and $X(t_2)$ are also statistically independent. Therefore,

$$R_{XX}(t_1, t_2) = E[X(t_1)X(t_2)] = E[X(t_1)]E[X(t_2)] = 0$$

since the autocorrelation is not a function of one variable $\tau = t_1 - t_2$, the process $X(t)$ is not stationary. This process is referred to as *semi-random binary transmission*.

Example 2.4

Consider the random process $Y(t) = X(t - \Theta)$ where $X(t)$ is the process of Example 2.3, and Θ is a random variable uniformly distributed over the interval 0 to T. Θ is statistically independent of $X(t)$. Is $Y(t)$ stationary in the wide-sense?

Solution.

A sample function of $Y(t)$ is shown in Figure 2.5.

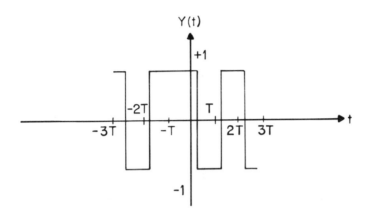

Figure 2.5 Sample function of $Y(t)$.

As in the previous example, the mean of $Y(t)$ is

$$E[Y(t)] = (1)P(H) + (-1)P(T) = 0$$

which is a constant. The autocorrelation is given by

$$R_{YY}(t_1, t_2) = E[Y(t_1)Y(t_2)]$$

where we have few possible cases.

Case 1 : $\tau = t_1 - t_2$ and $|\tau| > T$

In this case, t_1 and t_2 are in different tossing intervals for each sample function, and hence $Y(t_1)$ and $Y(t_2)$ are statistically independent. Thus,

$$R_{YY}(t_1, t_2) = E[Y(t_1)Y(t_2)] = E[Y(t_1)]E[Y(t_2)] = 0$$

74

Case 2 : $|\tau| \leq T$

In this case, t_1 and t_2 may or may not be in the same tossing interval. Let SI denote the event that t_1 and t_2 occur in the same interval and SI^C, complementary event of SI, be the event that t_1 and t_2 are not in the same interval. Thus,

$$
\begin{aligned}
R_{YY}(t_1, t_2) &= E[Y(t_1)Y(t_2)] \\
&= E[Y(t_1)Y(t_2)|SI]P(SI) + E[Y(t_1)Y(t_2)|SI^C]P(SI^C)
\end{aligned}
$$

We already have from Example 2.3 that

$$
E[Y(t_1)Y(t_2)|SI] = 1
$$

and

$$
E[Y(t_1)Y(t_2)|SI^C] = 0
$$

Hence, the autocorrelation is just the probability that the event SI occurs

$$
R_{YY}(t_1, t_2) = P(SI)
$$

The event SI occurs in two possible ways : $t_1 < t_2(\tau < 0)$ and $t_2 < t_1(\tau > 0)$.

When $t_1 < t_2$, $-T \leq \tau \leq 0$. The situation is best represented by the following diagram, Figure 2.6, representing one interval only :

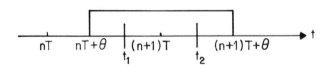

Figure 2.6 One interval for $-T \leq \tau \leq 0$.

t_1 and t_2 are in the same interval if $t_1 > nT + \theta$ and $t_2 < (n+1)T + \theta$, which yields

$$t_2 - (n+1)T < \theta < t_1 - nT$$

Since Θ is uniformly distributed between 0 and T, then the probability that t_1 and t_2 are in the same interval is

$$
\begin{aligned}
P(SI) &= \int_{t_2-(n+1)T}^{t_1-nT} \frac{1}{T} d\theta \\
&= 1 + \frac{\tau}{T} \quad \text{for} \quad -T \le \tau \le 0
\end{aligned}
$$

Similarly, when $t_2 < t_1$, and t_1 and t_2 are in the same interval, we have

$$t_1 < (n+1)T + \theta$$

and

$$t_2 > nT + \theta$$

which yields

$$t_1 - (n+1)T < \theta < t_2 - nT$$

and

$$P(SI) = 1 - \frac{\tau}{T} \quad \text{for } 0 \le \tau \le T$$

Therefore,

$$
R_{YY}(t_1, t_2) = \begin{cases} 1 - \frac{|\tau|}{T} & , |\tau| \le T \\ 0 & , |\tau| > T \end{cases}
$$

and is shown in Figure 2.7.

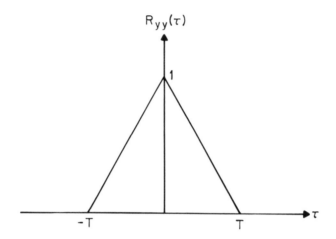

Figure 2.7 Autocorrelation of $Y(t)$.

Because both conditions, (the mean is constant and the autocorrelation is a function of τ only), are satisfied, the process $Y(t)$ is wide-sense stationary. $Y(t)$ is also referred to as *random binary transmission*.

Example 2.5

Let $I(t)$ and $Q(t)$ be two random processes such that

$$I(t) = X \cos\omega t + Y \sin\omega t$$
$$Q(t) = Y \cos\omega t - X \sin\omega t$$

where X and Y are zero mean and uncorrelated random variables. The mean square values of X and Y are $E[X^2] = E[Y^2] = \sigma^2$. Derive the cross-correlation function between the processes $I(t)$ and $Q(t)$.

Solution.

The cross-correlation between $I(t)$ and $Q(t)$ is

$$
\begin{aligned}
R_{IQ}(t+\tau,t) &= E[I(t+\tau)Q(t)] \\
&= E[\{X\cos(\omega t+\omega\tau)+Y\sin(\omega t+\omega\tau)\} \\
&\quad \cdot\{Y\cos\omega t-X\sin\omega t\}] \\
&= E[XY]\{\cos(\omega t+\omega\tau)\cos\omega t-\sin(\omega t+\tau)\sin\omega t\} \\
&\quad -E[X^2]\cos(\omega t+\omega\tau)\sin\omega t \\
&\quad +E[Y^2]\sin(\omega t+\omega\tau)\cos\omega t
\end{aligned}
$$

Using trigonometric identities and the fact that X and Y are uncorrelated and zero mean $(E[XY] = E[X]E[Y] = 0)$, we obtain

$$
R_{IQ}(t+\tau,t) = -\sigma^2\sin\omega_0\tau
$$

2.3 PROPERTIES OF CORRELATION FUNCTIONS

The autocorrelation and the cross-correlation functions introduced in the previous sections are very important concepts in understanding random processes. In this section, we study some of their properties that are most relevant without giving any formal proof.

2.3.1 Autocorrelation

(i)
$$
R_{XX}(t_2,t_1) = R^*_{XX}(t_1,t_2) \tag{2.17}
$$

If $X(t)$ is real, then the autocorrelation is symmetric about the line $t_1 = t_2$ in the (t_1,t_2) plane; that is,

$$
R_{XX}(t_2,t_1) = R_{XX}(t_1,t_2) \tag{2.18}
$$

(ii) The mean square value function of a random process $X(t)$ is always positive. That is,

$$R_{XX}(t_1, t_1) = E[X(t_1)X^*(t_1)] = E[|X(t_1)|^2] \geq 0 \qquad (2.19)$$

If $X(t)$ is real, the mean square value is $E[X^2(t_1)]$ and is always non-negative.

(iii) $\qquad |R_{XX}(t_1, t_2)| \leq \sqrt{R_{XX}(t_1, t_1)R_{XX}(t_2, t_2)} \qquad (2.20)$

This is known as *Schwarz Inequality* and can be written as

$$|R_{XX}(t_1, t_2)|^2 \leq E[|X(t_1)|^2]E[|X(t_2)|^2] \qquad (2.21)$$

2.3.2. Cross-Correlation

Consider $X(t)$ and $Y(t)$ to be two random processes, then,

(i) $\qquad R_{XY}(t_2, t_1) = R^*_{YX}(t_1, t_2) \qquad (2.22)$

If the random processes $X(t)$ and $Y(t)$ are real,

$$R_{XY}(t_2, t_1) = R_{YX}(t_1, t_2) \qquad (2.23)$$

In general, $R_{XY}(t_2, t_1)$ and $R_{YX}(t_1, t_2)$ are not equal.

(ii) $\quad |R_{XY}(t_1, t_2)| = |E[X(t_1)Y(t_2)]|$

$$\leq \sqrt{R_{XX}(t_1, t_1)R_{YY}(t_2, t_2)} = \sqrt{E[|X(t_1)|^2]E[|Y(t_2)|^2]}$$

$$(2.24)$$

2.3.3 Wide-Sense Stationary

We now consider the processes $X(t)$ and $Y(t)$ to be real and wide-sense stationary.

(i) The autocorrelation is an even function of τ, that is,

$$R_{XX}(-\tau) = R_{XX}(\tau) \tag{2.25}$$

(ii)
$$R_{XX}(0) = E[X^2(t)] \tag{2.26}$$

The autocorrelation at $\tau = 0$ is constant and is equal to the mean square value.

(iii)
$$|R_{XX}(\tau)| \le R_{XX}(0) \tag{2.27}$$

The maximum value of the autocorrelation occurs at $\tau = 0$ and it is nonnegative as shown in Figure 2.8.

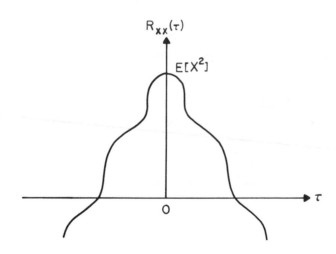

Figure 2.8 A possible autocorrelation function.

If $X(t)$ and $Y(t)$ are jointly stationary in the wide-sense, similar properties can be obtained. $\implies$

(iv) $$R_{XY}^*(-\tau) = R_{YX}(\tau) \qquad (2.28)$$

(v) $$|R_{XY}(\tau)|^2 \leq R_{XX}(0)R_{YY}(0) \qquad (2.29)$$

In many situations of interest, two observations of a wide-sense stationary process may become uncorrelated as τ approaches infinity. In this case, the covariance function goes to zero. That is,

$$
\begin{aligned}
\lim_{\tau \to \infty} C_{XX}(\tau) &= E[\{X(t+\tau) - m_X\}\{X(t) - m_X\}] \\
&= R_{XX}(\tau) - m_X^2 \\
&= 0 \qquad (2.30)
\end{aligned}
$$

The autocorrelation is equal to the square of the mean; $R_{XX}(\tau) = m_X^2$.

2.4 SOME RANDOM PROCESSES

In this section, we shall study certain types of random processes that may characterize some applications.

2.4.1 A Single Pulse of Known Shape but Random Amplitude and Arrival Time

In radar and sonar applications, a return signal may be characterized as a random process consisting of a pulse with known shape, but with random amplitude and random arrival time. The pulse may be expressed as

$$X(t) = A\, S(t - \Theta) \qquad (2.31)$$

where A and Θ are two statistically independent random variables, and $s(t)$ is a deterministic function. A sample function may be represented as shown in Figure 2.9.

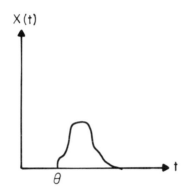

Figure 2.9 Pulse $X(t)$.

The mean value function of this particular random process is given by

$$E[X(t)] = E[A \ S(t - \Theta)] \tag{2.32}$$

because A and Θ are statistically independent, we have

$$\begin{aligned} E[X(t)] &= E[A]E[S(t - \Theta)] \\ &= E[A] \int_{-\infty}^{\infty} S(t - \theta)f_\Theta(\theta)d\theta \end{aligned} \tag{2.33}$$

The integral $\int_{-\infty}^{\infty} S(t - \theta)f_\Theta(\theta)d\theta$ is simply the convolution of the pulse $s(t)$ with the density function of Θ. Thus,

$$E[X(t)] = E[A]s(t) * f_\Theta(\theta) \tag{2.34}$$

similarly, the autocorrelation function is given by

$$R_{XX}(t_1, t_2) = E[A^2] \int_{-\infty}^{\infty} S(t_1 - \theta)S(t_2 - \theta)f_\Theta(\theta)d\theta \tag{2.35}$$

82

If the arrival time is known to be some fixed value θ_0, then the mean and the autocorrelation function of $X(t)$ become

$$E[X(t)] = E[A]s(t - \theta_0) \tag{2.36}$$

and

$$R_{XX}(t_1, t_2) = E[A^2]s(t_1 - \theta_0)s(t_2 - \theta_0) \tag{2.37}$$

Another special case is that the arrival time may be uniformly distributed over the interval from 0 to T. The mean and autocorrelation functions are in this case:

$$E[X(t)] = \frac{E[A]}{T} \int_0^T S(t - \theta)d\theta \tag{2.38}$$

and

$$R_{XX}(t_1, t_2) = \frac{E[A]^2}{T} \int_0^T S(t_1 - \theta)S(t_2 - \theta)d\theta \tag{2.39}$$

2.4.2 Multiple Pulses

We now assume that we have a multiple pulse situation. This may be the case in radar applications for a multiple target environment. The random process $X(t)$ can be expressed as

$$X(t) = \sum_{k=1}^{n} A_k \, S(t - \Theta_k) \tag{2.40}$$

where the $2n$ random variables A_k and Θ_k, $k = 1, 2, \cdots, n$, are mutually and statistically independent. In addition, the amplitudes are independent of the phase shifts, we assume that the A_k's are identically distributed with density function $f_A(a)$, while the Θ_k's are identically distributed with density function $f_\Theta(\theta)$. We can easily obtain the mean and autocorrelation functions to be

$$
\begin{aligned}
E[X(t)] &= E[\sum_{k=1}^{n} A_k\, S(t - \Theta_k)] \\
&= \sum_{k=1}^{n} E[A_k]E[S(t - \Theta_k)] \\
&= nE[A_k] \int_{-\infty}^{\infty} S(t - \theta)f_\Theta(\theta)d\theta \\
&= nE[A_k][s(t) * f_\Theta(\theta)]
\end{aligned}
\tag{2.41}
$$

and

$$
\begin{aligned}
R_{XX}(t_1, t_2) &= E[\sum_{k=1}^{n} A_k\, S(t_1 - \Theta_k) \sum_{j=1}^{n} A_j\, S(t_2 - \Theta_j)] \\
&= \sum_{k}\sum_{j} E[A_k A_j]E[S(t_1 - \Theta_k)S(t_2 - \Theta_j)] \\
&= nE[A_k^2] \int_{-\infty}^{\infty} S(t_1 - \theta)S(t_2 - \theta)f_\Theta(\theta)d\theta \\
&\quad +(n^2 - n)(E[A_k])^2 \int_{-\infty}^{\infty} S(t_1 - \theta)f_\Theta(\theta)d\theta \\
&\quad \cdot \int_{-\infty}^{\infty} S(t_2 - \theta)f_\Theta(\theta)d\theta
\end{aligned}
\tag{2.42}
$$

If the random variable Θ is uniformly distributed over the interval $(0, T)$, the mean and the autocorrelation functions of $X(t)$ become

$$
E[X(t)] = nE[A_k]\frac{1}{T} \int_{0}^{T} S(t - \theta)d\theta
\tag{2.43}
$$

and

$$
R_{XX}(t_1, t_2) = nE[A_k^2]\frac{1}{T} \int_{0}^{T} S(t_1 - \theta)S(t_2 - \theta)d\theta
$$
$$
+(n^2 - n)\frac{1}{T^2} \int_{0}^{T} S(t_1 - \theta)d\theta \int_{0}^{T} S(t_2 - \theta)d\theta
\tag{2.44}
$$

2.4.3 Periodic Random Processes

The random process $X(t)$ is said to be periodic with period T if all its sample functions are periodic with period T, except those sample functions that occur with probability zero.

Theorem: If the random process $X(t)$ is stationary in the wide sense, then the autocorrelation is periodic with period T, if and only if $X(t)$ is periodic with period T, and vice versa.

Proof: The first condition says that $R_{XX}(\tau + nT) = R_{XX}(\tau)$ if $X(t)$ is periodic. $X(t)$ periodic means that $X(t + \tau + nT) = X(t + \tau)$. Then,

$$
\begin{aligned}
R_{XX}(\tau + nT) &= E[X(t + \tau + nT)X(t)] \\
&= E[X(t + \tau)X(t)] \\
&= R_{XX}(\tau)
\end{aligned}
\tag{2.45}
$$

The second condition states that if the autocorrelation is periodic, then $X(t + nT) = X(t)$ where $X(t)$ is wide-sense stationary. Consider *Chebyschev's inequality* which states

$$
P[|Y(t) - m_Y| > k] \leq \frac{\sigma_Y^2}{k^2}
\tag{2.46}
$$

where m_Y and σ_Y^2 are the mean and variance of the process $Y(t)$, respectively, and k is a positive constant.

Let $Y(t) = X(t + T) - X(t)$. Then, the mean and variance of $Y(t)$ are

$$
\begin{aligned}
m_Y &= E[Y(t)] \\
&= E[X(t + T) - X(t)] \\
&= E[X(t + T)] - E[X(t)] \\
&= 0
\end{aligned}
\tag{2.47}
$$

because $X(t)$ is wide-sense stationary (mean is constant). Also,

$$
\begin{aligned}
\sigma_Y^2 &= E[Y^2(t)] \\
&= E[\{X(t+T) - X(t)\}^2] \\
&= E[X^2(t+T)] - 2E[X(t+T)X(t)] + E[X^2(t)] \\
&= R_{XX}(0) - 2R_{XX}(T) + R_{XX}(0) \\
&= 2[R_{XX}(0) - R_{XX}(T)] \tag{2.48}
\end{aligned}
$$

The variance σ_Y^2 is zero due to the fact that the autocorrelation is periodic with period T; $R_{XX}(0) = R_{XX}(T)$. Consequently, from Chebyschev's inequality, we have

$$
P[|X(t+T) - X(t)| > k] = 0 \qquad \text{for all } t \tag{2.49}
$$

Hence, $X(t)$ must be periodic.

Corollary: Let $s(t)$ be a deterministic function and periodic with period T. The random process $X(t)$ defined as $X(t) = S(t - \Theta)$, where Θ is a random variable uniformly distributed over the interval $(0, T)$, is stationary in the wide-sense.

Proof: For $X(t)$ to be wide-sense stationary, the mean $E[X(t)]$ must be constant and the autocorrelation must be a function of the time difference τ. The mean value function of $X(t)$ is

$$
\begin{aligned}
E[X(t)] &= \int_{-\infty}^{\infty} S(t - \theta) f_\Theta(\theta) d\theta \\
&= \frac{1}{T} \int_{\theta=0}^{\theta=T} S(t - \theta) d\theta \tag{2.50}
\end{aligned}
$$

We make a change of variable by letting $\lambda = t - \theta$. Then,

$$E[X(t)] = -\frac{1}{T}\int_t^{t-T} s(\lambda)d\lambda$$

$$= \frac{1}{T}\int_{t-T}^t s(\lambda)d\lambda$$

$$= constant \tag{2.51}$$

since we are integrating a periodic function, $s(t)$, over its period. Using the same reasoning, we can easily show that $R_{XX}(t+\tau,t) = R_{XX}(\tau)$.

The process $X(t)$ is *periodically stationary* or *cyclostationary* with period T if its statistics are not changed by a shift of nT, $n = \pm1, \pm2, \cdots$, from the time origin. That is,

$$f_{X_1,\cdots,X_m}(x_1\cdots x_m; t_1,\cdots,t_m) =$$

$$f_{X_1,\cdots,X_m}(x_1,\cdots,x_m; t_1 + nT,\cdots,t_m + nT) \tag{2.52}$$

for all integers n and m.

$X(t)$ is cyclostationary in the wide sense with period T, if its mean and autocorrelation functions are periodic with the same period T. That is,

$$m_X(t+kT) = m_X(t) \tag{2.53}$$

and

$$R_{XX}(t_1 + kT, t_2 + kT) = R_{XX}(t_1, t_2) \tag{2.54}$$

for all t, t_1, t_2 and any integer k.

Theorem: If $X(t)$ is a wide-sense cyclostationary process with period T, then the process $Y(t) = X(t - \Theta)$, where Θ is uniformly distributed over the interval $(0, T)$, is wide-sense stationary.

The proof of this theorem is straightforward and similar to that of previous theorem. Therefore, we will not show it.

2.4.4 Gaussian Process

A random process $X(t)$ is Gaussian if the random variables $X(t_1), X(t_2)$, $\cdots, X(t_n)$, are jointly Gaussian for all possible values of n and t_1, t_2, $\cdots, t_n$. Since the multivariate Gaussian random variable depends only on the mean vector, and the covariance matrix of the n random variables, we observe that if $X(t)$ is stationary in the wide-sense, it is also strictly stationary.

If $X(t)$ is a Gaussian random process applied to a linear time invariant system with impulse response $h(t)$, as shown in Figure 2.10, then the output process $Y(t)$ is also Gaussian. Hence, the output process $Y(t)$ will be completely specified given the input process $X(t)$ and the impulse response $h(t)$.

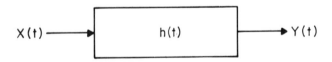

Figure 2.10 Impulse response h(t).

Let the processes $Y_1(t)$ and $Y_2(t)$ be the outputs of two linear time invariant systems with respective inputs $X_1(t)$ and $X_2(t)$. The processes $Y_1(t)$ and $Y_2(t)$ are jointly Gaussian, provided that $X_1(t)$ and $X_2(t)$ are jointly Gaussian.

2.4.5 Poisson Process

The Poisson process is used for modeling situations such as alpha particles emitted from a radioactive material, failure times of components of a system, people serviced at a post office, and telephone calls received in an office. In modeling the aforementioned cases, the concept of *Poisson points* is sometimes used. Let the instants at which the events occur be as depicted in Figure 2.11. We start observing the process at time $t = 0$.

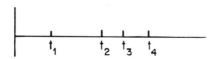

Figure 2.11 Possible occuring times of particular events.

We say that the points $\mathbf{t}_i$ are *Poisson points* with parameter λt provided the following properties are satisfied.

(i) the number of points $\mathbf{t}_i$ in an internal (t_1, t_2), denoted $\mathbf{n}(t_1, t_2)$, is a Poisson random variable. That is, the probability of k points in time $t = t_2 - t_1$ is

$$P[\mathbf{n}(t_1, t_2) = k] = \frac{e^{-\lambda t}(\lambda t)^k}{k!} \tag{2.55}$$

λ is called the *density* or average arrival rate of the Poisson process.

(ii) If the intervals (t_1, t_2) and (t_3, t_4) are nonoverlapping, then the corresponding random variables $\mathbf{n}(t_1, t_2)$ and $\mathbf{n}(t_3, t_4)$ are independent.

We define the Poisson process as

(i) $$X(t) = \mathbf{n}(0, t) \tag{2.56}$$

(ii) $$X(0) = 0 \tag{2.57}$$

A typical sample function is shown in Figure 2.12.

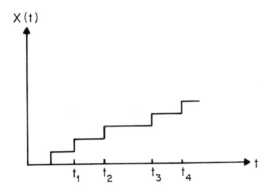

Figure 2.12 Sample function of a Poisson process.

The first-order distribution of $X(t)$ is

$$
\begin{aligned}
F_{X_1}(x_1, t_1) &= P[X(t_1) \le x_1] \\
&= P[\text{the number of points in } (0, t_1) \le x_1] \\
&= \sum_{k=0}^{x_1} e^{-\lambda t_1} \frac{(\lambda t_1)^k}{k!}
\end{aligned}
\tag{2.58}
$$

2.5 POWER SPECTRAL DENSITY

Given a deterministic signal $s(t)$, its Fourier transform (FT) is

$$
S(f) = \int_{-\infty}^{\infty} s(t) e^{-j2\pi ft} dt
\tag{2.59}
$$

which exists if the integral converges. The function $S(f)$, is sometimes called the *spectrum* of $s(t)$. In going from the time-domain description,

$s(t)$, to the frequency-domain, $S(f)$, no information about the signal is lost. In other words, $S(f)$ forms a complete description of $s(t)$ and vice versa. The signal $s(t)$ can be obtained from $S(f)$ by just taking the inverse Fourier transform (IFT) . That is,

$$s(t) = \int_{-\infty}^{\infty} S(f)e^{+j2\pi ft}df \tag{2.60}$$

In dealing with random processes, the ensemble is assumed to exist for all time t. In general, the sample functions are not absolutely integrable. However, since we are still interested in the notion of spectrum, we proceed in a manner similar to that of deterministic signals with infinite energy. We define $x_T(t)$ as the sample function $x(t)$, truncated between $-T$ and T, of the random process $X(t)$. That is,

$$x_T(t) = \begin{cases} x(t) & ,-T \le t \le T \\ 0 & ,\text{otherwise} \end{cases} \tag{2.61}$$

The truncated Fourier transform of the process $X(t)$ is

$$\begin{aligned} X_T(f) &= \int_{-T}^{T} x_T(t)e^{-j2\pi ft}dt \\ &= \int_{-\infty}^{\infty} x(t)e^{-j2\pi ft}dt \end{aligned} \tag{2.62}$$

The average power of $x_T(t)$ is

$$P_{ave} = \frac{1}{2T} \int_{-T}^{T} x_T^2(t)dt \tag{2.63}$$

Using Parseval's theorem, which says that

$$\int_{-\infty}^{\infty} x^2(t)dt = \int_{-\infty}^{\infty} |X_T(f)|^2 df \tag{2.64}$$

then, the average power of $x_T(t)$ is

$$P_T = \int_{-\infty}^{\infty} \frac{|X_T(f)|^2}{2T} df \tag{2.65}$$

the term $|X_T(f)|^2/2T$ is the power spectral density of $x_T(t)$. The ensemble average of P_T is given by

$$E[P_T] = \int_{-\infty}^{\infty} E[\frac{|X_T(f)|^2}{2T}]df \qquad (2.66)$$

The power spectral density of the random process $X(t)$ is defined to be

$$S_{XX}(f) = \lim_{T\to\infty} E[\frac{|X_T(f)|^2}{2T}] \qquad (2.67)$$

If $X(t)$ is stationary in the wide sense, the power spectral density $S_{XX}(f)$ can be expressed as the Fourier transform of the autocorrelation $R_{XX}(\tau)$. That is,

$$S_{XX}(f) = \int_{-\infty}^{\infty} R_{XX}(\tau)e^{-j2\pi f\tau}d\tau \qquad (2.68)$$

Proof: By definition:

$$
\begin{aligned}
E[\frac{|X_T(f)|^2}{2T}] &= E[\frac{1}{2T}\int_{-T}^{T} X_T(t_1)e^{-j2\pi ft_1}dt_1 \int_{-T}^{T} X_T(t_2)e^{j2\pi ft_2}dt_2]\\
&= E[\frac{1}{2T}\int_{-T}^{T} X(t_1)e^{-j2\pi ft_1}dt_1 \int_{-T}^{T} X(t_2)e^{j2\pi ft_2}dt_2]\\
&= \frac{1}{2T}\int_{-T}^{T}\int_{-T}^{T} E[X(t_1)X(t_2)]e^{-j2\pi f(t_1-t_2)}dt_1dt_2 \quad (2.69)
\end{aligned}
$$

where $E[X(t_1)X(t_2)] = R_{XX}(t_1,t_2)$. Since this is a wide-sense stationary process, we would like to express the autocorrelation in terms of the time difference $\tau = t_1 - t_2$ and consequently, replace the double integral in t_1 and t_2 to one integral in τ. Let $t_2 = t$ and $t_1 = t_2 + \tau = t + \tau$. The region of integration in the $t_1 - t_2$ plane and $t - \tau$ plane, are shown in Figures 2.13a and 2.13b .

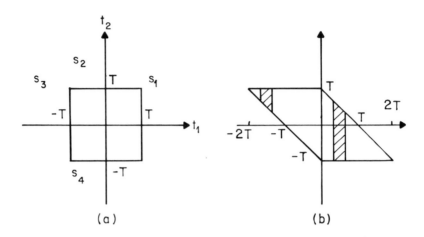

Figure 2.13 Regions of integration for autocorrelation.

Let s_1, s_2, s_3, s_4 denote the four sides of the square. From Figure 2.13 we see that the change of variables for the four sides will be

$$s_1 \rightsquigarrow \tau = T - t_2, \quad s_2 \rightsquigarrow \tau = t_1 - T, \quad s_3 \rightsquigarrow \tau = -T - t_2,$$
$$s_4 \rightsquigarrow \tau = t_1 + T$$

It follows that

$$
\begin{aligned}
E\left[\frac{|X_T(t)|^2}{2T} \right] &= \frac{1}{2T} \int_0^{2T} R_{XX}(\tau) e^{-j2\pi f \tau} \left[\int_{-T}^{T-\tau} R_{XX}(\tau) e^{-j2\pi f \tau} dt \right] d\tau \\
&\quad + \frac{1}{2T} \int_{-2T}^0 R_{XX}(\tau) e^{-j2\pi f \tau} \left[\int_{-T-\tau}^{T} R_{XX}(\tau) e^{-j2\pi f \tau} dt \right] d\tau \\
&= \int_{-2T}^{2T} R_{XX}(\tau) e^{-j2\pi f \tau} \left[1 - \frac{|\tau|}{2T} \right] d\tau
\end{aligned}
\tag{2.70}
$$

In the limit, as T approaches infinity, we conclude

$$S_{XX}(f) = \int_{-\infty}^{\infty} R_{XX}(\tau) e^{-j2\pi f \tau} d\tau \tag{2.71}$$

provided that $R_{XX}(\tau)$ approaches zero at least at the rate $1/|\tau|$ with increasing τ.

Thus, the power spectral density of a wide-sense stationary process is the Fourier transform of its autocorrelation. The inverse relationship using the inverse Fourier transform is

$$R_{XX}(\tau) = \int_{-\infty}^{\infty} S_{XX}(f)e^{j2\pi f\tau}df \qquad (2.72)$$

Equations (2.71) and (2.72) are sometimes called the *Wiener-Kinchin* relations. Note that the power spectral density is, from the definition, real, positive, and an even function of f. The autocorrelation is an even function of τ.

Example 2.6

Consider the random process $X(t) = A\cos(\omega_0 t + \Theta)$ where Θ is a random variable uniformly distributed over the interval $(0, 2\pi)$, and A and ω_0 are constants. Determine the power spectral density of this process.

Solution.

Since $X(t)$ is stationary in the wide sense with autocorrelation function $R_{XX}(\tau) = (A^2/2)\cos(2\pi f_0\tau)$ as shown in Example 2.2, then using Equation (2.71), the power spectral density is

$$
\begin{aligned}
S_{XX}(f) &= \int_{-\infty}^{\infty} \frac{A^2}{2}\cos(2\pi f_0\tau)e^{-j2\pi f\tau}d\tau \\
&= \frac{A^2}{4}[\delta(f - f_0) + \delta(f + f_0)]
\end{aligned}
$$

2.5.1 Cross Spectral Densities

Let $X(t)$ and $Y(t)$ be two jointly wide-sense stationary processes. Their cross-spectral densities are defined as

$$S_{XY}(f) = \int_{-\infty}^{\infty} R_{XY}(\tau)e^{-j2\pi f\tau}d\tau \qquad (2.73)$$

and

$$S_{YX}(f) = \int_{-\infty}^{\infty} R_{YX}(\tau)e^{-j2\pi f\tau}d\tau \qquad (2.74)$$

By the *Wiener-Kinchin* relations, the correlation functions $R_{XY}(\tau)$ and $R_{YX}(\tau)$ are just the respective inverse Fourier transforms of $S_{XY}(f)$ and $S_{YX}(f)$. From property (iv) of Section (2.3.3), we have

$$R_{YX}(\tau) = R_{XY}^*(-\tau) \qquad (2.75)$$

Consequently, their two cross-spectral densities are related by the following:

$$S_{YX}(f) = S_{XY}^*(f) \qquad (2.76)$$

It should be observed that while the power spectral densities $S_{XX}(f)$ and $S_{YY}(f)$ of the respective processes $X(t)$ and $Y(t)$ are always real, their cross-spectral densities $S_{XY}(f)$ and $S_{YX}(f)$ may be complex.

Example 2.7

Consider the process $Y(t) = X(t - T)$ where $X(t)$ is a real wide-sense stationary process with autocorrelation $R_{XX}(\tau)$ and power spectral density $S_{XX}(f)$. T is a constant. Express the power spectral density $S_{XY}(f)$ of the process $Y(t)$ in terms of $S_{XX}(f)$.

Solution.

The cross-correlation $R_{XY}(\tau)$ is

$$
\begin{aligned}
R_{XY}(\tau) &= E[X(t+\tau)Y(t)] \\
&= E[X(t+\tau)X(t-T)] \\
&= R_{XX}(\tau+T)
\end{aligned}
$$

Hence,

$$
S_{XY}(f) = S_{XX}(f)e^{j2\pi fT}
$$

That is, the delay T appears in the exponent as a phase angle scaled by $2\pi f$.

2.6 LINEAR TIME INVARIANT SYSTEMS

A linear-time invariant system is characterized by its *impulse response* $h(t)$, or by its *system function* $H(f)$ which is the Fourier transform of $h(t)$. That is,

$$
H(f) = \int_{-\infty}^{\infty} h(t)e^{-j2\pi ft}dt \tag{2.77}
$$

and

$$
h(t) = \int_{-\infty}^{\infty} H(f)e^{j2\pi ft}df \tag{2.78}
$$

If $x(t)$ the applied input signal to the linear time-invariant system is deterministic, as shown in Figure 2.14, the output signal is just the convolution of $x(t)$ and $h(t)$.

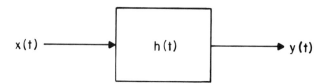

Figure 2.14 Impulse response $h(t)$.

$$
\begin{aligned}
y(t) &= x(t) * h(t) \\
&= \int_{-\infty}^{\infty} x(t-\tau)h(\tau)d\tau
\end{aligned} \tag{2.79}
$$

where $*$ denotes convolution. Taking the Fourier transform of (2.79), the convolution integral becomes a multiplication in the frequency domain. Specifically,

$$
Y(f) = X(f)H(f) \tag{2.80}
$$

where $X(f)$ and $Y(f)$ are the Fourier transforms of $x(t)$ and $y(t)$, respectively.

The system is realizable provided the impulse response is causal; that is, $h(t) = 0$ for $t < 0$. In this case, the convolution integral becomes

$$
\begin{aligned}
y(t) &= \int_{0}^{\infty} x(t-\tau)h(\tau)d\tau \\
&= \int_{-\infty}^{t} x(\tau)h(t-\tau)d\tau
\end{aligned} \tag{2.81}
$$

2.6.1 Stochastic Signals

Consider the linear time-invariant system shown in Figure 2.14 . The output signal is a sample function of the random process $Y(t)$ corresponding to the sample function of the input process $X(t)$. The time domain expression for the output is

$$
\begin{aligned}
Y(t) &= h(t) * X(t) \\
&= X(t) * h(t) \\
&= \int_{-\infty}^{\infty} X(t - \lambda)h(\lambda)d\lambda \\
&= \int_{-\infty}^{\infty} X(\lambda)h(t - \lambda)d\lambda
\end{aligned}
\tag{2.82}
$$

Mean Value Function

The mean value function of the output process is given by

$$
\begin{aligned}
E[Y(t)] &= \int_{-\infty}^{\infty} E[X(t - \lambda)]h(\lambda)d\lambda \\
&= \int_{-\infty}^{\infty} m_X(t - \lambda)h(\lambda)d\lambda
\end{aligned}
\tag{2.83}
$$

where $m_X(t)$ is the mean function of the process $X(t)$. If $X(t)$ is stationary in the wide-sense:

$$
\begin{aligned}
m_X(t - \lambda) &= m_X(t) \\
&= constant
\end{aligned}
\tag{2.84}
$$

Then, the mean function $m_Y(t)$ of the process $Y(t)$ is

$$
\begin{aligned}
m_Y(t) &= E[Y(t)] \\
&= m_X \int_{-\infty}^{\infty} h(\lambda)d\lambda
\end{aligned}
\tag{2.85}
$$

From (2.83), we recall that the system function evaluated at $f = 0$ is just the dc gain and $\int_{-\infty}^{\infty} h(\lambda)d\lambda = H(0)$. Hence,

$$
m_Y = m_X H(0)
\tag{2.86}
$$

The Mean-Square Value

The mean square value of the output process signal is

$$E[Y^2(t)] = E[\int_{-\infty}^{\infty} \int_{-\infty}^{\infty} X(t - t_1)X(t - t_2)h(t_1)h(t_2)dt_1 dt_2] \quad (2.87)$$

Simplifying Equation (2.87), the mean-square value function becomes

$$\begin{aligned} E[Y^2(t)] &= \int_{-\infty}^{\infty} \int_{-\infty}^{\infty} R_{XX}(t - t_1, t - t_2)h(t_1)h(t_2)dt_1 dt_2 \\ &= \int_{-\infty}^{\infty} \int_{-\infty}^{\infty} R_{XX}(t_1, t_2)h(t - t_1)h(t - t_2)dt_1 dt_2 \quad (2.88) \end{aligned}$$

Assuming $X(t)$ is stationary in the wide-sense and making the following change of variables $\alpha = t - t_1$ and $\beta = t - t_2$, the above results reduces to

$$E[Y^2(t)] = \int_{-\infty}^{\infty} \int_{-\infty}^{\infty} R_{XX}(\alpha - \beta)h(\alpha)h(\beta)d\alpha d\beta \quad (2.89)$$

which is independent of the time t.

Cross-Correlation Function between Input and Output

Assume that the input process $X(t)$ is wide-sense stationary. The cross-correlation function between the input and output is

$$R_{YX}(t + \tau, t) = E[Y(t + \tau)X^*(t)] \quad (2.90)$$

Using (2.82) in (2.90) and making a change of variables, the cross-correlation function can be rewritten as

$$\begin{aligned} R_{YX}(t + \tau, t) &= \int_{-\infty}^{\infty} R_{XX}(\tau - \lambda)h(\lambda)d\lambda \\ &= R_{XX}(\tau) * h(\tau) \quad (2.91) \end{aligned}$$

Observe that this result does not depend on t, and hence $R_{YX}(t+\tau, t) = R_{YX}(\tau)$. Similarly, it can be shown that the cross-correlation function between the input and output process signals is

$$R_{XY}(\tau) = R_{XX}(\tau) * h(\tau) \tag{2.92}$$

If the processes $X(t)$ and $Y(t)$ are jointly wide-sense stationary, their cross-spectral density is the Fourier transform of their cross-correlation function. Since a convolution in time domain is equivalent to a multiplication in frequency domain, taking the Fourier transform of (2.91) and (2.92), we obtain

$$S_{YX}(f) = S_{XX}(f)H(f) \tag{2.93}$$

and

$$S_{XY}(f) = S_{XX}(f)H^*(f) \tag{2.94}$$

Autocorrelation and Spectrum of Output

The autocorrelation of the output process is

$$R_{YY}(t + \tau, t) = E[Y(t + \tau)Y(t)] \tag{2.95}$$

Using the fact that

$$Y(t + \tau) = \int_{-\infty}^{\infty} X(t + \tau - \alpha)h(\alpha)d\alpha \tag{2.96}$$

and

$$Y(t) = \int_{-\infty}^{\infty} X(t - \beta)h(\beta)d\beta \tag{2.97}$$

substituting in (2.95) and making a change of variables ($\beta = -\alpha$), we obtain

$$
\begin{aligned}
R_{YY}(\tau) &= R_{YX}(\tau) * h(-\tau) \\
&= R_{XY}(\tau) * h(\tau) \\
&= R_{XX}(\tau) * h(\tau) * h(-\tau)
\end{aligned} \tag{2.98}
$$

Taking the Fourier transforms of the above equations results in

$$
\begin{aligned}
S_{YY}(f) &= S_{YX}(f)H^*(f) \\
&= S_{XX}(f)H(f)H^*(f) \\
&= S_{XX}(f)|H(f)|^2
\end{aligned}
\tag{2.99}
$$

Example 2.8

A white noise process with autocorrelation function $R_{XX}(\tau) = (N_0/2)\delta(\tau)$ is applied to a filter with impulse response

$$
h(t) = \begin{cases} \alpha e^{-\alpha t} & , \ t \geq 0 \text{ and } \alpha > 0 \\ 0 & , \ t < 0 \end{cases}
$$

Determine the autocorrelation function, $R_{YY}(\tau)$, of the output process.

Solution.
The problem can be solved in two ways. We can directly solve the convolution integral of Equation (2.98), or obtain the power spectral density $S_{YY}(f)$ using Equation (2.99) and then take the inverse Fourier transform of $S_{YY}(f)$. We shall solve this problem using both methods.

Method 1:   For $\tau < 0$, we have from Figure 2.15:

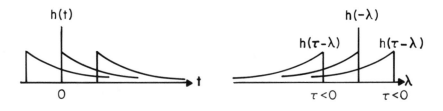

Figure 2.15 Impulse response with τ as a parameter.

$$h(\tau) * h(-\tau) = \int_{-\infty}^{\lambda} \alpha e^{-\alpha(\tau-\lambda)} \alpha e^{\alpha\lambda} d\lambda$$

$$= \alpha^2 e^{-\alpha\tau} \int_{-\infty}^{\tau} e^{2\alpha\lambda} d\lambda$$

$$= \frac{\alpha}{2} e^{\alpha\tau}$$

For $\tau > 0$, we have

$$h(\tau) * h(-\tau) = \int_{-\infty}^{0} \alpha e^{-\alpha(\tau-\lambda)} \alpha e^{\alpha\lambda} d\lambda$$

$$= \frac{\alpha}{2} e^{-\alpha\tau}$$

Hence,

$$g(\tau) = h(\tau) * h(-\tau) = \begin{cases} \frac{\alpha}{2} e^{\alpha\tau} & , \tau \leq 0 \\ \frac{\alpha}{2} e^{-\alpha\tau} & , \tau \geq 0 \end{cases}$$

Consequently,

$$R_{YY}(\tau) = R_{XX}(\tau) * g(\tau) = \begin{cases} \frac{N_0\alpha}{4} e^{\alpha\tau} & , \tau \leq 0 \\ \frac{N_0\alpha}{4} e^{-\alpha\tau} & , \tau \geq 0 \end{cases}$$

or

$$R_{YY}(\tau) = \frac{N_0\alpha}{4} e^{-\alpha|\tau|}$$

Method 2:   From Equation (2.99), we first need to determine the Fourier transform $H(f)$ of the impulse response $h(t)$. Thus,

$$H(f) = \int_{0}^{\infty} \alpha e^{-\alpha t} e^{-j2\pi f t} dt$$

$$= \alpha \int_{0}^{\infty} e^{-(j2\pi f + \alpha)t} dt$$

$$= \frac{\alpha}{j2\pi f + \alpha}$$

The magnitude of $H(f)$ squared is

$$|H(f)|^2 = \frac{\alpha^2}{4\pi^2 f^2 + \alpha^2}$$

while the output power spectral density is

$$S_{YY}(f) = S_{XX}(f)|H(f)|^2 = \frac{N_0\alpha}{4} \cdot \frac{2\alpha^2}{\omega^2 + \alpha^2}$$

where $\omega = 2\pi f$. Taking the inverse Fourier transform of $S_{YY}(f)$, we obtain the autocorrelation, shown in Figure 2.16, to be

$$R_{YY}(\tau) = \frac{N_0\alpha}{4} e^{-\alpha|\tau|}$$

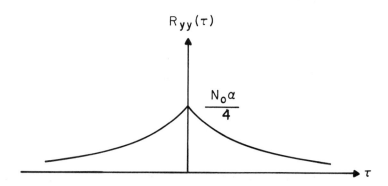

Figure 2.16 Autocorrelation of $Y(t)$.

Both results of Method 1 and Method 2 agree.

2.6.2 Systems with Multiple Terminals

Linear time-invariant systems may have more than one input and/or output. A simple case would be a system with one input and two outputs as shown in Figure 2.17 .

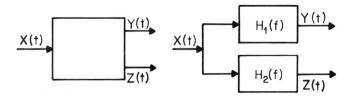

Figure 2.17 System with multiple terminals.

The relationship between the output processes $Y(t)$ and $Z(t)$ may be obtained from (2.99) as

$$S_{YY}(f) = |H_1(f)|^2 S_{XX}(f) \qquad (2.100)$$

and

$$
\begin{aligned}
S_{ZZ}(f) &= |H_2(f)|^2 S_{XX}(f) \\
&= S_{YY}(f) \left| \frac{H_2(f)}{H_1(f)} \right|^2 \qquad H_1(f) \neq 0
\end{aligned}
\qquad (2.101)
$$

In a similar manner, we can obtain the cross-spectral densities of the random processes $Y(t)$ and $Z(t)$ to be

$$S_{ZY}(f) = S_{XY}(f)H_2(f) = S_{XX}(f)H_1^*(f)H_2(f) \qquad (2.102)$$

and

$$S_{YZ}(f) = S_{YX}(f)H_2^*(f) = S_{XX}(f)H_1(f)H_2^*(f) \qquad (2.103)$$

in time domain, we have

$$R_{ZY}(\tau) = R_{XX}(\tau) * h_1(-\tau) * h_2(\tau) \qquad (2.104)$$

and

$$R_{YZ}(\tau) = R_{XX}(\tau) * h_1(\tau) * h_2(-\tau) \qquad (2.105)$$

If $Y(t)$ and $Z(t)$ are orthogonal, then $R_{ZY}(\tau) = R_{YZ}(\tau) = 0$. In this case, the system is said to be *disjoint* because their transfer functions do not overlap; that is, $H_1(f)H_2(f) = 0$.

2.7 ERGODICITY

A random process $X(t)$ is ergodic if all of its statistics can be determined (with probability one) from a sample function of the process. That is, the ensemble averages equal the corresponding time averages with probability one.

Usually, we are not interested in estimating all the ansemble average of a random process, but rather we are concerned with a weaker form of ergodicity, such as ergodicity in the mean, and ergodicity in the autocorrelation.

2.7.1 Ergodicity in the Mean

A random process $X(t)$ is *ergodic in the mean* if the time-averaged mean value of a sample function $x(t)$ is equal to the ensemble-averaged mean value function. That is,

$$E[X(t)] = < x(t) > \qquad (2.106)$$

where the symbol $< \cdot >$ denotes time-average and $< x(t) >$ is defined to be

$$< x(t) > = \lim_{T \to \infty} \frac{1}{2T} \int_{-T}^{T} x(t)dt \qquad (2.107)$$

The necessary and sufficient condition, under which the process $X(t)$ is ergodic in the mean, is

$$\lim_{T \to \infty} \frac{1}{2T} \int_{-T}^{T} R_{XX}(\tau)d\tau = m_X^2 \qquad (2.108)$$

where m_X is the mean value of $X(t)$.

2.7.2 Ergodicity in the Autocorrelation

The random process $X(t)$ is *ergodic in the autocorrelation* if

$$R_{XX}(\tau) = <x(t+\tau)x(t)> \tag{2.109}$$

$<x(t+\tau)x(t)>$ denotes the time-averaged autocorrelation function of the sample function $x(t)$, and is defined as

$$<x(t+\tau)x(t)> = \lim_{T\to\infty} \frac{1}{2T} \int_{-T}^{T} x(t+\tau)x(t)dt \tag{2.110}$$

The necessary and sufficient condition for ergodicity in the autocorrelation is that the random variables $X(t+\tau)X(t)$ and $X(t+\tau+\lambda)X(t+\lambda)$ become uncorrelated as λ approaches infinity, and for each τ.

Example 2.9

Consider the random process $X(t) = A\cos(2\pi f_c t + \Theta)$ where A and f_c are constants, and Θ is a random variable uniformly distributed over the interval $[0, 2\pi]$.

Solution.
It was shown in Example 2.2 that the mean and autocorrelation functions of $X(t)$ are $E[X(t)] = 0$ and $R_{XX}(\tau) = (A^2/2)\cos(2\pi f_c \tau)$. Let the sample function of the process $X(t)$ be

$$x(t) = A\cos(2\pi f_c t + \theta)$$

The time-averaged mean and the time-averaged autocorrelation are

$$<x(t)> = \lim_{T\to\infty} \frac{A}{2T} \int_{-T}^{T} \cos(2\pi f_c t + \theta)dt = 0$$

and

$$
\begin{aligned}
<x(t+\tau)x(t)> &= \lim_{T\to\infty} \frac{A^2}{2T} \int_{-T}^{T} \cos[2\pi f_c(t+\tau)+\theta]\cos(2\pi f_c t+\theta)dt \\
&= \frac{A^2}{2}\cos(2\pi f_c \tau)
\end{aligned}
$$

Hence, the process $X(t)$ is ergodic in the mean and in the autocorrelation.

2.7.3 Ergodicity of the First-Order Distribution

Let $X(t)$ be a stationary random process. Define the random process $Y(t)$ as

$$Y(t) = \begin{cases} 1 & , \ X(t) \leq x_t \\ 0 & , \ X(t) > x_t \end{cases} \tag{2.111}$$

We say that the random process $X(t)$ is *ergodic in the first-order distribution* if

$$F_X(x;t) = \lim_{T \to \infty} \frac{1}{2T} \int_{-T}^{T} y(t) dt \tag{2.112}$$

where $F_X(x;t) = P\{X(t) \leq x(t)\}$ and $y(t)$ is a sample function of the process $Y(t)$.

The necessary and sufficient condition, under which the process is ergodic in the first order distribution, is that $X(t+\tau)$ and $X(t)$ become statistically independent as τ approaches infinity.

2.7.4 Ergodicity of Power Spectral Density

A wide-sense stationary process $X(t)$ is ergodic in the power spectral density if, for any sample function $x(t)$,

$$S_{XX}(f) = \lim_{T \to \infty} \frac{1}{2T} \left| \int_{-T}^{T} x(t) e^{-j2\pi f t} dt \right|^2 \tag{2.113}$$

except for a set of sample functions which occur with zero probability.

2.8 SAMPLING THEOREM

The sampling theorem for deterministic bandlimited signals of finite energy states that:

a bandlimited signal of finite energy with no frequency higher than f_m hertz may be completely recovered from its samples taken at the rate of $2f_m$ per second.

Specifically, if $x(t)$ is a deterministic signal and bandlimited to the frequency range $(-f_m, f_m)$, then

$$x(t) = \sum_{n=-\infty}^{\infty} x(nT) \, \text{sinc}(\frac{t}{T} - n) \qquad (2.114)$$

where $T = 1/2f_m$.

An analogous sampling theorem may be stated for random processes.

Theorem: Let $X(t)$ be a wide-sense stationary random process bandlimited to the frequency $(-f_m, f_m)$; that is, $S_{XX}(f) = 0$ for $|f| > f_m$ Then,

$$X(t) = \sum_{n=-\infty}^{\infty} X(nT) \, \text{sinc}(\frac{t}{T} - n) \qquad (2.115)$$

where $T = 1/2f_m$.

PROBLEMS

2.1 Consider a random process $X(t)$ defined by

$$X(t) = A \cos(\omega_0 t + \Theta)$$

where A and ω_0 are constants, and Θ is a random variable with probability density function

$$f_\Theta(\theta) = \begin{cases} \frac{4}{\pi} & , |\theta| \leq \frac{\pi}{8} \\ 0 & , \text{otherwise} \end{cases}$$

(a) Find the mean and autocorrelation functions.
(b) Is the process stationary ?

2.2 Let $s(t) = \text{rect}(t)$. Define the process $X(t) = S(t - T_0)$ where T_0 is a discrete random variable taking values 0 and 1 with equal probability.
(a) Determine and sketch the distribution function $F_{X_t}(x_t, 0)$.
(b) Determine and sketch the autocorrelation function $R_{XX}(t_1, t_2)$.

2.3 Consider the random process defined in Problem 2.1 .
(a) Is the process ergodic in the mean ?
(b) Is the process ergodic in the autocorrelation ?

2.4 Consider the random process defined by

$$X(t) = A \cos(\omega_0 t + \Theta)$$

where A and ω_0 are constants and Θ is a random variable uniformly distributed over the interval $(0, 2\pi)$. Let $Y(t)$ be a random process defined as $Y(t) = X^2(t)$.
(a) Find the autocorrelation function of $Y(t)$.
(b) Is $Y(t)$ a stationary process ?

2.5 Consider the random process defined by

$$X(t) = A \, e^{j(\omega t + \Theta)}$$

where A is a random variable with density function

$$f_A(a) = \begin{cases} \frac{a}{\sigma^2} \, e^{-\frac{a^2}{2\sigma^2}} & , a \geq 0 \\ 0 & , \text{otherwise} \end{cases}$$

σ^2 is constant and Θ is a random variable uniformly distributed over the interval $(0, 2\pi)$. A and Θ are statistically independent. Determine
(a) $E[X(t)]$
(b) $R_{XX}(t_1, t_2)$

2.6 Let $X(t)$ be the random process shown in Figure $P2.6$. The square wave is periodic with period T. The amplitude A is random with zero mean and variance σ^2. t_0 is governed by a random variable T_0 which is uniformly distributed over the interval $(0, T)$. A and T_0 are statistically independent. Determine the autocorrelation function $R_{XX}(t_1, t_2)$.

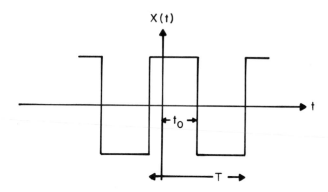

Figure P2.6 Random process $X(t)$.

2.7 Let $s(t)$ be the periodic deterministic waveform shown in Figure P2.7. Define the random process $X(t) = S(t - T_0)$ where T_0 is a random variable uniformly distributed over the interval $(0, T)$.
(a) Find the autocorrelation function $R_{XX}(t_1, t_2)$. Is the process $X(t)$ stationary in the wide sense?
(b) Determine and sketch the distribution function $F_{X_t}(x_t)$.
(c) Determine and sketch the density function $f_{X_t}(x_t)$.
(d) Find $E[X(t)], E[X^2(t)]$, and $\sigma_{X_t}^2$.
(e) Find $< x(t) >$ and $< x^2(t) >$.

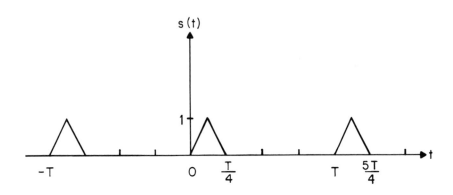

Figure P2.7 Deterministic signal $s(t)$.

2.8 Let $X(t)$ be a wide-sense stationary process with autocorrelation function

$$R_{XX}(\tau) = \begin{cases} 1 - |\tau| & , |\tau| < 1 \\ 0 & , \text{otherwise} \end{cases}$$

be applied to the system shown in Figure P2.8. Determine and sketch the output autocorrelation $R_{YY}(\tau)$.

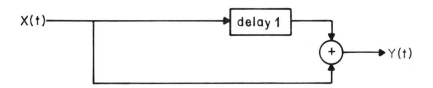

Figure P2.8 System function for X(t).

2.9 Consider two orthogonal processes $X(t)$ and $Y(t)$ with power spectral densities

$$S_{XX}(f) = S_{YY}(f) = \begin{cases} 1 - |f| & , |f| < 1 \\ 0 & , \text{otherwise} \end{cases}$$

Define a new process $Z(t) = Y(t) - X(t-1)$. Determine and sketch the power spectral density $S_{ZZ}(f)$.

2.10 Let $X(t)$ be the input process to a linear system with impulse response

$$h(t) = \begin{cases} e^{-t} & , t \geq 0 \\ 0 & , \text{otherwise} \end{cases}$$

$X(t)$ is stationary in the wide sense with autocorrelation

$$R_{XX}(\tau) = \begin{cases} 1 - |\tau| & , |\tau| \leq 1 \\ 0 & , \text{otherwise} \end{cases}$$

Determine the autocorrelation function of the output process $R_{YY}(\tau)$.

2.11 The random process $X(t)$ with autocorrelation function

$$R_{XX}(\tau) = e^{-\alpha|\tau|} \quad \alpha \text{ is constant}$$

is applied to the RC filter shown in Figure $P2.11$. Determine the output power spectral density $S_{YY}(f)$.

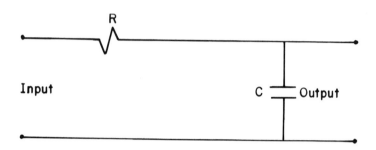

Figure P2.11 RC filter.

2.12 Let $X(t)$ be the input process to the RLC network shown in Figure P2.12. $X(t)$ is a wide-sense stationary process with mean $m_X(t) = 2$ and autocorrelation function

$$R_{XX}(\tau) = 4 + e^{-2|\tau|}$$

Find the mean $m_Y(t)$ of the output process and the output power spectral density $S_{YY}(f)$.

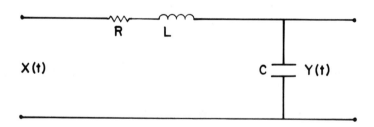

Figure P2.12 RLC filter.

2.13 Let $X(t)$ be a stationary random process with autocorrelation function $R_{XX}(\tau)$. Define the stochastic process

$$Y(t) = \int_0^t X(\tau)d\tau$$

Is the process $Y(t)$ stationary given that

$$R_{XX}(\tau) = 4 + e^{-2|\tau|}$$

2.14 Let $Y(t)$ be the process defined in Problem 2.13. Obtain the autocorrelation of $Y(t)$ when $R_{XX}(t_1, t_2) = 2\delta(t_1 - t_2)$.

2.15 Let $X(t)$ be a wide-sense stationary process with power spectral density

$$S_{XX}(f) = \begin{cases} 1 - |f| & , |f| < 1 \\ 0 & , \text{otherwise} \end{cases}$$

and sampled at the Nyquist rate.
(a) What is the interval between samples ?
(b) Determine the correlation coefficient between the samples $X(nT)$ and $X\{(n+1)T\}$; n arbitrary.

REFERENCES

1. Gray, R.M., and L.D. Davisson, *Random Processes: A Mathematical Approach for Engineers,* Prentice-Hall, Englewood Cliffs, NJ, 1986.

2. Haykin, S., *Communication Systems,* John Wiley and Sons, New York, 1983.

3. Helstron, C.W., *Probability and Stochastic Processes for Engineers,* Macmillan, New York, 1984.

4. Papoulis, A., *Probability, Random Variables, and Stochastic Processes,* McGraw-Hill, New York, 1986.

5. Peebles, P.Z., *Probability, Random Variables, and Random Signal Principles,* McGraw-Hill, New York, 1980.

6. Stark, H., and J.W. Woods, *Probability, Random Processes, and Estimation Theory for Engineers,* Prentice-Hall, Englewood Cliffs, NJ, 1986.

7. Urkowitz, H., *Signal Theory and Random Processes,* Artech House, Norwood, MA, 1983.

8. Wong, E., *Introduction to Random Processes,* Springer-Verlag, New York, 1983.

9. Wozencraft, J.M., and I.M. Jacobs, *Principles of Communication Engineering,* John Wiley and Sons, New York, 1965.

Chapter 3

Signal Detection

3.1 INTRODUCTION

In our daily life, we are constantly making decisions. Given some hypotheses, a criterion is selected upon which a decision has to be made. For example, in engineering, when there is a radar signal detection problem, the returned signal is observed and a decision is made as to whether a target is present or absent. In a digital communication system, a string of zeros and ones may be transmitted over some medium. At the receiver, the received signals representing the zeros and ones are corrupted in the medium by some additive noise and by the receiver noise. The receiver does not know which signal represents a zero and which signal represents a one, but must make a decision as to whether the received signals represent zeros or ones. The process that the receiver undertakes in selecting a decision rule falls under the theory of signal detection.

The situation above may be described by a source emitting two possible outputs at various instants of time. The outputs are referred to as *hypotheses*. The *null hypothesis* H_0 represents a zero (target not present) while the *alternate hypothesis* H_1 represents a one (target present) as shown in Figure 3.1.

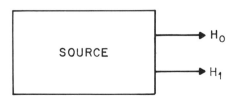

Figure 3.1 Source for binary hypothesis.

Each hypothesis corresponds to one or more observations which are represented by random variables. Based on the observed values of these random variables, the receiver decides as to which hypothesis (H_0 or H_1) is true. Assume that the receiver is to make a decision based on a single observation of the received signal. The range of values that the random variable Y takes constitute the observation space Z. The observation space is partitioned into two regions Z_0 and Z_1 such that if Y lies in Z_0 the receiver decides in favor of H_0 while if Y lies in Z_1 the receiver decides in favor of H_1 as shown in Figure 3.2 . The observation space Z is the union of Z_0 and Z_1; that is,

$$Z = Z_0 \cup Z_1 \qquad (3.1)$$

The probability density functions of Y corresponding to each hypothesis are $f_{Y|H_0}(y|H_0)$ and $f_{Y|H_1}(y|H_1)$ where y is a particular value of the random variable Y.

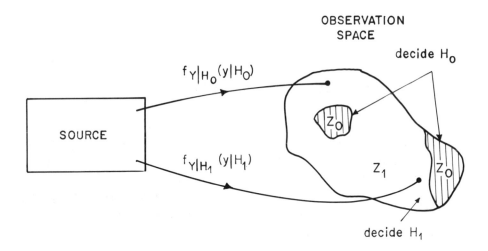

Figure 3.2 Decision region.

Each time a decision is made, based on some criterion, for this binary hypothesis testing problem, four possible cases can occur:

(i)   Decide H_0 when H_0 is true.
(ii) Decide H_0 when H_1 is true.
(iii) Decide H_1 when H_0 is true.
(iv) Decide H_1 when H_1 is true.

Observe that for cases (i) and (iv) the receiver makes a correct decision, while for cases (ii) and (iii) the receiver makes an error. From radar nomenclature, case (ii) is called *miss*, case (iii) a *false alarm*, and case (iv) a *detection*.

In this chapter, we develop the basic and necessary principles needed for solving decision problems. The observations are represented by random variables. Extension of these results to time varying waveforms will be studied in later chapters.

In the next sections, we study some of the criteria that are used in decision theory, and the conditions under which these criteria are useful.

3.2 BAYES CRITERION

3.2.1 Binary Hypothesis Testing

In using Bayes criterion, two assumptions are made. First, the probability of occurrence of the two source outputs is known. They are the *a priori* probabilities $P(H_0)$ and $P(H_1)$. $P(H_0)$ is the probability of occurrence of hypothesis H_0 while $P(H_1)$ is the probability of occurrence of hypothesis H_1. Denoting the *a priori* probabilities $P(H_0)$ and $P(H_1)$ by P_0 and P_1 respectively, and since either hypothesis H_0 or H_1 will always occur, we have

$$P_0 + P_1 = 1 \tag{3.2}$$

The second assumption is that a cost is assigned to each possible decision. The consequences of one decision are different from the consequences of another. For example, in a radar detection problem the consequences of miss are not the same as the consequences of a false alarm. If we let D_i, $i = 0, 1$, where D_0 denotes "decide H_0" and D_1 denotes "decide H_1", we can define C_{ij}, $i, j = 0, 1$, as the cost associated with the decision D_i given that the true hypothesis is H_j.

In particular, the cost for this binary hypothesis testing problem are C_{00} for case (i), C_{01} for case (ii), C_{10} for case (iii), and C_{11} for case (iv). The goal in the Bayes criterion is to determine the decision rule so that the average cost $E[\mathcal{C}]$, also known as *risk R*, is minimized. The operation $E[\mathcal{C}]$ denotes expected value. It is also assumed that the cost of making a wrong decision is greater than the cost of making a correct decision. That is,

$$C_{01} > C_{11} \tag{3.3a}$$

and

$$C_{10} > C_{00} \qquad (3.3b)$$

Given $P(D_i, H_j)$, the joint probability that we decide D_i and that the hypothesis H_j is true, the average cost is

$$
\begin{aligned}
R &= E[\mathcal{C}] \\
&= C_{00}P(D_0, H_0) + C_{01}P(D_0, H_1) + C_{10}P(D_1, H_0) + C_{11}P(D_1, H_1)
\end{aligned}
$$
$$(3.4)$$

From Bayes rule, we have

$$P(D_i, H_j) = P(D_i|H_j)P(H_j) \qquad (3.5)$$

The conditional density functions $P(D_i|H_j)$, $i, j = 0, 1$, in terms of the regions shown in Figure 3.2 are

$$P(D_0|H_0) \equiv P(\text{decide } H_0|H_0 \text{ true}) = \int_{Z_0} f_{Y|H_0}(y|H_0)dy \qquad (3.6)$$

$$P(D_0|H_1) \equiv P(\text{decide } H_0|H_1 \text{ true}) = \int_{Z_0} f_{Y|H_1}(y|H_1)dy \qquad (3.7)$$

$$P(D_1|H_0) \equiv P(\text{decide } H_1|H_0 \text{ true}) = \int_{Z_1} f_{Y|H_0}(y|H_0)dy \qquad (3.8)$$

and

$$P(D_1|H_1) \equiv P(\text{decide } H_1|H_1 \text{ true}) = \int_{Z_1} f_{Y|H_1}(y|H_1)dy \qquad (3.9)$$

The probabilities $P(D_0|H_1)$, $P(D_1|H_0)$ and $P(D_1|H_1)$ represent the probability of miss, P_M, the probability of false alarm, P_F, and the probability of detection, P_D, respectively. We also observe that

$$P_M = 1 - P_D \qquad (3.10)$$

and

$$P(D_0|H_0) = 1 - P_F \qquad (3.11)$$

Consequently, the probability of a correct decision is given by

$$
\begin{aligned}
P(\text{correct decision}) = P(C) &= P(D_0, H_0) + P(D_1, H_1) \\
&= P(D_0|H_0)P(H_0) + P(D_1|H_1)P(H_1) \\
&= (1 - P_F)P_0 + P_D P_1 \tag{3.12}
\end{aligned}
$$

and the probability of error is given by

$$
\begin{aligned}
P(error) = P(\varepsilon) &= P(D_0, H_1) + P(D_1, H_0) \\
&= P(D_0|H_1)P(H_1) + P(D_1|H_0)P(H_0) \\
&= P_M P_1 + P_F P_0 \tag{3.13}
\end{aligned}
$$

The average cost now becomes

$$
R = E[\mathcal{C}] = C_{00}(1-P_F)P_0 + C_{01}(1-P_D)P_1 + C_{10}P_F P_0 + C_{11}P_D P_1 \tag{3.14}
$$

In terms of the decision regions defined in Equations (3.6) to (3.9), the average cost is

$$
\begin{aligned}
R =\ & P_0 C_{00} \int_{Z_0} f_{Y|H_0}(y|H_0)dy \\
& + P_1 C_{01} \int_{Z_0} f_{Y|H_1}(y|H_1)dy \\
& + P_0 C_{10} \int_{Z_1} f_{Y|H_0}(y|H_0)dy \\
& + P_1 C_{11} \int_{Z_1} f_{Y|H_1}(y|H_1)dy \tag{3.15}
\end{aligned}
$$

Using Equation (3.1) and the fact that

$$
\int_Z f_{Y|H_0}(y|H_0)dy = \int_Z f_{Y|H_1}(y|H_1)dy = 1 \tag{3.16}
$$

it follows that

$$
\int_{Z_1} f_{Y|H_j}(y|H_j)dy = 1 - \int_{Z_0} f_{Y|H_j}(y|H_j)dy \ ; \ j = 0,1 \tag{3.17}
$$

where $f_{Y|H_j}(y|H_j)$, $j = 0,1$, is the probability density function of Y corresponding to each hypothesis. Substituting for Equation (3.17) in Equation (3.15), we obtain

$$
\begin{aligned}
R = \ & P_0 C_{10} + P_1 C_{11} \\
& + \int_{Z_0} \{ [P_1(C_{01} - C_{11}) f_{Y|H_1}(y|H_1)] \\
& - [P_0(C_{10} - C_{00}) f_{Y|H_0}(y|H_0)] \} dy
\end{aligned}
$$
(3.18)

We observe that the quantity $P_0 C_{10} + P_1 C_{11}$ is constant, independent of how we assign points in the observation space, and that the only variable quantity is the region of integration Z_0. From Equations (3.3a) and (3.3b), the terms inside the brackets of Equation (3.18), $P_1(C_{01} - C_{11}) f_{Y|H_1}(y|H_1)$ and $P_0(C_{10} - C_{00}) f_{Y|H_0}(y|H_0)$, are both positive. Consequently, the risk is minimized by selecting the decision region Z_0 to include only those points of Y for which the second term is larger, and hence, the integrand is negative. Specifically, we assign to the region Z_0 those points for which:

$$
P_1(C_{01} - C_{11}) f_{Y|H_1}(y|H_1) < P_0(C_{10} - C_{00}) f_{Y|H_0}(y|H_0)
$$
(3.19)

All values for which the second term is greater will be excluded from Z_0 and assigned to Z_1. The values for which the two terms are equal do not affect the risk and can be assigned to either Z_0 or Z_1. Consequently, we say if

$$
P_1(C_{01} - C_{11}) f_{Y|H_1}(y|H_1) > P_0(C_{10} - C_{00}) f_{Y|H_0}(y|H_0)
$$
(3.20)

decide H_1. Otherwise, decide H_0. Hence, the decision rule resulting from the Bayes criterion is

$$
\frac{f_{Y|H_1}(y|H_1)}{f_{Y|H_0}(y|H_0)} \underset{H_0}{\overset{H_1}{\gtrless}} \frac{P_0(C_{10} - C_{00})}{P_1(C_{01} - C_{11})}
$$
(3.21)

The ratio of $f_{Y|H_1}(y|H_1)$ over $f_{Y|H_0}(y|H_0)$ is called the *likelihood ratio* and is denoted by $\Lambda(y)$. That is,

$$\Lambda(y) = \frac{f_{Y|H_1}(y|H_1)}{f_{Y|H_0}(y|H_0)} \tag{3.22}$$

It should be noted that if we have K observations, say K samples of a received waveform, $Y_1, Y_2, \cdots, Y_K$, based on which we make the decision, the likelihood ratio can be expressed as

$$\Lambda(\mathbf{y}) = \frac{f_{\mathbf{Y}|H_1}(\mathbf{y}|H_1)}{f_{\mathbf{Y}|H_0}(\mathbf{y}|H_0)} \tag{3.23}$$

where $\mathbf{Y}$, the received vector, is

$$\mathbf{Y} = \begin{bmatrix} Y_1 \\ Y_2 \\ \vdots \\ Y_K \end{bmatrix} \tag{3.24}$$

The *likelihood statistic* $\Lambda(\mathbf{Y})$ is a random variable since it is a function of the random variable $\mathbf{Y}$.

The threshold is

$$\eta = \frac{P_0(C_{10} - C_{00})}{P_1(C_{01} - C_{11})} \tag{3.25}$$

Therefore, Bayes criterion, which minimizes the average cost results in the likelihood ratio test

$$\Lambda(\mathbf{y}) \underset{H_0}{\overset{H_1}{\underset{<}{\overset{>}{}}}} \eta \tag{3.26}$$

An important observation is that the likelihood ratio test is performed by simply processing the receiving vector to yield the likelihood ratio and comparing it with the threshold. Thus, in practical situations where the *a priori* probabilities and the cost may change, only the threshold changes but the computation of the likelihood ratio is not affected.

Because the natural logarithm is a monotonically increasing function as shown in Figure 3.3, and the likelihood ratio $\Lambda(\mathbf{y})$ and the threshold η are nonnegative, an equivalent decision rule to (3.26) is

$$\ln\Lambda(\mathbf{y}) \underset{H_0}{\overset{H_1}{\underset{<}{\gtrless}}} \ln\eta \qquad (3.27)$$

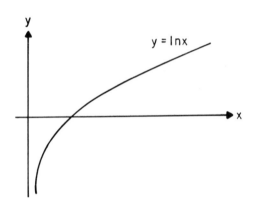

Figure 3.3 Natural logarithmic function.

We note that if we select the cost of an error to be one and the cost of a correct decision to be zero; that is,

$$C_{01} = C_{10} = 1 \qquad (3.28a)$$

and

$$C_{00} = C_{11} = 0 \qquad (3.28b)$$

The risk function of (3.14) reduces to

$$R = P_M P_1 + P_F P_0 = P(\varepsilon) \tag{3.29}$$

Thus, in this case, minimizing the average cost is equivalent to minimizing the probability of error. Receivers for such cost assignment are called *minimum probability of error receivers*. Also, the threshold reduces to

$$\eta = \frac{P_0}{P_1} \tag{3.30}$$

If the *a priori* probabilities are equal, η is equal to one and the log likelihood ratio test uses a zero threshold.

Example 3.1

In a digital communication system, consider a source whose output under hypothesis H_1 is a constant voltage of value m while its output under hypothesis H_0 is zero. The received signal is corrupted by N an additive white Gaussian noise of zero mean and variance σ^2. (a) Set up the likelihood ratio test and determine the decision regions. (b) Calculate the probability of false alarm and the probability of detection.

Solution.
The received signal under each hypothesis is

$$H_1 : Y = m \quad + \quad N$$
$$H_0 : Y = \qquad\quad N$$

where the noise N is Gaussian with zero mean and variance σ^2. Under hypothesis H_0,

$$f_{Y|H_0}(y|H_0) = f_N(n) = \frac{1}{\sqrt{2\pi}\sigma} e^{-\frac{y^2}{2\sigma^2}}$$

Under hypothesis H_1, the mean of Y is $E[Y] = E[m+N] = m+E[N] = m$ since $E[N] = 0$. The variance of Y is

$$\text{var}(Y) = \text{var}(m+N) = E[(m+N)^2] - (E[m+N])^2$$
$$= E[N^2] = \sigma^2$$

Hence,

$$f_{Y|H_1}(y|H_1) = \frac{1}{\sqrt{2\pi}\sigma} e^{-\frac{1}{2}\frac{(y-m)^2}{\sigma^2}}$$

The likelihood ratio test is

$$\Lambda(y) = \frac{f_{Y|H_1}(y|H_1)}{f_{Y|H_0}(y|H_0)} = e^{-\frac{m^2 - 2ym}{2\sigma^2}}$$

Taking the natural logarithm on both sides of the above equation, the likelihood ratio test becomes

$$\ln\Lambda(y) = \frac{m}{\sigma^2}y - \frac{m^2}{2\sigma^2} \mathop{\gtrless}\limits_{H_0}^{H_1} \ln\eta$$

rearranging terms, an equivalent test is

$$y \mathop{\gtrless}\limits_{H_0}^{H_1} \frac{\sigma^2}{m}\ln\eta + \frac{m}{2} = \gamma$$

That is, the received observation is compared with the threshold γ. The decision regions are as shown in Figure 3.4.

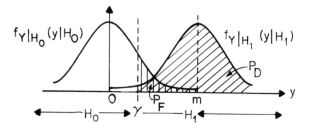

Figure 3.4 Decision regions.

(b) The probabilities of false alarm and detection are

$$P_F = P(\text{decide } H_1 | H_0 \text{ true}) = \int_\gamma^\infty \frac{1}{\sqrt{2\pi}\sigma} e^{-\frac{y^2}{2\sigma^2}} dy$$

$$= Q(\frac{\gamma}{\sigma})$$

where

$$Q(\alpha) = \int_\alpha^\infty \frac{1}{\sqrt{2\pi}} e^{-\frac{u^2}{2}} du$$

$$P_D = P(\text{decide } H_1 | H_1 \text{ true}) = \int_\gamma^\infty \frac{1}{\sqrt{2\pi}\sigma} e^{-\frac{(y-m)^2}{2\sigma^2}} dy$$

$$= Q(\frac{\gamma - m}{\sigma})$$

Example 3.2

Suppose that the receiver of Example 3.1 takes K samples, $Y_1, Y_2, \cdots,$ Y_K. The noise samples are independent Gaussian random variables, each with mean zero and variance σ^2. Obtain the optimum decision rule.

Solution.

The received signal under hypotheses H_0 and H_1 is

$$H_1 : Y_k = m \quad + \quad N_k, \quad k = 1, 2, \cdots, K$$
$$H_0 : Y_k = \qquad\quad N_k, \quad k = 1, 2, \cdots, K$$

Under hypothesis H_0

$$f_{Y_k|H_0}(y_k|H_0) = f_{N_k}(y_k) = \frac{1}{\sqrt{2\pi}\sigma} e^{-\frac{y_k^2}{2\sigma^2}}$$

Under hypothesis H_1, the kth received sample is a Gaussian random variable with mean m and variance σ^2. Thus,

$$f_{Y_k|H_1}(y_k|H_1) = f_{N_k}(y_k - m) = \frac{1}{\sqrt{2\pi}\sigma} e^{-\frac{(y_k-m)^2}{2\sigma^2}}$$

From (3.23) we need $f_{\mathbf{Y}|H_1}(\mathbf{y}|H_1)$ and $f_{\mathbf{Y}|H_0}(\mathbf{y}|H_0)$. Since the noise samples are statistically independent, the joint density function of the K samples is just the product of the individual density functions. This yields

$$f_{\mathbf{Y}|H_0}(\mathbf{y}|H_0) = \prod_{k=1}^{K} \frac{1}{\sqrt{2\pi}\sigma} e^{-\frac{y_k^2}{2\sigma^2}}$$

and

$$f_{\mathbf{Y}|H_1}(\mathbf{y}|H_1) = \prod_{k=1}^{K} \frac{1}{\sqrt{2\pi}\sigma} e^{-\frac{(y_k-m)^2}{2\sigma^2}}$$

where $\prod$ denotes product.

Using the fact that $\prod_k e^{x_k} = e^{\sum_k x_k}$, the likelihood ratio test is

$$\Lambda(\mathbf{y}) = e^{[\sum_{k=1}^{K} \frac{y_k^2}{2\sigma^2} - \sum_{k=1}^{K} \frac{(y_k-m)^2}{2\sigma^2}]}$$
$$= e^{[\frac{m}{\sigma^2} \sum_{k=1}^{K} y_k - \frac{Km^2}{2\sigma^2}]}$$

Taking the natural logarithm of both sides, the likelihood ratio test becomes

$$\ln\Lambda(\mathbf{y}) = \frac{m}{\sigma^2}\sum_{k=1}^{K} y_k - \frac{Km^2}{2\sigma^2} \underset{H_0}{\overset{H_1}{\gtrless}} \ln\eta$$

Rearranging terms, an equivalent is

$$\sum_{k=1}^{K} y_k \underset{H_0}{\overset{H_1}{\gtrless}} \frac{\sigma^2}{m}\ln\eta + \frac{Km}{2}$$

That is, the receiver adds the K samples and compares them to the threshold $(\sigma^2/m)\ln\eta + (Km/2)$.

Sufficient Statistics

A statistic is any random variable which can be computed from observed data. Let T be the value of a statistic given by $T = t(x)$. Let T' be the value of another statistic, and that T and T' have a joint density function given by $f(x, y|\theta)$. Then,

$$f(x,y|\theta) = f_T(x|\theta)f_{T'}(y|x,\theta) \qquad (3.31)$$

where $f_T(x|\theta)$ is the probability density function of T, and $f_{T'}(y|x,\theta)$ is the conditional density function of T' given $T = x$. Note that in (3.31) we have used the fact that $P(A \text{ and } B) = P(A)P(B|A)$. Assume the conditional density function $f_{T'}(y|x,\theta)$ does not involve θ. Then, if T is known, the conditional density function of T' does not depend on

θ and T' is not relevant in the decision making problem. This can be shown to be the case of all T' for all data. Consequently, T summarizes all the data of the experiment relevant to θ and is called a *sufficient statistic*.

Example 3.3

In Example 3.2, we observe that only knowledge of sum $\sum_{k=1}^{K} y_k$ is relevant in making a decision about **Y**. Hence, $T(\mathbf{y}) = \sum_{k=1}^{K} y_k$ is a sufficient statistic.

Example 3.4

Consider the situation where the samples $Y_1, Y_2, \cdots, Y_K$ are independent random variables each having a Bernoulli distribution with parameter p. Assume that the test statistic is

$$T(y) = \sum_{k=1}^{K} Y_k$$

Is $T(y)$ a sufficient statistic ?

Solution.
A random variable Y is said to have a Bernoulli distribution with parameter p if

$$f_Y(y; p) = p^y (1 - p)^{1-y} \; ; y = 0, 1$$

where $0 \le p \le 1$. Since the random variables $Y_1, Y_2, \cdots, Y_k$ are statistically independent, the joint density function

$$
\begin{aligned}
f_{\mathbf{Y}}(\mathbf{y}; p) &= [p^{y_1}(1 - p)^{1-y_1}][p^{y_2}(1 - p)^{1-y_2}] \cdots [p^{y_K}(1 - p)^{1-y_K}] \\
&= p^{\sum_{k=1}^{K} y_k} (1 - p)^{K - \sum_{k=1}^{K} y_k}
\end{aligned}
$$

That is, the joint density function of the sample values does not involve the parameter p and depends only through the sum $T(y) = \sum_{k=1}^{K} y_k$. Hence, $T(y)$ is a sufficient statistic.

Example 3.5

Consider the problem where the conditional density functions under each hypothesis are

$$f_{Y|H_0}(y|H_0) = \frac{1}{\sqrt{2\pi}\sigma_0} e^{-\frac{y^2}{2\sigma_0^2}}$$

$$f_{Y|H_1}(y|H_1) = \frac{1}{\sqrt{2\pi}\sigma_1} e^{-\frac{y^2}{2\sigma_1^2}}$$

where $\sigma_1^2 > \sigma_0^2$. Determine the decision rule.

Solution.

Applying the likelihood ratio test given in Equation (3.22), we obtain

$$\Lambda(y) = \frac{\frac{1}{\sqrt{2\pi}\sigma_1} e^{-\frac{y^2}{2\sigma_1^2}}}{\frac{1}{\sqrt{2\pi}\sigma_0} e^{-\frac{y^2}{2\sigma_0^2}}} \underset{H_0}{\overset{H_1}{\gtrless}} \eta$$

or

$$\frac{\sigma_0}{\sigma_1} e^{\frac{y^2}{2}\left(\frac{1}{\sigma_0^2} - \frac{1}{\sigma_1^2}\right)} \underset{H_0}{\overset{H_1}{\gtrless}} \eta$$

Taking the logarithm on both sides, we have

$$\ln\frac{\sigma_0}{\sigma_1} + \frac{y^2}{2}\left(\frac{1}{\sigma_0^2} - \frac{1}{\sigma_1^2}\right) \underset{H_0}{\overset{H_1}{\gtrless}} \ln\eta$$

or

$$y^2 \begin{array}{c} H_1 \\ \gtrless \\ H_0 \end{array} \frac{2\sigma_0^2\sigma_1^2}{\sigma_1^2 - \sigma_0^2} \ln\frac{\eta\sigma_1}{\sigma_0} = \gamma$$

$T(y) = y^2$ is the sufficient statistic and hence, the test can be written as

$$T(y) \begin{array}{c} H_1 \\ \gtrless \\ H_0 \end{array} \gamma$$

Example 3.6

Redo Example 3.5 assuming that we have K independent observations.

Solution.
Since the random variables $Y_1, Y_2, \cdots, Y_K$, are independent, the joint density function is simply the product of the individual densities. That is,

$$f_{\mathbf{Y}|H_0}(\mathbf{Y}|H_0) = \prod_{k=1}^{K} \frac{1}{\sqrt{2\pi}\sigma_0} e^{-\frac{y_k^2}{2\sigma_0^2}}$$

$$f_{\mathbf{Y}|H_1}(\mathbf{y}|H_1) = \prod_{k=1}^{K} \frac{1}{\sqrt{2\pi}\sigma_1} e^{-\frac{y_k^2}{2\sigma_1^2}}$$

Substituting in Equation (3.23) and taking the logarithm, we have

$$\frac{1}{2}(\frac{1}{\sigma_0^2} - \frac{1}{\sigma_1^2}) \sum_{k=1}^{K} y_k^2 + K \ln\frac{\sigma_0}{\sigma_1} \begin{array}{c} H_1 \\ \gtrless \\ H_0 \end{array} \ln\eta$$

or

$$\sum_{k=1}^{K} y_k^2 \underset{H_1}{\overset{H_1}{\gtrless}} \frac{2\sigma_0^2\sigma_1^2}{\sigma_1^2 - \sigma_0^2}\left(\ln\eta - K \ln\frac{\sigma_0}{\sigma_1}\right) = \gamma$$

The sufficient statistic is $T(\mathbf{y}) = \sum_{k=1}^{K} y_k^2$ and the test can be written as

$$T(\mathbf{y}) = \sum_{k=1}^{K} y_k^2 \underset{H_0}{\overset{H_1}{\gtrless}} \gamma$$

Note that if $\sigma_1^2 < \sigma_0^2$, $\sigma_1^2 - \sigma_0^2$ is negative and the inequality is reversed; that is,

$$T(\mathbf{y}) \underset{H_1}{\overset{H_0}{\gtrless}} \gamma$$

3.2.2 M-ary Hypothesis Testing

In the previous subsection, we considered the case where we had to choose between two hypotheses, H_0 and H_1. We now consider the case of choosing one hypothesis among M hypotheses, $H_0, H_1, \cdots, H_{M-1}$, each time an experiment is conducted. Since any one of M decisions can be made, there are M^2 possible alternatives. Bayes criterion assigns a cost to each alternative. To the ijth alternative, which is decide D_i given hypothesis H_j, the cost C_{ij}, $i, j = 0, 1, \cdots, (M - 1)$, is assigned. In addition to the hypotheses $H_0, H_1, \cdots, H_{M-1}$, we assign the a priori probabilities $P_0, P_1, \cdots, P_{M-1}$; respectively. The goal is to minimize the risk defined as

$$R = \sum_{i=1}^{M-1} \sum_{j=1}^{M-1} P_j C_{ij} P(D_i|H_j) \qquad (3.32)$$

Using the fact that

$$P(D_i|H_j) = \int_{Z_i} f_{Y|H_j}(y|H_j)dy \qquad (3.33)$$

The average cost becomes

$$R = \sum_{i=1}^{M-1} \sum_{j=1}^{M-1} P_j C_{ij} \int_{Z_i} f_{Y|H_j}(y|H_j)dy \qquad (3.34)$$

The observation space Z is now divided into M subspaces, $Z_0, Z_1, \cdots,$ Z_{M-1}, such that

$$Z = Z_0 \cup Z_1 \cup \cdots \cup Z_{M-1} \qquad (3.35)$$

In order to find the decision surfaces so that R is minimized, we rewrite Equation (3.34) as

$$R = \sum_{i=0}^{M-1} \sum_{\substack{j=0 \\ j \neq i}}^{M-1} P_j C_{ij} \int_{Z_i} f_{Y|H_j}(y|H_j)dy + \sum_{i=0}^{M-1} P_i C_{ii} \int_{Z_i} f_{Y|H_i}(y|H_i)dy$$

$$(3.36)$$

because $\int_Z f_{Y|H_j}(y|H_j)dy = 1$ and the surface $Z_i = Z - \bigcup_{j=0}^{M-1} Z_j$, substituting in (3.36), the risk becomes

$$R = \sum_{i=1}^{M-1} \int_{Z_i} \sum_{\substack{j=0 \\ j \neq i}}^{M-1} P_j(C_{ij} - C_{jj}) f_{Y|H_j}(y|H_j)dy + \sum_{i=0}^{M-1} P_i C_{ii} \qquad (3.37)$$

Using the same reasoning as before, we observe that the second term of Equation (3.37) is fixed while the first term determines the cost for the selected decision regions. Hence, the smallest integral value yields selection of the hypothesis for which

$$I_i(y) = \sum_{\substack{j=0 \\ j \neq i}}^{M-1} P_j(C_{ij} - C_{jj}) f_{Y|H_j}(y|H_j) \qquad (3.38)$$

is minimum.

Defining the likelihood ratio $\Lambda_i(y)$, $i = 1, 2, \cdots, M-1$, as

$$\Lambda_i(y) = \frac{f_{Y|H_i}(y|H_i)}{f_{Y|H_0}(y|H_0)}, i = 1, 2, \cdots, M-1 \qquad (3.39)$$

and the term $J_i(y)$ as

$$J_i(y) = \frac{I_i(y)}{f_{Y|H_0}(y|H_0)} = \sum_{j=1}^{M-1} P_j(C_{ij} - C_{jj})\Lambda_j(y) \qquad (3.40)$$

the decision rule is to choose the hypothesis for which Equation (3.40) is minimum.

For simplicity, let $M = 3$ and the observation space be $Z = Z_0 \cup Z_1 \cup Z_2$. From Equation (3.34), we obtain

$$\begin{aligned}
R = {} & P_0 C_{00} \int_{Z_0} f_{Y|H_0}(y|H_0) dy + P_0 C_{10} \int_{Z_1} f_{Y|H_0}(y|H_0) dy \\
& + P_0 C_{20} \int_{Z_2} f_{Y|H_0}(y|H_0) dy + P_1 C_{11} \int_{Z_1} f_{Y|H_1}(y|H_1) dy \\
& + P_1 C_{01} \int_{Z_0} f_{Y|H_1}(y|H_1) dy + P_1 C_{21} \int_{Z_2} f_{Y|H_1}(y|H_1) dy \\
& + P_2 C_{22} \int_{Z_2} f_{Y|H_2}(y|H_2) dy + P_2 C_{02} \int_{Z_0} f_{Y|H_2}(y|H_2) dy \\
& + P_2 C_{12} \int_{Z_1} f_{Y|H_2}(y|H_2) dy \qquad (3.41)
\end{aligned}$$

Note that

$$\int_{Z_i} f_{Y|H_i}(y|H_i)dy = \int_{Z-\sum_j Z_j} f_{Y|H_i}(y|H_i)dy$$
$$j \neq i$$

$$= 1 - \sum_j \int_{Z_j} f_{Y|H_i}(y|H_i)dy \;,\; i,j = 0,1,2 \quad (3.42)$$
$$j \neq i$$

substituting (3.42) for the terms involving C_{00}, C_{11} and C_{22} of (3.41), we obtain

$$\begin{aligned}
R = \; & P_0 C_{00} + P_1 C_{11} + P_2 C_{22} \\
& + \int_{Z_0} P_2(C_{02} - C_{22})f_{Y|H_2}(y|H_2) + P_1(C_{01} - C_{11})f_{Y|H_1}(y|H_1)dy \\
& + \int_{Z_1} P_0(C_{10} - C_{00})f_{Y|H_0}(y|H_0) + P_2(C_{12} - C_{22})f_{Y|H_2}(y|H_2)dy \\
& + \int_{Z_2} P_2(C_{20} - C_{00})f_{Y|H_0}(y|H_0) + P_1(C_{21} - C_{11})f_{Y|H_1}(y|H_1)dy
\end{aligned}$$
$$(3.43)$$

We define $I_0(y), I_1(y)$ and $I_2(y)$ as

$$I_0(y) = P_2(C_{02} - C_{22})f_{Y|H_2}(y|H_2) + P_1(C_{01} - C_{11})f_{Y|H_1}(y|H_1) \quad (3.44a)$$
$$I_1(y) = P_0(C_{10} - C_{00})f_{Y|H_0}(y|H_0) + P_2(C_{12} - C_{22})f_{Y|H_2}(y|H_2) \quad (3.44b)$$

and

$$I_2(y) = P_0(C_{20} - C_{00})f_{Y|H_0}(y|H_0) + P_1(C_{21} - C_{11})f_{Y|H_1}(y|H_1) \quad (3.44c)$$

To minimize R we assign values of Y to the region having the smallest integrands in (3.43), since $I_0(y), I_1(y)$ and $I_2(y)$ are nonnegative. Consequently,

$$Z_0 = \{y | I_0(y) < I_1(y) \text{ and } I_2(y)\}$$
$$Z_1 = \{y | I_1(y) < I_0(y) \text{ and } I_2(y)\}$$

and

$$Z_2 = \{y | I_2(y) < I_0(y) \text{ and } I_1(y)\}$$

where $|$ denotes "such that."

From Equation (3.39), the likelihood ratios $\Lambda_1(y)$ and $\Lambda_2(y)$ are

$$\Lambda_1(y) = \frac{f_{Y|H_1}(y|H_1)}{f_{Y|H_0}(y|H_0)} \qquad (3.45a)$$

and

$$\Lambda_2(y) = \frac{f_{Y|H_2}(y|H_2)}{f_{Y|H_0}(y|H_0)} \qquad (3.45b)$$

In order to incorporate the likelihood ratios into the decision rule, we use the following equivalent test:

$$I_0(y) \underset{H_0 \text{ or } H_2}{\overset{H_1 \text{ or } H_2}{\underset{<}{\gtrless}}} I_1(y) \qquad (3.46a)$$

$$I_0(y) \underset{H_0 \text{ or } H_1}{\overset{H_1 \text{ or } H_2}{\underset{<}{\gtrless}}} I_2(y) \qquad (3.46b)$$

and

$$I_1(y) \quad \overset{H_0 \text{ or } H_2}{\underset{H_0 \text{ or } H_1}{\gtrless}} \quad I_2(y) \tag{3.46c}$$

Substituting Equations (3.44) and (3.45) into (3.46), we obtain the test:

$$P_1(C_{01} - C_{11})\Lambda_1(y) \quad \overset{H_1 \text{ or } H_2}{\underset{H_0 \text{ or } H_2}{\gtrless}} \quad P_0(C_{10} - C_{00}) + P_2(C_{12} - C_{02})\Lambda_2(y)$$

$$\tag{3.47a}$$

$$P_2(C_{02} - C_{22})\Lambda_2(y) \quad \overset{H_1 \text{ or } H_2}{\underset{H_0 \text{ or } H_1}{\gtrless}} \quad P_0(C_{20} - C_{00}) + P_1(C_{21} - C_{01})\Lambda_1(y)$$

$$\tag{3.47b}$$

and

$$P_2(C_{12} - C_{22})\Lambda_2(y) \quad \overset{H_0 \text{ or } H_2}{\underset{H_0 \text{ or } H_1}{\gtrless}} \quad P_0(C_{20} - C_{10}) + P_1(C_{21} - C_{11})\Lambda_1(y)$$

$$\tag{3.47c}$$

Because $M = 3$, there are only two likelihood ratios and the decision space is two-dimensional and may be as shown in Figure 3.5.

138

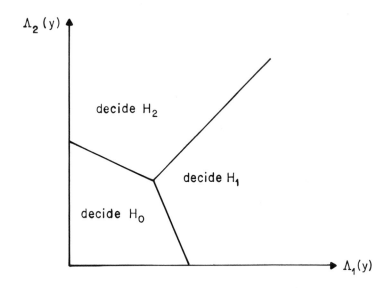

Figure 3.5 Decision space for M=3.

MAP Criterion

In communication problems, it is common to have the costs

$$C_{ii} = 0 \ , \ i = 0, 1, 2 \qquad (3.48a)$$

and

$$C_{ij} = 1 \ , \ i \neq j \text{ and } i, j = 0, 1, 2 \qquad (3.48b)$$

In this case minimizing the risk is equivalent to minimizing the probability of error. After substitution of Equation (3.48) into Equation (3.47), the decision rule reduces to

$$\Lambda_1(y) \quad \overset{H_1 \text{ or } H_2}{\underset{H_0 \text{ or } H_2}{\gtrless}} \quad \frac{P_0}{P_1} \qquad (3.49a)$$

$$\Lambda_2(y) \quad \overset{H_1 \text{ or } H_2}{\underset{H_0 \text{ or } H_1}{\gtrless}} \quad \frac{P_0}{P_2} \qquad (3.49b)$$

and

$$\Lambda_2(y) \quad \overset{H_0 \text{ or } H_2}{\underset{H_0 \text{ or } H_1}{\gtrless}} \quad \frac{P_1}{P_2}\Lambda_1(y) \qquad (3.49c)$$

Substituting Equation (3.45) in Equation (3.49), dividing by $f_Y(y)$ and using $P(A|B)/P(B)$, we obtain the following decision rule

$$P(H_1|Y) \quad \overset{H_1 \text{ or } H_2}{\underset{H_0 \text{ or } H_2}{\gtrless}} \quad P(H_0|Y) \qquad (3.50a)$$

$$P(H_2|Y) \quad \overset{H_1 \text{ or } H_2}{\underset{H_0 \text{ or } H_1}{\gtrless}} \quad P(H_0|Y) \qquad (3.50b)$$

and

$$P(H_2|Y) \quad \overset{H_0 \text{ or } H_2}{\underset{H_0 \text{ or } H_1}{\gtrless}} \quad P(H_1|Y) \qquad (3.50c)$$

We observe that in this decision rule, the receiver computes the *a posteriori* probabilities $P(H_i|Y)$, $i = 0, 1, 2$, and decides in favor of the hypothesis corresponding to the largest *a posteriori* probability. Such a minimum probability of error receiver is also referred to as the *Maximum a posteriori probability* (MAP) receiver.

3.3 MINIMAX CRITERION

Bayes criterion assigns costs to decisions and assumes knowledge of the *a priori* probabilities. In many situations, we may not have enough information about the *a priori* probabilities and consequently, Bayes criterion cannot be used. One approach would be to select a value of P_1, the *a priori* probability of H_1, for which the risk is maximum and then minimize that risk function. This principle of minimizing the maximum average cost for the selected P_1 is referred to as the *minimax criterion*.

From Equation (3.2), we have

$$P_0 = 1 - P_1 \tag{3.51}$$

substituting Equation (3.2) in Equation (3.14) we obtain the risk function in terms of P_1 as

$$R = C_{00}(1-P_F)+C_{10}P_F+P_1[(C_{11}-C_{00})+(C_{01}-C_{11})P_M-(C_{10}-C_{00})P_F] \tag{3.52}$$

Assuming a fixed value of P_1, $P_1\epsilon[0,1]$, we can design a Bayes test. These decision regions are then determined and so are the probabilities of false alarm, P_F, and miss, P_M. The test results in

$$\Lambda(y) \underset{H_0}{\overset{H_1}{\gtrless}} \frac{(1 - P_1)(C_{10} - C_{00})}{P_1(C_{01} - C_{11})} \tag{3.53}$$

As P_1 varies, the decision regions change resulting in a nonoptimum decision rule. This in turn causes a variation in the average cost which would be larger than the Bayes cost. The two extreme cases are when P_1 is zero or one. If P_1 is zero, the threshold is infinity and the decision rule is

$$\Lambda(y) \underset{H_0}{\overset{H_1}{\gtrless}} \infty \tag{3.54}$$

H_0 is always true. The observation is Z_0 and the resulting probability of false alarm and probability of miss are

$$P_F = \int_{Z_1} f_{Y|H_0}(y|H_0)dy = 0 \tag{3.55}$$

and

$$P_M = \int_{Z_0} f_{Y|H_1}(y|H_1)dy = 1 \tag{3.56}$$

Substituting for the values of P_1, P_F, and P_M in Equation (3.52), we obtain that the risk is

$$R = C_{00} \tag{3.57}$$

Similarly, when $P_1 = 1$, the threshold of Equation (3.53) is zero and the new decision rule is

$$\Lambda(y) \underset{H_0}{\overset{H_1}{\gtrless}} 0 \tag{3.58}$$

Since $\Lambda(y)$ is nonnegative, we always decide H_1. Hence, $P_F = 1$ and $P_M = 0$. The resulting risk is

$$R = C_{11} \tag{3.59}$$

If $P_1 = P_1^*$ such that $P_1^* \epsilon (0,1)$, the risk as a function of P_1 is as shown in Figure 3.6.

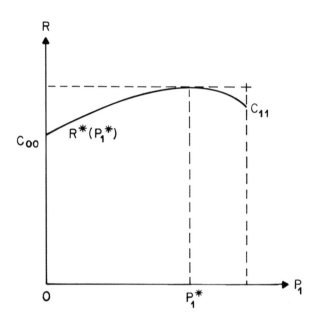

Figure 3.6 Risk as a function of P_1.

From Equation (3.52), we see that the risk R is linear in terms of P_1 and the Bayes test for $P_1 = P_1^*$ gives the minimum risk R_{min}. The tangent to R_{min} is horizontal and that $R^*(P_1)$ at $P_1 = P_1^*$ represents the maximum cost. Thus, the average cost will not exceed $R^*(P_1^*)$. Taking the derivative of R with respect to P_1 and setting it equal to zero, we obtain the *minimax equation* to be

$$(C_{11} - C_{00}) + (C_{01} - C_{11})P_M - (C_{10} - C_{00})P_F = 0 \qquad (3.60)$$

If the cost of a correct decision is zero $(C_{00} = C_{11} = 0)$ the minimax equation reduces to

$$C_{01}P_M = P_{10}P_F \tag{3.61}$$

Furthermore, if the cost of a wrong decision is one ($C_{01} = C_{10} = 1$), the probability of false alarm equals the probability of miss. That is,

$$P_F = P_M \tag{3.62}$$

and the minimax cost is

$$
\begin{aligned}
R &= P_F(1 - P_1) + P_1 P_M \\
&= P_0 P_F + P_1 P_M
\end{aligned} \tag{3.63}
$$

which is the average probability of error.

Example 3.7

Consider the problem of Example 3.1. Calculate the minimum probability of error.
(a) $P_0 = P_1$
(b) P_0 and P_1 unknown.

Solution.
(a) From Example 3.1, we found that the decision rule is

$$y \underset{H_0}{\overset{H_1}{\gtrless}} \frac{\sigma^2}{m} \ln\eta + \frac{m}{2} = \gamma$$

Given $P_0 = P_1 = 1/2$, the probability of error is

$$P(\varepsilon) = \frac{1}{2}(P_F + P_M)$$

where

$$P_F = Q(\frac{\gamma}{\sigma})$$

and

$$P_M = 1 - P_D$$
$$= 1 - Q(\frac{\gamma - m}{\sigma})$$
$$= Q(\frac{m - \gamma}{\sigma})$$

(b) In this case, the optimum threshold γ^* is obtained when $P_F = P_M$ as given in Equation (3.62). Hence,

$$Q(\frac{\gamma^*}{\sigma}) = Q(\frac{m - \gamma^*}{\sigma})$$

or, the threshold γ^* is

$$\gamma^* = \frac{m}{2}$$

Consequently, the average probability of error is

$$P(\varepsilon) = P_0 P_F + P_1 P_M$$
$$= (P_0 + P_1) P_M$$
$$= Q(\frac{m}{2\sigma})$$

In order to compare the results of (b) and (a), we normalize the standard deviation of the observation in (a) to one. Let $y' = y/\sigma$ and since $\eta = 1$, the decision rule becomes

$$y' \underset{H_0}{\overset{H_1}{\underset{<}{>}}} \frac{m}{2\sigma} = \gamma$$

Let $\alpha = m/\sigma$, the decision rule reduces to

$$y' \underset{H_0}{\overset{H_1}{\underset{<}{>}}} \frac{\alpha}{2}$$

The probability of false alarm and the probability of detection are given by

$$P_F = \int_{\frac{\alpha}{2}}^{\infty} \frac{1}{\sqrt{2\pi}} e^{-\frac{y'^2}{2}} dy' = Q(\frac{\alpha}{2})$$

and

$$P_D = \int_{\frac{\alpha}{2}}^{\infty} \frac{1}{\sqrt{2\pi}} e^{-\frac{(y'-\alpha)^2}{2}} dy' = Q(\frac{\alpha}{2} - \alpha)$$
$$= Q(-\frac{\alpha}{2})$$

$$\implies P_M = 1 - P_D = 1 - Q(-\frac{\alpha}{2})$$
$$= Q(\frac{\alpha}{2})$$

The average probability of error is

$$P(\varepsilon) = \frac{1}{2}[Q(\frac{\alpha}{2}) + Q(\frac{\alpha}{2})]$$
$$= Q(\frac{\alpha}{2}) = Q(\frac{m}{2\sigma})$$

Therefore, both results obtained in (a) and (b) are the same.

3.4 NEYMAN – PEARSON CRITERION

In the previous sections, we have seen that for Bayes criterion we require knowledge of the a priori probabilities and cost assignments for each possible decision. Then, we have studied the minimax criterion which is useful in situations where knowledge of the *a priori* probabilities is not possible. In many other physical situations, such as radar detection, it is very difficult to assign realistic costs and *a priori* probabilities. To overcome this difficulty, we use the conditional probabilities of false alarm, P_F, and detection, P_D. The *Neyman–Pearson test* requires that P_F be fixed to some value α while P_D is maximized. Since $P_M = 1 - P_D$, maximizing P_D is equivalent to minimizing P_M.

In order to minimize P_M (maximize P_D) subject to the constraint that $P_F = \alpha$, we use the calculus of extrema and form the objective function J to be

$$J = P_M + \lambda[P_F - \alpha] \tag{3.64}$$

where $\lambda(\lambda \geq 0)$ is the Lagrange multiplier. We note that given the observation space Z, there are many decision regions Z_1 for which $P_F = \alpha$. The question is to determine those decision regions for which P_M is minimum. Consequently, we rewrite the objective function J in terms of the decision region to obtain

$$J = \int_{Z_0} f_{Y|H_1}(y|H_1)dy + \lambda[\int_{Z_1} f_{Y|H_0}(y|H_0)dy - \alpha] \tag{3.65}$$

Using Equation (3.1), Equation (3.65) can be rewritten as

$$
\begin{aligned}
J &= \int_{Z_0} f_{Y|H_1}(y|H_1)dy + \lambda[1 - \int_{Z_0} f_{Y|H_0}(y|H_0)dy - \alpha] \\
&= \lambda(1 - \alpha) + \int_{Z_0} [f_{Y|H_1}(y|H_1) - \lambda f_{Y|H_0}(y|H_0)]dy
\end{aligned} \tag{3.66}
$$

Hence, J is minimized when values for which $f_{Y|H_1}(y|H_1) > \lambda f_{Y|H_0}(y|H_0)$ are assigned to the decision region Z_1. The decision rule is, therefore,

$$\Lambda(y) = \frac{f_{Y|H_1}(y|H_1)}{f_{Y|H_0}(y|H_0)} \overset{H_1}{\underset{H_0}{\gtrless}} \lambda \tag{3.67}$$

The threshold η derived from Bayes criterion is equivalent to λ the Lagrange multiplier in the Neyman–Pearson (N–P) test for which the probability of false alarm is fixed to the value α. If we define the conditional density of Λ given that H_0 is true as $f_{\Lambda|H_0}(\lambda|H_0)$, then $P_F = \alpha$ may be rewritten as

$$P_F = \int_{Z_1} f_{Y|H_0}(y|H_0)dy = \int_{\lambda}^{\infty} f_{\Lambda(Y)|H_0}[\Lambda(y)|H_0]d\lambda \tag{3.68}$$

Example 3.8

Consider the binary hypothesis problem with received conditional probabilities:

$$f_{Y|H_0}(y|H_0) = \frac{1}{2(1 - e^{-1})} \, e^{-|y|} \text{ for } |y| \leq 1$$

and

$$f_{Y|H_1}(y|H_1) = \frac{1}{2} \, \text{rect}(\frac{1}{2})$$

The hypotheses H_0 and H_1 are equally likely.
(a) Find the decision regions for which the probability of error is minimum.
(b) Calculate the minimum probability of error.
(c) Find the decision rule based on the Neyman–Pearson criterion such that the probability of false alarm is constrained to be $P_F = 0.5$.
(d) Calculate the probability of detection for the given constraint of P_F in (b).

Solution.
(a) The minimum probability of error receiver requires that $C_{00} = C_{11} = 0$ and $C_{01} = C_{10} = 1$. Since the *a priori* probabilities are equal, the likelihood ratio test reduces to

$$\Lambda(y) = \frac{f_{Y|H_1}(y|H_1)}{f_{Y|H_0}(y|H_0)} \mathop{\gtrless}_{H_0}^{H_1} 1$$

That is, we choose the hypothesis for which $f_{Y|H_j}(y|H_j)$, $j = 0, 1$, is maximum. The decision regions are as shown in Figure 3.7.

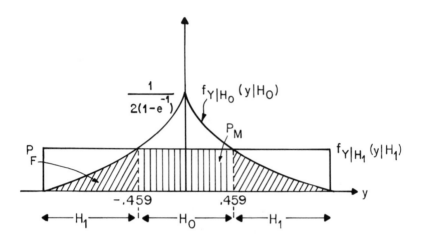

Figure 3.7 Decision region for Example 3.8.

Decide H_1 for $-1 \leq y \leq 0.459$ and $0.459 \leq y \leq 1$; decide H_0 for $-0.459 < y < 0.459$.

(b) The probability of error is $P(\varepsilon) = P_0 P_F + P_1 P_M$ where

$$
\begin{aligned}
P_F &= P(\text{decide } H_1 | H_0 \text{ true}) \\
&= \frac{1}{2(1 - e^{-1})} \left(\int_{-1}^{-0.459} e^y dy + \int_{0.459}^{1} e^{-y} dy \right) = 0.418
\end{aligned}
$$

$$
\begin{aligned}
P_M &= P(\text{decide } H_0 | H_1 \text{ true}) \\
&= 2\{(0.459)(\frac{1}{2})\} = 0.459
\end{aligned}
$$

$$
\Longrightarrow P(\varepsilon) = \frac{1}{2}(0.418 + 0.459) = 0.4385
$$

(c) In using the Neyman–Pearson Criterion, we have

$$\Lambda(y) = \frac{\frac{1}{2}}{\frac{1}{2(1-e^{-1})}\, e^{-|y|}} \mathop{\gtrless}_{H_0}^{H_1} \eta$$

$$\implies \frac{1-e^{-1}}{e^{-|y|}} \mathop{\gtrless}_{H_0}^{H_1} \eta \quad \text{or} \quad |y| \mathop{\gtrless}_{H_0}^{H_1} - \ln\frac{1-e^{-1}}{\eta} = \gamma$$

P_F is as shown in Figure 3.8.

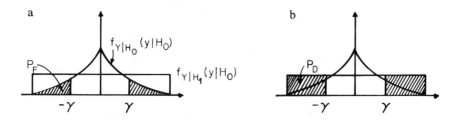

Figure 3.8 Regions showing: (a) P_F; (b) P_D.

Hence, $P_F = P(D_1|H_0) = \frac{1}{2(1-e^{-1})}\{\int_{-1}^{-\gamma} e^y dy + \int_{\gamma}^{1} e^{-y} dy\} = 0.5$

$$\implies \gamma = 0.38$$

is the threshold.

(d) The probability of detection is

$$P_D = 2\{(1 - 0.38)\frac{1}{2}\} = 0.62$$

150

Receiver Operating Characteristic - ROC

A plot of the probability of detection, P_D, versus the probability of false alarm with the threshold as a parameter is referred to as *receiver operating characteristic (ROC)* curves. We note that the *ROC* depends on the conditional density function of the observed signal under each hypothesis, that is, $f_{Y|H_j}(y|H_j)$, $j = 0, 1$, and not on the assigned costs, or the *a priori* probabilities. From Example 3.1, the conditional density functions $f_{Y|H_j}(y|H_j)$, $j = 0, 1$, and the probabilities of detection and false alarm are as shown in Figure 3.9.

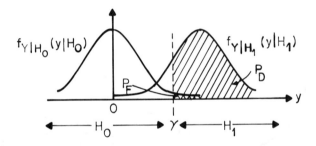

Figure 3.9 Decision regions showing P_D and P_F.

Varying the threshold γ, the areas representing P_D and P_F vary. The corresponding *ROC* curves are shown in Figure 3.10.

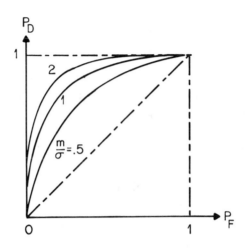

Figure 3.10 *ROC* with m/σ as a parameter.

The two extreme points on the *ROC* for $P_F = P_D = 1$ and $P_F = P_D = 0$ are easily verified. Because both the Neyman–Pearson receiver and the Bayes receiver employ the likelihood ratio test, and that $\Lambda(y)$ is a random variable, P_D and P_F may be rewritten as

$$
\begin{aligned}
P_D &= P(\text{decide } H_1 | H_1 \text{ true}) \\
&= \int_\eta^\infty f_{\Lambda|H_1}(\lambda|H_1)d\lambda
\end{aligned}
\tag{3.69}
$$

and

$$
\begin{aligned}
P_F &= P(\text{decide } H_1 | H_0 \text{ true}) \\
&= \int_\eta^\infty f_{\Lambda|H_0}(\lambda|H_0)d\lambda
\end{aligned}
\tag{3.70}
$$

$\Lambda(y)$ is a ratio of two nonnegative quantities, $f_{Y|H_1}(y|H_1)$ and $f_{Y|H_0}(y|H_0)$, and thus takes values from zero to infinity. When the threshold η is zero ($\eta = 0$ corresponds to $P_0 = 0$), hypothesis H_1 is always true and thus $P_D = P_F = 1$. When the threshold η is infinity ($\eta \to \infty$ corresponds to $P_1 = 0$), hypothesis H_0 is always true and thus $P_D = P_F = 0$. This is clearly depicted in Figure 3.10.

The slope of the ROC at a particular point on the curve represents the threshold η for the Neyman–Pearson test to achieve the P_D and P_F at that point. Taking the derivative of Equation (3.69) and (3.70) with respect to η, we have

$$\begin{aligned}
\frac{dP_D}{d\eta} &= \frac{d}{d\eta} \int_\eta^\infty f_{\Lambda|H_1}(\lambda|H_1)d\lambda \\
&= -f_{\Lambda|H_1}(\eta|H_1)
\end{aligned} \tag{3.71}$$

and

$$\begin{aligned}
\frac{dP_F}{d\eta} &= \frac{d}{d\eta} \int_\eta^\infty f_{\Lambda|H_0}(\lambda|H_0)d\lambda \\
&= -f_{\Lambda|H_0}(\eta|H_0)
\end{aligned} \tag{3.72}$$

Also,

$$\begin{aligned}
P_D(\eta) &= P[\Lambda(y) \geq \eta|H_1] \\
&= \int_\eta^\infty f_{\Lambda|H_1}[\lambda(y)|H_1]d\lambda \\
&= \int_\eta^\infty \Lambda(y)f_{\Lambda|H_0}[\lambda(y)|H_0]d\lambda
\end{aligned} \tag{3.73}$$

Taking the derivative of the above equation with respect to η, we obtain

$$\frac{dP_D}{d\eta} = -\eta f_{\Lambda|H_1}(\eta|H_0) \tag{3.74}$$

Combining Equations (3.71), (3.72), and (3.74), results in

$$\frac{f_{\Lambda|H_1}(\eta|H_1)}{f_{\Lambda|H0}(\eta|H_0)} = \eta \tag{3.75}$$

and

$$\frac{dP_D}{dP_F} = \eta \qquad (3.76)$$

In the Bayes criterion, the threshold η is determined by the *a priori* probabilities and the costs. Consequently, the probability of detection P_D and the probability of false alarm, P_F, are determined on the point of the *ROC* curve at which the tangent has a slope of η.

The minimax equation represents a straight line in the P_D–P_F plane starting at the point $P_D = 0$ and $P_F = 1$, and crosses the *ROC* curve. The slope of the tangent of the intersection with the *ROC* is the threshold η.

Example 3.9

Consider a problem with the following conditional density functions:

$$f_{Y|H_0}(y|H_0) = \begin{cases} e^{-y} & , y \geq 0 \\ 0 & , \text{ otherwise} \end{cases}$$

$$f_{Y|H_1}(y|H_1) = \begin{cases} \alpha e^{-\alpha y} & , y \geq 0, \alpha > 1 \\ 0 & , \text{ otherwise} \end{cases}$$

Plot the *ROC*.

Solution.

The *ROC* is a plot of P_D, the probability of detection, *versus* P_F, the probability of false alarm, with the threshold η as a parameter. The likelihood ratio is

$$\Lambda(y) \quad = \quad \frac{\alpha \, e^{-\alpha y}}{e^{-y}}$$

$$= \quad \alpha \, e^{-(\alpha-1)y} \quad \underset{H_0}{\overset{H_1}{\underset{<}{\gtrless}}} \quad \eta$$

Taking the logarithm and rearranging terms, the decision rule becomes

$$y \quad \underset{H_1}{\overset{H_0}{\underset{<}{\gtrless}}} \quad \frac{1}{1-\alpha} \ln\frac{\eta}{\alpha} = \gamma$$

Also, from the Neyman–Pearson test, the probability of detection, and the probability of false alarm are

$$P_D = P(D_1|H_1) \quad = \quad \int_0^\gamma \alpha \, e^{-\alpha y} dy$$

$$= \quad 1 - e^{-\alpha\gamma}$$

and

$$P_F = P(D_1|H_0) \quad = \quad \int_0^\gamma e^{-y} dy$$

$$= \quad 1 - e^{-\gamma}$$

Note that taking the derivative of P_D and P_F with respect to the threshold γ, and substituting in Equation (3.76), we obtain the threshold η; i.e.,

$$\frac{dP_D}{dP_F} \quad = \quad \gamma \, e^{-(\alpha-1)\gamma}$$

$$= \quad \eta$$

A plot of the ROC with α as a parameter is shown in Figure 3.11.

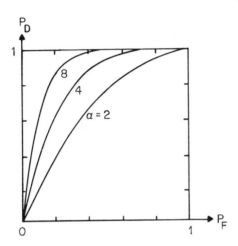

Figure 3.11 ROC of Example 3.9.

3.5 COMPOSITE HYPOTHESIS TESTING

In the simple hypothesis testing problem previously considered, the parameters characterizing a hypothesis were all known. In many situations, the parameters characterizing a hypothesis may not be known. In this case, the hypothesis is called a *composite hypothesis.*

Example 3.10

Consider the situation where the observations under each hypothesis are given by

$$H_1 : Y = m + N$$
$$H_0 : Y = \quad\quad N$$

where N denotes a white Gaussian noise of zero mean, and variance σ^2, and m is unknown. Then, we say that H_0 is a simple hypothesis, and H_1 a composite hypothesis.

In the previous sections, we developed the theory of designing good tests for simple hypotheses. We now consider tests for composite hypothesis. The situation may be best described by the following block diagram of Figure 3.12.

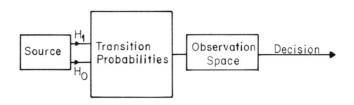

Figure 3.12 Block diagram showing composite hypothesis.

Each hypothesis is characterized by a set of K parameters such that

$$\Theta = \begin{bmatrix} \theta_1 \\ \theta_2 \\ \vdots \\ \theta_K \end{bmatrix} \tag{3.77}$$

Two cases will be considered. First, Θ may be a random variable with known density functions $f_{\Theta|H_1}(\theta|H_1)$ and $f_{\Theta|H_0}(\theta|H_0)$. Second, Θ may not be a random variable but still an unknown constant.

3.5.1 Θ a Random Variable

If Θ is a random variable with known density functions, $f_{\Theta|H_1}(\theta|H_1)$ and $f_{\Theta|H_0}(\theta|H_0)$, then the optimum decision is obtained by using the Bayes criterion and minimizing the risk. The analysis is as before. In order to apply the likelihood ratio test, we need $f_{Y|H_1}(y|H_1)$ and

$f_{\mathbf{Y}|H_0}(\mathbf{y}|H_0)$. They are readily obtained by averaging over all possible values of Θ. That is,

$$f_{\mathbf{Y}|H_j}(\mathbf{y}|H_j) = \int f_{\mathbf{Y}|\Theta,H_j}(\mathbf{y}|\theta, H_j) f_{\Theta|H_j}(\theta|H_j) d\theta \qquad ; j = 0, 1 \quad (3.78)$$

The likelihood ratio becomes

$$\begin{aligned}
\Lambda(\mathbf{y}) &= \frac{f_{\mathbf{Y}|H_1}(\mathbf{y}|H_1)}{f_{\mathbf{Y}|H_0}(\mathbf{y}|H_0)} \\[2mm]
&= \frac{\int f_{\mathbf{Y}|\Theta,H_1}(\mathbf{y}|\theta, H_1) f_{\Theta|H_1}(\theta|H_1) d\theta}{\int f_{\mathbf{Y}|\Theta,H_0}(\mathbf{y}|\theta, H_0) f_{\Theta|H_0}(\theta|H_0) d\theta}
\end{aligned} \qquad (3.79)$$

Example 3.11

Consider the problem of Example 3.10 where the constant m, now denoted M, is a Gaussian random variable with mean zero and variance σ_M^2. Determine the optimum decision rule.

Solution.

Using Equation (3.79), the optimum decision rule can be directly obtained from the likelihood ratio test. Hence,

$$\Lambda(y) = \frac{\int_{-\infty}^{\infty} f_{Y|M,H_1}(y|m, H_1) f_{M|H_1}(m|H_1) dm}{f_{Y|H_0}(y|H_0)}$$

Note that only H_1 is a composite hypothesis and consequently, the numerator of $\Lambda(y)$ is integrated over M. Since the actual value of M is not important, M is referred to as the "unwanted parameter." The numerator of $\Lambda(y)$, denoted $Num(y)$, is

$$\begin{aligned}
Num(y) &= \frac{1}{2\pi\sigma\sigma_M} \int_{-\infty}^{\infty} e^{-\frac{(y-m)^2}{2\sigma^2} - \frac{m^2}{2\sigma_M^2}} \, dm \\[2mm]
&= \frac{1}{2\pi\sigma\sigma_M} \int_{-\infty}^{\infty} e^{-\frac{\sigma_M^2+\sigma^2}{2\sigma^2\sigma_M^2}[m^2 - \frac{2\sigma_M^2 y}{\sigma_M^2+\sigma^2} m] - \frac{y^2}{2\sigma^2}} \, dm
\end{aligned}$$

Completing the square in the exponent, $Num(y)$ becomes

$$Num(y) = \frac{1}{2\pi\sigma\sigma_M} \int_{-\infty}^{\infty} e^{-\frac{\sigma_M^2+\sigma^2}{2\sigma^2\sigma_M^2}(m-\frac{\sigma_M^2 y}{\sigma_M^2+\sigma^2})^2 + \frac{\sigma_M^2 y^2}{2\sigma^2(\sigma_M^2+\sigma^2)} - \frac{y^2}{2\sigma^2}} \, dm$$

$$= \frac{1}{2\pi\sigma\sigma_M} e^{-\frac{y^2}{2(\sigma_M^2+\sigma^2)}} \int_{-\infty}^{\infty} e^{-\frac{\sigma_M^2+\sigma^2}{2\sigma^2\sigma_M^2}(m-\frac{\sigma_M^2 y}{\sigma_M^2+\sigma^2})^2} \, dm$$

Because the integral

$$\int_{-\infty}^{\infty} e^{-\frac{\sigma_M^2+\sigma^2}{2\sigma^2\sigma_M^2}(m-\frac{\sigma_M^2 y}{\sigma_M^2+\sigma^2})^2} \, dm = \sqrt{2\pi}\,\frac{\sigma\sigma_M}{\sqrt{\sigma_M^2+\sigma}}$$

$Num(y)$ becomes

$$f_{Y|H_1}(y|H_1) = \frac{1}{\sqrt{\pi}\sqrt{\sigma_M^2+\sigma^2}} e^{-\frac{y^2}{2(\sigma_M^2+\sigma^2)}}$$

The likelihood ratio test reduces to

$$\Lambda(y) = \frac{\frac{1}{\sqrt{\pi}\sqrt{\sigma_M^2+\sigma^2}} e^{-\frac{y^2}{2(\sigma_M^2+\sigma^2)}}}{\frac{1}{\sqrt{2\pi}\sigma} e^{-\frac{y^2}{2\sigma^2}}} \underset{H_0}{\overset{H_1}{\gtrless}} \eta$$

Taking the natural logarithm on both sides and simplifying the expression, we obtain

$$y^2 \underset{H_0}{\overset{H_1}{\gtrless}} \frac{2\sigma^2(\sigma_M^2+\sigma^2)}{\sigma_M^2} [\,\ln\eta + \frac{1}{2}\ln(1+\frac{\sigma_M^2}{\sigma^2})]$$

We observe that exact knowledge of m "the unwanted parameter" is not important because it does not appear in the decision rule.

3.5.2 Θ Nonrandom; Unknown

If Θ is not a random variable but still unknown, Bayes test is no longer applicable since Θ does not have a probability density function over which $f_{Y|\Theta,H_j}(y|\theta, H_j)$, $j = 0, 1$, can be averaged, and consequently, the risk cannot be determined. Instead, we use the Neyman–Pearson test. In this case, we maximize the probability of detection, P_D, while the probability of false alarm, P_F, is fixed, given that the assumed value θ is the true value.

Performing this test for several values of θ results in a plot of P_D versus Θ known as the *power function*. A test that maximizes the probability of detection as mentioned above for *all* possible values of Θ is referred to as a *Uniformly Most Powerful* (UMP) test. Hence, a UMP test maximizes the probability of detection irrespect of the values of Θ.

Example 3.12

Consider the problem of Example 3.10 where m is a positive constant. Determine the optimum decision rule.

Solution.
The conditional density functions under hypotheses H_0 and H_1 are

$$H_0 : f_{Y|H_0}(y|H_0) = \frac{1}{\sqrt{2\pi}\sigma} e^{-\frac{y^2}{2\sigma^2}}$$
$$H_1 : f_{Y|H_1}(y|H_1) = \frac{1}{\sqrt{2\pi}\sigma} e^{-\frac{(y-m)^2}{2\sigma^2}}$$

The exact value of m is not known, but it is known to be positive. Assuming a value of m, the likelihood ratio test is given by

$$\Lambda(y) = \frac{\frac{1}{\sqrt{2\pi}\sigma} e^{-\left(\frac{y^2 - my + m^2}{2\sigma^2}\right)}}{\frac{1}{\sqrt{2\pi}\sigma} e^{-\frac{y^2}{2\sigma^2}}} \overset{H_1}{\underset{H_0}{\gtrless}} \eta$$

simplifying the likelihood ratio test, and taking the natural logarithm, we obtain

$$y \overset{H_1}{\underset{H_0}{\gtrless}} \frac{\sigma^2}{m} \ln\eta + \frac{m}{2} = \gamma_1$$

Note that the threshold η is determined from the specified value of the probability of false alarm P_F. In fact, knowledge of η is not necessary to determine γ_1. Assuming γ_1, as shown in Figure 3.13, we have

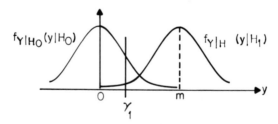

Figure 3.13 Threshold γ_1 for composite hypothesis.

$$P_F = \int_{\gamma_1}^{\infty} f_{Y|H_0}(y|H_0)dy = \int_{\gamma_1}^{\infty} \frac{1}{\sqrt{2\pi}\sigma} e^{-\frac{y^2}{2\sigma^2}} dy$$

Once γ_1 is determined, the application of the likelihood ratio test:

$$y \overset{H_1}{\underset{H_0}{\gtrless}} \gamma_1$$

does not require any knowledge of m. That is, a best test can be completely designed independently of m. Hence, a UMP test exists.

Similarly if m were unknown but negative, the likelihood ratio test reduces to

$$y \overset{H_0}{\underset{H_1}{\gtrless}} \frac{\sigma^2}{m} \ln\eta + \frac{m}{2} = \gamma_2$$

γ_2 is determined from the specified probability of false alarm to be

$$P_F = \int_{-\infty}^{\gamma_2} f_{Y|H_0}(y|H_0)dy = \int_{-\infty}^{\gamma_2} \frac{1}{\sqrt{2\pi}\sigma} e^{-\frac{y^2}{2\sigma^2}} dy$$

Again, a UMP test exists since application of the likelihood ratio test is independent of m. It should be noted that the probability of dectection for both cases, $m < 0$ and $m > 0$, cannot be evaluated because the exact value of m is not known. Nevertheless, the test is optimum for all possible values of m positive or negative.

Note that the test designed for m positive is not the same for m negative. Consequently, if m were unknown and takes all possible values, positive and negative, a UMP test does not exist. Because we know from the definition that a UMP test exists if it is optimum for all possible values of m. In this case, the test designed for m positive (negative) is not optimum for m negative (positive). This requires that different tests are to be used which will be discussed in the coming chapter after we cover *Maximum Likelihood Estimation* (MLE).

3.6 SEQUENTIAL DETECTION

In the previous sections, we considered the theory of hypothesis testing such that the number of observations, on which the test was based, was fixed. In many practical situations, observations may be taken in a sequential manner so that a test is performed after each observation. Each time an observation is taken, one of the three possible decisions is made.

(i) Decide H_1

(ii) Decide H_0

(iii) Not enough information to decide in favor of either H_1 or H_0.

If decision (i) or (ii) is made, the hypothesis testing procedure stops. Otherwise, an additional observation is taken, and the test is performed again. This process continues until a decision is made either in favor of H_1 or H_0. Note that the number of observations K is not fixed but a random variable.

The test to be performed after each observation is to compute the likelihood ratio and compare it to two threshold η_0 and η_1. Such a test that makes one of the three possible decisions mentioned above after the kth observation is referred to as *sequential likelihood ratio test*.

Let Y_k, $k = 1, 2, \cdots, K$, represent the kth observation sample of the vector $\mathbf{Y}_K$ defined as

$$\mathbf{Y}_K = \begin{bmatrix} Y_1 \\ Y_2 \\ \vdots \\ Y_K \end{bmatrix} \tag{3.80}$$

The likelihood ratio based on the first K observations is

$$\Lambda(\mathbf{y}_K) = \frac{f_{\mathbf{Y}_K|H_1}(\mathbf{y}_K|H_1)}{f_{\mathbf{Y}_K|H_0}(\mathbf{y}_K|H_0)} \tag{3.81}$$

To compute the likelihood ratio of Equation (3.81), we need to know the joint density function of these K observations. For simplicity, we assume that the observations are identically distributed, and are taken independently. The likelihood ratio can be written as a product of K likelihood ratios to obtain

$$\Lambda(\mathbf{y}_K) = \frac{f_{\mathbf{Y}_K|H_1}(\mathbf{y}_K|H_1)}{f_{\mathbf{Y}_K|H_0}(\mathbf{y}_K|H_0)} = \prod_{k=1}^{K} \frac{f_{Y_k|H_1}(y_k|H_1)}{f_{Y_k|H_0}(y_k|H_0)} \tag{3.82}$$

The goal is to determine η_0 and η_1 in terms of P_F, the probability of false alarm, and P_M, the probability of miss. We set

$$P_F = \alpha \tag{3.83}$$

and

$$P_M = \beta \tag{3.84}$$

and perform the following test. If

$$\Lambda(\mathbf{y}_K) \geq \eta_1 \tag{3.85}$$

we decide in favor of H_1. If

$$\Lambda(\mathbf{y}_K) \leq \eta_0 \tag{3.86}$$

we decide in favor of H_0. Otherwise, if

$$\eta_0 < \Lambda(\mathbf{y}_K) < \eta_1 \tag{3.87}$$

we take an additional observation and perform another test. The probability of detection, P_D, in terms of the integral over the observation space is

$$P_D = P(\text{decide } H_1|H_1 \text{ true}) = \int_{Z_1} f_{\mathbf{Y}_K|H_1}(\mathbf{y}_K|H_1) d\mathbf{y}_K \tag{3.88}$$

Using Equation (3.70), P_D can be written as

$$P_D = \int_{Z_1} \Lambda(\mathbf{y}_K) f_{\mathbf{Y}_K|H_0}(\mathbf{y}_K|H_0) d\mathbf{y}_K \tag{3.89}$$

The decision in favor of H_1 means that $\Lambda(\mathbf{y}_K) \geq \eta_1$. Hence, substituting for Equation (3.85) into (3.89) we obtain the inequality:

$$P_D \geq \eta_1 \int_{Z_1} f_{\mathbf{Y}_K|H_0}(\mathbf{y}_K|H_0)d\mathbf{y}_K \qquad (3.90)$$

Note that the integral:

$$\int_{Z_1} f_{\mathbf{Y}_K|H_0}(\mathbf{y}_K|H_0)d\mathbf{y}_K = P_F = \alpha \qquad (3.91)$$

and since $P_D = 1 - P_M = 1 - \beta$, Equation (3.90) reduces to

$$1 - \beta \geq \eta_1 \alpha \qquad (3.92)$$

or, the threshold η_1 is

$$\eta_1 \leq \frac{1 - \beta}{\alpha} \qquad (3.93)$$

Similarly, it can be shown that the threshold η_0 is

$$\eta_0 \geq \frac{\beta}{1 - \alpha} \qquad (3.94)$$

At this stage, some important questions need to be investigated and answered. What is the probability that the procedure never terminates? What are some of the properties of the distribution of the random variable K? In particular, what is the expected value of this sample size K?

To answer such questions, it is much easier to use the log-likelihood function. Taking the natural logarithm of Equation (3.87), we obtain

$$\ln\eta_0 < \ln\frac{f_{Y_1|H_1}(y_1|H_1)}{f_{Y_1|H_0}(y_1|H_0)} + \cdots + \ln\frac{f_{Y_K|H_1}(y_K|H_1)}{f_{Y_K|H_0}(y_K|H_0)} < \ln\eta_0 \qquad (3.95)$$

Let the kth term, $k = 1, 2, \cdots, K$, of the above sum be denoted as

$$L(y_k) = \ln\frac{f_{Y_k|H_1}(y_k|H_1)}{f_{Y_k|H_0}(y_k|H_0)} \qquad (3.96)$$

then, Equation (3.95) becomes

$$\ln\eta_0 < L(y_1) + \cdots + L(y_k) + \cdots + L(y_K) < \ln\eta_1 \qquad (3.97)$$

The sum may be written in a recursive relation as

$$L(\mathbf{y}_K) = L(\mathbf{y}_{K-1}) + L(y_K) \qquad (3.98a)$$

where

$$
\begin{aligned}
L(\mathbf{y}_{K-1}) &= L(y_1) + L(y_2) + \cdots + L(y_{K-1}) \\
&= \sum_{k=1}^{K-1} L(y_k) \qquad (3.98b)
\end{aligned}
$$

In order to calculate $E[K]$ the average number of observations under each hypothesis, we assume that the test terminates in K observations with probability one. This assumption implies that $L(\mathbf{y}_K)$ takes on two possible values $\ln\eta_0$ and $\ln\eta_1$. If hypothesis H_1 is true, a detection is declared when $L(\mathbf{y}_K) \geq \ln\eta_1$, with probability $P_D = 1 - P_M = 1 - \beta$. Also, a miss occurs when $L(\mathbf{y}_K) \leq \ln\eta_0$, with probability $P_M = \beta$. Hence, the expected value of $L(\mathbf{y}_K)$ under hypothesis H_1 is

$$E[L(\mathbf{y}_K)|H_1] = \beta \ln\eta_0 + (1 - \beta) \ln\eta_1 \qquad (3.99)$$

Following the same reasoning, the expected value of $L(\mathbf{y}_K)$ under hypothesis H_0 is

$$E[L(\mathbf{y}_K)|H_0] = \alpha \ln\eta_1 + (1 - \alpha) \ln\eta_0 \qquad (3.100)$$

Let B be a random variable taking binary numbers zero and one such that

$$
B_k = \begin{cases} 1 & , \text{ no decision made up to } (k-1) \text{ sample} \\ 0 & , \text{ decision made at an earlier sample} \end{cases} \qquad (3.101)
$$

that is, B_k depends on the observations Y_k, $k = 1, 2, \cdots, K - 1$, and not Y_K. Rewriting the log-likelihood ratio in terms of the variable B, we obtain

$$L(\mathbf{y}_K) = \sum_{k=1}^{K} L(y_K) = \sum_{k=1}^{\infty} B_k L(y_k) \qquad (3.102)$$

Because the observations are independent and identically distributed, we have

$$E[L(\mathbf{y}_K)|H_j] = E[L(y)|H_j] \sum_{k=1}^{\infty} E[B_k] \;, j = 0,1 \qquad (3.103a)$$

where

$$E[L(y)|H_j] = E[L(y_1)|H_j] = \cdots = E[L(y_K)|H_j] \qquad (3.103b)$$

The sum in Equation (3.103a) is just

$$\begin{aligned}
\sum_{k=1}^{\infty} E[B_k] &= \sum_{k=1}^{\infty} P(K \geq k) \\
&= \sum_{k=1}^{\infty} kP(K = k) = E[K] \qquad (3.104)
\end{aligned}$$

Substituting Equation (3.104) and Equation (3.103b) into Equation (3.99), we obtain

$$E[L(y)|H_1]E[K|H_1] = \alpha \ln\eta_1 + (1 - \alpha) \ln\eta_0 \qquad (3.105)$$

or

$$E[K|H_1] = \frac{(1 - \beta) \ln\eta_1 + \beta \ln\eta_0}{E[L(y)|H_1]} \qquad (3.106)$$

Similarly, the expected value of K under hypothesis H_0 can be expected to be

$$E[K|H_0] = \frac{\alpha \ln\eta_1 + (1 - \alpha) \ln\eta_0}{E[L(y)|H_0]} \qquad (3.107)$$

To answer the question that the process terminates with probability one, we need to show that

$$\lim_{k \to \infty} P(K \geq k) = 0 \qquad (3.108)$$

which is straightforward. Furthermore, it can be shown that the expected value of the number of observations K under each hypothesis is minimum for the specified values of P_F and P_M.

Example 3.13

Suppose that the receiver of Example 3.2 takes the K observations sequentially. Let the variance $\sigma^2 = 1$ and the mean $m = 1$. Determine (a) the decision rule such that $P_F = \alpha = 0.1 = P_M = \beta$. (b) the expected value of K under each hypothesis.

Solution.

(a) The definition of the decision rule is expressed in Equations (3.85), (3.86) and (3.87). Consequently, we need to solve for the likelihood ratio at the kth stage and for the thresholds η_0 and η_1. Substituting for $\sigma^2 = 1$ and $m = 1$ in the likelihood ratio of Example 3.2, we obtain the likelihood ratio at the kth stage to be

$$\Lambda(\mathbf{y}_K) = e^{\sum_{k=1}^{K} y_k - \frac{K}{2}}$$

The log likelihood ratio is just

$$L(\mathbf{y}_K) = \ln\Lambda(\mathbf{y}_K) = \sum_{k=1}^{K} y_k - \frac{K}{2}$$

From Equations (3.93) and (3.94), the two thresholds are

$$\ln\eta_1 = 2.197 \text{ and } \ln\eta_0 = -2.197$$

Hence, the decision rule in terms of the log-likelihood ratio is:

If $L(\mathbf{y}_K) \geq 2.197$ decide H_1

If $L(\mathbf{y}_K) \leq -2.197$ decide H_0

If $-2.197 \leq L(\mathbf{y}_K) \leq 2.197$ take an additional observation $K + 1$ and perform another test.

(b) The expected value of K under hypotheses H_1 and H_0 is given by Equations (3.106) and (3.107), respectively. We observe that we need to obtain $E[L(y)|H_1]$ and $E[L(y)|H_0]$. Again, assuming that the observations are identical, we have

$$E[L(y)|H_1] = 1 - \frac{1}{2} = \frac{1}{2}$$

and

$$E[L(Y)|H_0] = 0 - \frac{1}{2} = -\frac{1}{2}$$

Substituting for the values of $E[L(Y)|H_1]$ and $E[L(Y)|H_0]$, in Equations (3.106) and (3.107), we obtain

$$E[K|H_1] = 3.515$$

$$E[K|H_0] = 3.515$$

That is, we need four samples to obtain the performance specified by $P_F = P_M = 0.1$

3.7 SUMMARY

In this chapter, we have developed the basic concepts of hypothesis testing. First, we studied Bayes criterion which assumes knowledge of the *a priori* probability of each hypothesis, and the cost assignment of each possible decision. The average cost, known as the risk function, was minimized to obtain the optimum decision rule. Bayes criterion was considered for the simple binary hypothesis testing and the M-ary hypothesis testing. The minimax criterion, which minimizes the average cost for a selected *a priori* probability, P_1, was studied in Section 3.3. The minimax criterion applies to situations where the *a priori* probabilities are not known, even though realistic cost assignments to the various decisions are possible. In cases, where realistic cost assignments are not possible and the *a priori* probabilities are not known, we considered the Neyman–Pearson approach. In the Neyman–Pearson criterion, the probability of detection (miss) is maximized (minimized), while the probability of false alarm is fixed to a designed value. The receiver operating characteristic, which is a plot of the probability of detection versus the probability of false alarm, was useful in analyzing the performance of detectors, based on the Neyman–Pearson approach.

In Section 3.5, we studied the composite hypothesis testing problem. A composite hypothesis is characterized by an unknown parameter. When the parameter was a random variable with known density function, we applied the likelihood ratio test by averaging the conditional density function corresponding to the hypotheses, over all possible values of the parameter. However, if the parameter were not random but still unknown, Bayes test was no longer applicable, and instead we used the Neyman–Pearson test. Furthermore, when it was possible to apply the Neyman–Pearson test to all possible values of the parameter, a *uniformly most powerful* test was said to exist. Otherwise, a different approach which estimates the parameter should be considered. This will be described in the next chapter. We concluded this chapter with a brief section on sequential detection.

PROBLEMS

3.1 Consider the hypothesis testing problem in which

$$f_{Y|H_1}(y|H_1) = \frac{1}{2} \operatorname{rect}(\frac{y-1}{2})$$

and

$$f_{Y|H_0}(y|H_0) = e^{-y} \text{ for } y > 0$$

(a) Set up the likelihood ratio test and determine the decision regions.
(b) Find the minimum probability of error when
   (i) $P_0 = \frac{1}{2}$ (ii) $P_0 = \frac{2}{3}$ (iii) $P_0 = \frac{1}{3}$

3.2 Consider the hypothesis testing problem in which

$$f_{Y|H_0}(y|H_0) = \operatorname{rect}(y - \frac{1}{2})$$

and

$$f_{Y|H_1}(y|H_1) = \frac{1}{2} \, \text{rect}(\frac{y-1}{2})$$

(a) Set up the likelihood ratio test and determine the decision regions.
(b) Calculate P_F the probability of false alarm and P_M the probability of miss.

3.3 Consider the following binary hypothesis testing problem

$$H_1 : Y = S \; + \; N$$
$$H_0 : Y = \qquad N$$

where S and N are statistically independent random variables with probability density functions:

$$f_S(s) = \begin{cases} \frac{1}{2} & , \; -1 < s < 1 \\ 0 & , \; \text{otherwise} \end{cases}$$

and

$$f_N(n) = \begin{cases} \frac{1}{4} & , \; -2 < n < 2 \\ 0 & , \; \text{otherwise} \end{cases}$$

(a) Set up the likelihood ratio test and determine the decision regions when
  (i) $\eta = \frac{1}{4}$ (ii) $\eta = 1$ (iii) $\eta = 2$
(b) Find the probability of false alarm and the probability of detection for the three values of η in part (a).
(c) Sketch the ROC.

3.4 The conditional density functions corresponding to the hypotheses H_1 and H_0 are given by

$$f_{Y|H_0}(y|H_0) = \frac{1}{\sqrt{2\pi}} \, e^{-\frac{y^2}{2}}$$

$$f_{Y|H_1}(y|H_1) = \frac{1}{2} \, e^{-|y|}$$

(a) Find the likelihood ratio and determine the decision regions.
(b) Find the probability of false alarm and the probability of detection assuming minimum probability of error and $P_0 = 2/3$.
(c) Discuss the performance of the minimax test for the cost assignments as in Part (b).
(d) Determine the decision rule based on the Neyman–Pearson test for a probability of false alarm of 0.2.

3.5 In a binary hypothesis problem, the observed random variable under each hypothesis is

$$f_{Y|H_j}(y|H_j) = \frac{1}{\sqrt{2\pi}} e^{-\frac{(y-m_j)^2}{2}}, j = 0, 1$$

where $m_0 = 0$ and $m_1 = 1$.
(a) Find the decision rule for minimum probability of error and $P_0 = P_1$.
(b) Find the decision rule for a Neyman–Pearson if $P_F = 0.005$.
(c) Find P_D based on the test of (b).

3.6 Consider the hypothesis testing problem where we are given K independent observation.

$$H_1 : Y_k = m \;+\; N_k \quad, k = 1, 2, \cdots, K$$
$$H_0 : Y_k = N_k \quad, k = 1, 2, \cdots, K$$

where m is a constant and N_k is a zero mean Gaussian random variable with variance σ^2.
(a) Compute the likelihood ratio.
(b) Obtain the decision rule in terms of the sufficient statistic and the threshold γ.

3.7 Repeat Problem 3.6 assuming that m is zero and the variance of N_k, $k = 1, 2, \cdots, K$, under H_1 and H_0 are σ_1^2 and σ_0^2 $(\sigma_1 > \sigma_0)$, respectively.

3.8 Consider Problem 3.7.

(a) Obtain an expression for the probability of false alarm and the probability of miss for $K = 1$.

(b) Plot the *ROC* if $\sigma_1^2 = 2\sigma_0^2 = 2$.

(c) Determine the threshold for the minimax criterion assuming $C_{00} = C_{11} = 0$ and $C_{01} = C_{10}$.

3.9 The conditional density function of the observed random variable under each hypothesis is

$$f_{Y|H_j}(y|H_j) = \frac{1}{\sqrt{2\pi}\sigma} e^{-\frac{(y-m_j)^2}{2\sigma^2}} \ , \ j = 0, 1, 2$$

where $m_0 = 0$, $m_1 = -m$, and $m_2 = m$.

(a) Find the decision rule (draw the decision regions) assuming minimum probability of error criterion, and equal *a priori* probabilities.

(b) Calculate the minimum probability of error.

3.10 Repeat Problem 3.9 assuming

$$H_0 : m_0 = 0 \ , \ \sigma_0 = 1$$
$$H_1 : m_1 = 1 \ , \ \sigma_1 = 1$$
$$H_2 : m_2 = 0 \ , \ \sigma_2 = 2$$

3.11 Consider the Problem of 3.6 where m, now denoted M, is not a constant, but a zero mean Gaussian random variable with variance σ_M^2. M and N_k, $k = 1, \cdots, K$ are independent. Determine the optimum decision rule.

3.12 Consider the following hypothesis testing problem

$$H_1 : Y_k = M_k \ + \ N_k \ , \ k = 1, 2, \cdots, K$$
$$H_0 : Y_k = \qquad\quad N_k \ , \ k = 1, 2, \cdots, K$$

where M_k and N_k, $k = 1, 2, \cdots, K$, are statistically independent zero mean Gaussian random variables. Their respective variances are σ_M^2 and σ_N^2, where σ_N^2 is normalized to one, but σ_M^2 is unknown. Does a UMP test exist?

3.13 Consider the following composite hypothesis testing problem. The observations are $\mathbf{Y} = [Y_1, Y_2, \cdots, Y_K]^T$ where Y_k, $k = 1, 2, \cdots, K$, are independent Gaussian random variables with a known variance $\sigma^2 = 1$. The mean m_j, $j = 0, 1$, under each hypothesis is

$$H_1 : m_1 = m \quad , \quad m > 0$$
$$H_0 : m_0 = 0$$

(a) Does a UMP test exist?
(b) If $P_F = 0.05$ and $m_1 = 1$, using a most powerful test, find the smallest value of K which will guarantee a power greater than 0.9

REFERENCES

1. Chernoff, H., and L.E. Moses, *Elementary Decision Theory*, John Wiley and Sons, New York, 1959.

2. Dudewicz, E.J., *Introduction to Statistics and Probability*, Holt, Rinehart and Winston, New York, 1976.

3. Ehrenfeld, S., and S.B. Littauer, *Introduction to Statistical Method*, McGraw-Hill, New York, 1964.

4. Freeman, H., *Introduction to Statistical Inference*, Addison-Wesley, Reading, MA, 1963.

5. Hoel, P.G., *Introduction to Mathematical Statistics*, John Wiley and Sons, New York, 1962.

6. Hogg, R.V., and A.T. Craig, *Introduction to Mathematical Statistics*, Macmillan, New York, 1978.

7. Melsa, J.L., and D.L. Cohn, *Decision and Estimation Theory*, McGraw-Hill, New York, 1978.

8. Mohanty, N., *Signal Processing: Signals, Filtering, and Detection*, Van Nostrand Reinhold, New York, 1987.

9. Schwartz, M., *Information Transmission, Modulation, and Noise*, McGraw-Hill, New York, 1980.

10. Shanmugan, K.S., and A.M. Breipohl, *Random Signals: Detection, Estimation and Data Analysis*, John Wiley and Sons, New York, 1988.

11. Srinath, M.D., and P.K. Rajasekaran, *An Introduction to Statistical Signal Processing with Applications*, John Wiley and Sons, New York, 1979.

12. Urkowitz, H., *Signal Theory and Random Processes*, Artech House, Norwood, MA, 1983.

13. Van Trees, H.L., *Detection, Estimation, and Modulation Theory*, Part I, John Wiley and Sons, New York, 1968.

14. Wald, A., *Sequential Analysis*, Dover, New York, 1973.

15. Whalen, A.D., *Detection of Signals in Noise*, Academic, New York, 1971.

16. Wozencraft, J.M., and I.M. Jacobs, *Principles of Communication Engineering*, John Wiley and Sons, 1965.

Chapter 4

Parameter Estimation

4.1 INTRODUCTION

In the previous chapter, we considered the problem of detection theory where the receiver receives a noisy version of a signal and has to make a decision as to which hypothesis is true among the M possible hypotheses. In the binary case, the receiver had to decide between the null hypothesis H_0 and the alternate hypothesis H_1.

In this chapter, we assume that the receiver has made a decision in favor of the true hypothesis but some parameters associated with the signal may not be known. The goal is to estimate those parameters in an optimum fashion based on a finite number of samples of the signal.

Let $Y_1, Y_2, \cdots, Y_K$ be K independent, and identically distributed samples of a random variable Y with some density function depending on an unknown parameter θ. Let $y_1, y_2, \cdots, y_K$ be the corresponding values of the samples $Y_1, Y_2, \cdots, Y_K$ and $g(Y_1, \cdots, Y_K)$ a function (a statistic) of the samples used to estimate the parameter θ. We call

$$\hat{\theta} = g(Y_1, \cdots, Y_K) \tag{4.1}$$

the *estimator* of θ. The value that the statistic assumes is called the *estimate* of θ and is equal to $\hat{\theta} = g(y_1, \cdots, y_K)$. In order to avoid any confusion between a random variable and its value, it should be noted that $\hat{\theta}$, the estimate of θ, is actually $g(Y_1, \cdots, Y_K)$. Consequently, when we speak of the mean of $\hat{\theta}$, $E[\hat{\theta}]$, we are actually referring to $E[g(Y_1, \cdots, Y_K)]$.

The parameters to be estimated may be random or nonrandom. The estimation of random parameters is known as *Bayes estimation*, while the estimation of nonrandom parameters is referred to as *maximum likelihood estimation (MLE)*.

4.2 MAXIMUM LIKELIHOOD ESTIMATION

As mentioned in the previous section, the procedure commonly used to estimate nonrandom parameters is the *Maximum Likelihood (ML)* estimation. Let $Y_1, Y_2, \cdots, Y_K$ be K observations of the random variable Y, with sample values $y_1, y_2, \cdots, y_K$. These random variables are independent, and identically distributed. Let $f_{Y|\Theta}(y|\theta)$ denote the density function of the random variable Y. Note that the density function of Y depends on the parameter θ, $\theta \epsilon \Theta$, which needs to be estimated. The *likelihood function*, $L(\theta)$, is

$$
\begin{aligned}
L(\theta) &= f_{Y_1,\cdots,Y_K|\Theta}(y_1, \cdots, y_k|\theta) \\
&= f_{\mathbf{Y}|\Theta}(\mathbf{y}|\theta) \\
&= \prod_{k=1}^{K} f_{Y_k|\Theta}(y_k|\theta)
\end{aligned}
\tag{4.2}
$$

The value $\hat{\theta}$ which maximizes the likelihood function is called the *maximum likelihood estimator* of θ. In order to maximize the likelihood function, standard techniques of calculus may be used. Because the logarithmic function $\ln x$ is a monotonically increasing function of x, as was shown in Chapter 3, maximizing $L(\theta)$ is equivalent to maximizing $\ln L(\theta)$. Hence, it can be shown that a necessary but not sufficient condition to obtain the ML estimate $\hat{\theta}$ is to solve the *likelihood equation*

$$
\frac{\partial}{\partial \theta} \ln f_{Y|\Theta}(y|\theta) = 0
\tag{4.3}
$$

Invariance Property: Let $L(\theta)$ be a likelihood function of θ and $g(\theta)$ be a one-to-one function of θ (that is, if $g(\theta_1) = g(\theta_2) \Leftrightarrow \theta_1 = \theta_2$). If $\hat{\theta}$ is an MLE of θ, then $g(\hat{\theta})$ is an MLE of $g(\theta)$.

In many situations, we need to estimate more than one parameter at the same time. Let $\boldsymbol{\theta}$ be a vector such that

$$\boldsymbol{\theta} = \begin{bmatrix} \theta_1 \\ \theta_2 \\ \vdots \\ \theta_M \end{bmatrix} \tag{4.4}$$

Then, Equation (4.3) becomes the following set of simultaneous likelihood equations :

$$\frac{\partial}{\partial \theta_1} ln f_{\mathbf{Y}|\Theta}(y_1, \cdots, y_K | \theta_1, \cdots, \theta_M)$$

$$\frac{\partial}{\partial \theta_2} ln f_{\mathbf{Y}|\Theta}(y_1, \cdots, y_K | \theta_1, \cdots, \theta_M)$$

$$\vdots$$

$$\frac{\partial}{\partial \theta_M} ln f_{\mathbf{Y}|\Theta}(y_1, \cdots, y_K | \theta_1, \cdots, \theta_M) \tag{4.5}$$

Example 4.1

In Example 3.2, the received signal under hypotheses H_1 and H_0 was

$$H_1 : Y_k = \quad m \quad + \quad N_k \, , \, k = 1, 2, \cdots, K$$
$$H_0 : Y_k = \qquad\qquad N_k \, , \, k = 1, 2, \cdots, K$$

Detection theory (Chapter 3) was used to determine which of the two hypotheses was true. In this chapter of estimation theory, we assume that H_1 is true. However, the constant m is not known and the problem is to find the ML estimate $\hat{m}_{ml}$.

Solution.

The parameter $\hat{\theta}$ to be determined in this example is $\hat{m}_{ml}$ where the mean $m \epsilon M$. Since the samples are independent and identically distributed, the likelihood function, using Equation (4.2), is

$$f_{\mathbf{Y}|M}(\mathbf{y}|m) = \prod_{k=1}^{K} \frac{1}{\sqrt{2\pi}\sigma} e^{-\frac{(y_k-m)^2}{2\sigma^2}}$$

$$= \frac{1}{(2\pi)^{\frac{K}{2}}\sigma^K} e^{-\sum_{k=1}^{K}\frac{(y_k-m)^2}{2\sigma^2}}$$

Taking the logarithm on both sides, we obtain

$$\ln f_{\mathbf{Y}|M}(\mathbf{y}|m) = \ln(\frac{1}{(2\pi)^{\frac{K}{2}}\sigma^K}) - \sum_{k=1}^{K} \frac{(y_k-m)^2}{2\sigma^2}$$

The ML estimate is obtained by solving the likelihood equation as shown in Equation (4.3). Hence,

$$\frac{\partial \ln f_{\mathbf{Y}|M}(\mathbf{y}|m)}{\partial m} = \sum_{k=1}^{K} \frac{y_k - m}{\sigma^2}$$

$$= \sum_{k=1}^{K} \frac{y_k}{\sigma^2} - \frac{Km}{\sigma^2}$$

$$= \frac{K}{\sigma^2}(\frac{1}{K}\sum_{k=1}^{K} y_k - m) = 0$$

or

$$m = \frac{1}{K}\sum_{k=1}^{K} y_k$$

Thus, the ML estimator is

$$\hat{m}_{ml} = \frac{1}{K}\sum_{k=1}^{K} Y_k$$

Example 4.2

Obtain the *MLE* of $\theta = \sigma^2$ in Example 4.1, assuming that the mean m is known, but the variance σ^2 is unknown.

Solution.
The likelihood function is

$$L(\sigma^2) = \frac{1}{(2\pi)^{\frac{K}{2}} \sigma^K} e^{-\sum_{k=1}^{K} \frac{(y_k - m)^2}{2\sigma^2}}$$

Taking the logarithm, we obtain

$$\ln L(\sigma^2) = -\frac{K}{2} \ln 2\pi - K \ln \sigma - \sum_{k=1}^{K} \frac{(y_k - m)^2}{2\sigma^2}$$

Observe that maximizing $\ln L(\sigma^2)$ with respect to σ^2 is equivalent to minimizing

$$g(\sigma^2) = K \ln \sigma + \sum_{k=1}^{K} \frac{(y_k - m)^2}{2\sigma^2}$$

Using the invariance property, it is easier to differentiate $g(\sigma^2)$, with respect to σ, to obtain $\hat{\sigma}_{ml}$ the *MLE* of σ, instead of $\hat{\sigma}^2_{ml}$ the *MLE* of σ^2. Hence,

$$\frac{dg(\sigma^2)}{d\sigma} = \frac{K}{\sigma} - \sum_{k=1}^{K} \frac{(y_k - m)^2}{\sigma^3} = 0$$

or

$$\hat{\sigma} = \sqrt{\frac{1}{K} \sum_{k=1}^{K} (Y_k - m)^2}$$

Consequently, the *MLE* of σ^2 is

$$\hat{\sigma}^2_{ml} = \frac{1}{K} \sum_{k=1}^{K} (Y_k - m)^2$$

4.3 SOME CRITERIA FOR GOOD ESTIMATORS

Since the estimator $\hat{\theta}$ is a random variable and may assume more than one value, some characteristics of a "good" estimate need to be determined.

Unbiased Estimate: We say $\hat{\theta}$ is an *unbiased* estimator for θ if $E[\hat{\theta}] = \theta$ for all θ.

Bias of Estimator: Let $E[\hat{\theta}] = \theta + b(\theta)$.

(i) If $b(\theta)$ does not depend on θ $\{b(\theta) = b\}$, we say that the estimator $\hat{\theta}$ has a *known bias*. That is, $(\hat{\theta} - b)$ is an unbias estimate.
(ii) When $b(\theta) \neq b$, an unbiased estimate cannot be obtained, since θ is unknown. In this case, we say that the estimator has an *unknown bias*.

The fact that an estimator is unbiased, which means that the average value of the estimate is close to the true value, does not necessarily guarantee that the estimator is "good". This is easily seen by the conditional density function of the estimator shown in Figure 4.1.

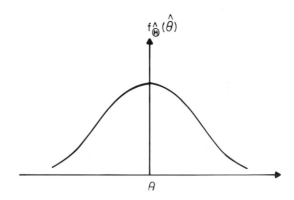

Figure 4.1 Density function of the unbiased estimator $\hat{\theta}$.

We observe that even though the estimate is unbiased, sizable errors are likely to occur since the variance of the estimate is large. However, if the variance is small, the variability of the estimator about its expected value is also small. Consequently, the variability of the estimator is close to the true value, since the estimate is unbiased, which is a desired feature. Hence, we say that the second measure of quality of the estimate is to have a small variance.

Unbiased Minimum Variance: $\hat{\theta}$ is an unbiased minimum variance estimate of θ if for all estimates θ' such that $E[\theta'] = \theta$, we have $\text{var}(\hat{\theta}) \leq \text{var}(\theta')$ for all θ'. That is, $\hat{\theta}$ has the smallest variance among all unbiased estimates of θ.

Consistent Estimate: $\hat{\theta}$ is a consistent estimate of the parameter θ if

$$\lim_{K \to \infty} P(|\hat{\theta} - \theta| > \epsilon) = 0 \qquad \text{for all } \epsilon > 0$$

where $P(\cdot)$ denotes probability.

Applying the above definition to verify the consistency of an estimate is not simple. The following theorem is used instead.

Theorem :   Let $\hat{\theta}$ be an unbiased estimator of θ based on K observed samples. If

$$\lim_{K \to \infty} E[\hat{\theta}] = \theta$$

and if

$$\lim_{K \to \infty} \text{var}[\hat{\theta}] = 0$$

then $\hat{\theta}$ is a consistent estimator of θ.

Example 4.3

Verify if the estimator of Example 4.1 is an unbiased estimate of m.

Solution.
The estimator $\hat{m}_{ml}$ is unbiased if $E[\hat{m}_{ml}] = m$. After substitution, we obtain

$$
\begin{aligned}
E[\hat{m}_{ml}] &= E[\frac{1}{K}\sum_{k=1}^{K} Y_k] \\
&= \frac{1}{K} E[\sum_{k=1}^{K} Y_k] = \frac{1}{K}Km = m
\end{aligned}
$$

Hence, $\hat{m}_{ml}$ is unbiased.

Example 4.4

Is the estimator $\hat{\sigma}_{ml}^2$ of Example 4.2 unbiased ?

Solution.
The estimator $\hat{\sigma}_{ml}^2$ is unbiased if $E[\hat{\sigma}_{ml}^2] = \sigma^2$. That is

$$
E[\frac{1}{K}\sum_{k=1}^{K}(Y_k - m)^2] = \frac{1}{K} E[Km^2 + \sum_{k=1}^{K}Y_k^2 - 2m\sum_{k=1}^{K}Y_k] = \sigma^2
$$

Hence, $\hat{\sigma}_{ml}^2$ is unbiased.

4.4 BAYES ESTIMATION

In Bayes estimation, we assign a cost $c(\theta, \hat{\theta})$ to all pairs $(\theta, \hat{\theta})$. The cost is a nonnegative real valued function of the two random variables θ and $\hat{\theta}$. As in Bayes detection, the risk function is defined to be the average value of the cost; that is,

$$
R = E[c(\theta, \hat{\theta})] \tag{4.6}
$$

The goal is to minimize the risk function in order to obtain $\hat{\theta}$, which is the optimum estimate. In many problems, only the error $\tilde{\theta}$ between the estimate and the true value is of interest

$$\tilde{\theta} = \theta - \hat{\theta} \tag{4.7}$$

Consequently, we will only consider costs which are a function of the error. The following three cases will be studied, and their corresponding sketches are shown in Figure 4.2.

(i) Squared error

$$c(\theta, \hat{\theta}) = (\theta - \hat{\theta})^2 \tag{4.8}$$

(ii) Absolute value of error

$$c(\theta, \hat{\theta}) = |\theta - \hat{\theta}| \tag{4.9}$$

(iii) Uniform cost function

$$c(\theta, \hat{\theta}) = \begin{cases} 1 & , |\theta - \hat{\theta}| \geq \frac{\Delta}{2} \\ 0 & , |\theta - \hat{\theta}| < \frac{\Delta}{2} \end{cases} \tag{4.10}$$

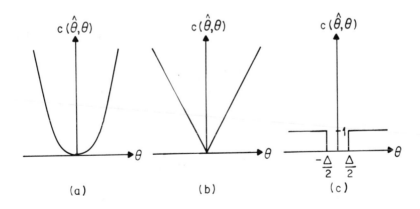

Figure 4.2 Cost functions: (a) squared error; (b) absolute value of error; (c) uniform.

The unknown parameter is assumed to be a continuous random variable with density function $f_\Theta(\theta)$. The risk function can then be expressed as

$$
\begin{aligned}
R &= E[c(\theta, \hat{\theta})] \\
&= \int_{-\infty}^{\infty} \int_{-\infty}^{\infty} c(\theta, \hat{\theta}) f_{\Theta, \mathbf{Y}}(\theta, \mathbf{y}) d\theta d\mathbf{y}
\end{aligned}
\tag{4.11}
$$

Note that we take the cost average over all possible values of θ and $\mathbf{Y}$ where $\mathbf{Y}$ is the vector $[Y_1, Y_2, \cdots, Y_K]^T$. We now find the estimator for the three cost functions considered.

Case (i): Squared error

The estimator which minimizes the risk function for the cost given in Equation (4.8) is referred to as the *Minimum Mean Square Estimate* (MMSE). The corresponding risk function is deneted by R_{ms}. We have

$$
\begin{aligned}
R_{ms} &= \int_{-\infty}^{\infty} \int_{-\infty}^{\infty} (\theta - \hat{\theta})^2 f_{\Theta, \mathbf{Y}}(\theta, \mathbf{y}) d\theta d\mathbf{y} \\
&= \int_{-\infty}^{\infty} d\theta \int_{-\infty}^{\infty} (\theta - \hat{\theta})^2 f_{\Theta, \mathbf{Y}}(\theta, \mathbf{y}) d\theta d\mathbf{y}
\end{aligned}
\tag{4.12}
$$

Using Equation (1.90), the risk function can be rewritten as

$$
R_{ms} = \int_{-\infty}^{\infty} d\mathbf{y} f_{\mathbf{Y}}(\mathbf{y}) \left[\int_{-\infty}^{\infty} (\theta - \hat{\theta})^2 f_{\Theta|\mathbf{Y}}(\theta|\mathbf{y}) d\theta \right]
\tag{4.13}
$$

Since the density function $f_{\mathbf{Y}}(\mathbf{y})$ is nonnegative, minimizing R_{ms} is equivalent to minimizing the expression in brackets of the above equation. Hence, taking the derivative with respect to $\hat{\theta}$ and setting it equal to zero we have

$$
\frac{d}{d\hat{\theta}} \int_{-\infty}^{\infty} (\theta - \hat{\theta})^2 f_{\Theta|\mathbf{Y}}(\theta|\mathbf{y}) d\theta = 0
\tag{4.14}
$$

Using Leibnitz's rule given in Equation (1.27), we obtain

$$
\begin{aligned}
\hat{\theta}_{ms} &= \int_{-\infty}^{\infty} \theta f_{\Theta|\mathbf{Y}}(\theta|\mathbf{y}) d\theta \\
&= E[\theta|\mathbf{Y}]
\end{aligned}
\tag{4.15}
$$

That is, the minimum mean-square estimate $\hat{\theta}_{ms}$ represents the conditional mean of θ given $\mathbf{Y}$. It can easily be shown that the second derivative with respect to $\hat{\theta}_{ms}$ is positive-definite, which corresponds to a unique minimum of R_{ms}, and is given by

$$
\begin{aligned}
R_{ms} &= \int_{-\infty}^{\infty} d\mathbf{y}\, f_{\mathbf{Y}}(\mathbf{y}) \int_{-\infty}^{\infty} (\theta - \hat{\theta}_{ms})^2 f_{\Theta|\mathbf{Y}}(\theta, \mathbf{y}) d\theta \\
&= \int_{-\infty}^{\infty} d\mathbf{y}\, f_{\mathbf{Y}}(\mathbf{y}) \int_{-\infty}^{\infty} (\theta - E[\theta|\mathbf{Y}])^2 f_{\Theta|\mathbf{Y}}(\theta|\mathbf{y}) d\theta
\end{aligned}
\tag{4.16}
$$

The conditional variance of θ given $\mathbf{Y}$ is

$$
\mathrm{var}(\theta|\mathbf{Y}) = \int_{-\infty}^{\infty} (\theta - E[\theta|\mathbf{Y}])^2 f_{\Theta|\mathbf{Y}}(\theta|\mathbf{y}) d\theta
\tag{4.17}
$$

Hence, R_{ms} is just the conditional variance of θ given $\mathbf{Y}$, averaged over all possible values of $\mathbf{Y}$. This estimation procedure using the squared error criterion is sometimes referred to as the *minimum variance of error estimation*.

Case (ii): Absolute error

In this case, the risk function is

$$
R_{abs} = \int_{-\infty}^{\infty} f_{\mathbf{Y}}(\mathbf{y}) \Big[\int_{-\infty}^{\infty} |\theta - \hat{\theta}| f_{\Theta|\mathbf{Y}}(\theta|\mathbf{y}) d\theta \Big] d\mathbf{y}
\tag{4.18}
$$

Using the same arguments as in case (i), the risk can be minimized by minimizing the integral in the brackets, which is given by

$$
\int_{-\infty}^{\hat{\theta}} (\hat{\theta} - \theta) f_{\Theta|\mathbf{Y}}(\theta|\mathbf{y}) d\theta + \int_{\hat{\theta}}^{\infty} (\theta - \hat{\theta}) f_{\Theta|\mathbf{Y}}(\theta|\mathbf{y}) d\theta
\tag{4.19}
$$

Differentiating Equation (4.19) with respect to $\hat{\theta}$, and setting the result equal to zero, we obtain

$$
\int_{-\infty}^{\hat{\theta}_{abs}} f_{\Theta|\mathbf{Y}}(\theta|\mathbf{y}) d\theta = \int_{\hat{\theta}_{abs}}^{\infty} f_{\Theta|\mathbf{Y}}(\theta|\mathbf{y}) d\theta
\tag{4.20}
$$

That is, the estimate $\hat{\theta}_{abs}$ is just the *median* of the conditional density function $f_{\Theta|\mathbf{Y}}(\theta|\mathbf{y})$.

Case (iii) : Uniform cost function

For the uniform cost function, the Bayes risk becomes

$$
\begin{aligned}
R_{unf} &= \int_{-\infty}^{\infty} f_{\mathbf{Y}}(\mathbf{y})[\int_{-\infty}^{\hat{\theta}-\frac{\Delta}{2}} f_{\Theta|\mathbf{Y}}(\theta|\mathbf{y})d\theta + \int_{\hat{\theta}+\frac{\Delta}{2}}^{\infty} f_{\Theta|\mathbf{Y}}(\theta|\mathbf{y})d\theta]dy \\
&= \int_{-\infty}^{\infty} f_{\mathbf{Y}}(\mathbf{y})[1 - \int_{\hat{\theta}-\frac{\Delta}{2}}^{\hat{\theta}+\frac{\Delta}{2}} f_{\Theta|\mathbf{Y}}(\theta|\mathbf{y})d\theta] \qquad (4.21)
\end{aligned}
$$

where

$$
\int_{\hat{\theta}-\frac{\Delta}{2}}^{\hat{\theta}+\frac{\Delta}{2}} f_{\Theta|\mathbf{Y}}(\theta|\mathbf{y})d\theta = P[(\hat{\theta} - \frac{\Delta}{2} \le \Theta \le \hat{\theta} + \frac{\Delta}{2})|\mathbf{y}] \qquad (4.22)
$$

$P[\cdot]$ denotes probability. Hence, the risk R_{unf} is minimized by maximizing Equation (4.22). Note that in maximizing Equation (4.22) (minimizing R_{unf}), we are searching for the estimate $\hat{\theta}$, which maximizes $f_{\Theta|\mathbf{Y}}(\theta|\mathbf{y})$. This is called the *maximum a posteriori estimate*, $\hat{\theta}_{MAP}$, which is defined as

$$
\frac{\partial f_{\Theta|\mathbf{Y}}(\theta|\mathbf{y})}{\partial \theta}\bigg|_{\theta=\hat{\theta}_{MAP}} = 0 \qquad (4.23)
$$

Using the logarithm, which is a monotonically increasing function, Equation (4.23) becomes

$$
\frac{\partial \ln f_{\Theta|\mathbf{Y}}(\theta|\mathbf{y})}{\partial \theta} = 0 \qquad (4.24)
$$

Equation (4.24) is called the *MAP* equation. This is a necessary but not sufficient condition because $f_{\Theta|\mathbf{Y}}(\theta|\mathbf{y})$ may have several local maxima. Using Bayes rule:

$$
f_{\Theta|\mathbf{Y}}(\theta|\mathbf{y}) = \frac{f_{\mathbf{Y}|\Theta}(\mathbf{y}|\theta)f_{\Theta}(\theta)}{f_{\mathbf{Y}}(\mathbf{y})} \qquad (4.25)
$$

and the fact that

$$
\ln f_{\Theta|\mathbf{Y}}(\theta|\mathbf{y}) = \ln f_{\mathbf{Y}|\Theta}(\mathbf{y}|\theta) + \ln f_{\Theta}(\theta) - \ln f_{\mathbf{Y}}(\mathbf{y}) \qquad (4.26)
$$

the MAP equation may be rewritten as

$$\frac{\partial \ln f_{\Theta|\mathbf{Y}}(\theta|\mathbf{y})}{\partial \theta} = \frac{\partial \ln f_{\mathbf{Y}|\Theta}(\mathbf{y}|\theta)}{\partial \theta} + \frac{\partial \ln f_{\Theta}(\theta)}{\partial \theta} = 0 \qquad (4.27)$$

We always assume that Δ is sufficiently small so that the estimate $\hat{\theta}_{MAP}$ is given by the MAP equation.

Example 4.5

Consider the problem where the observed samples are

$$Y_k = M + N_k \ , \ k = 1, 2, \cdots, K$$

M and N_k are statistically independent Gaussian random variables, with mean zero and variances σ^2. Find $\hat{m}_{ms}$, $\hat{m}_{MAP}$ and $\hat{m}_{abs}$.

Solution.
From Equation (4.15), the estimate $\hat{m}_{ms}$ is the conditional mean of m given $\mathbf{Y}$. The density function $f_{M|\mathbf{Y}}(m|\mathbf{y})$ is expressed as

$$f_{M|\mathbf{Y}}(m|\mathbf{y}) = \frac{f_{\mathbf{Y}|M}(\mathbf{y}|m)f_M(m)}{f_{\mathbf{Y}}(\mathbf{y})}$$

where

$$f_M(m) = \frac{1}{\sqrt{2\pi}\sigma} e^{-\frac{m^2}{2\sigma^2}}$$

$$f_{\mathbf{Y}|M}(\mathbf{y}|m) = \prod_{k=1}^{K} \frac{1}{\sqrt{2\pi}\sigma} e^{-\frac{(y_k-m)^2}{2\sigma^2}}$$

and the marginal density function $f_{\mathbf{Y}}(\mathbf{y})$ is

$$f_{\mathbf{Y}}(\mathbf{y}) = \int_{-\infty}^{\infty} f_{M,\mathbf{Y}}(m, \mathbf{y})dm$$

$$= \int_{-\infty}^{\infty} f_{\mathbf{Y}|M}(\mathbf{y}|m)f_M(m)dm$$

Note that $f_{M|\mathbf{Y}}(m|\mathbf{y})$ is a function of m, but $f_{\mathbf{Y}}(\mathbf{y})$ is the constant with $\mathbf{y}$ as a parameter needed to maintain the area under the conditional density function equal to one. That is,

$$f_{M|\mathbf{Y}}(m|\mathbf{y}) = \frac{\frac{1}{(\sqrt{2\pi}\sigma)^{K+1}}}{f_{\mathbf{Y}}(\mathbf{y})} e^{[-\frac{1}{2\sigma^2}\{\sum_{k=1}^{K}(y_k-m)^2+m^2\}]}$$

Expanding the exponent, we have

$$\sum_{k=1}^{K}(y_k^2 - 2y_k m + m^2) + m^2$$

$$= m^2(K+1) - 2m\sum_{k=1}^{K} y_k + \sum_{k=1}^{K} y_k^2$$

$$= (K+1)[m^2 - \frac{2m}{K+1}\sum_{k=1}^{K} y_k] + \sum_{k=1}^{K} y_k^2$$

$$= (K+1)(m - \frac{1}{k+1}\sum_{k=1}^{K} y_k)^2 - \frac{1}{K+1}(\sum_{k=1}^{K} y_k)^2 + \sum_{k=1}^{K} y_k^2$$

The last two terms in the exponent do not involve m, and can be absorbed in the multiplicative constant to obtain

$$f_{M|\mathbf{Y}}(m|\mathbf{y}) = c(\mathbf{y})\, e^{[-\frac{1}{2\sigma_m^2}(m - \frac{1}{K+1}\sum_{k=1}^{K} y_k^2)^2]}$$

where

$$\sigma_m = \frac{\sigma}{\sqrt{K+1}}$$

By inspection, the conditional mean is

$$\hat{m}_{ms} = E[M|\mathbf{Y}]$$

$$= \frac{1}{K+1}\sum_{k=1}^{K} Y_k$$

According to Equation (4.13), R_{ms} is given by

$$R_{ms} = \int_{-\infty}^{\infty} \text{var}(M|\mathbf{y})f_{\mathbf{Y}}(\mathbf{y})dy$$

$$= \sigma_m^2$$

Hence,

$$R_{ms} = \sigma_m^2 \int_{-\infty}^{\infty} f_{\mathbf{Y}}(\mathbf{y})dy = \sigma_m^2$$

since $\int_{-\infty}^{\infty} f_{\mathbf{Y}}(\mathbf{y})dy = 1$.

The MAP estimate is obtained using the MAP Equation of (4.24) and Equation (4.25). Taking the logarithm of $f_{M|\mathbf{Y}}(m|\mathbf{y})$, we have

$$\ln f_{M|\mathbf{Y}}(m|\mathbf{y}) = \ln c(\mathbf{y}) - \frac{1}{2\sigma_m^2}(m - \frac{1}{K+1}\sum_{k=1}^{K} y_k)^2$$

$$\frac{\partial \ln f_{M|\mathbf{Y}}(m|\mathbf{y})}{\partial m} = -\frac{1}{\sigma_m^2}(m - \frac{1}{K+1}\sum_{k=1}^{K} y_k) = 0$$

$$\Longrightarrow \hat{m}_{MAP} = \frac{1}{K+1}\sum_{k=1}^{K} Y_k$$

That is, $\hat{m}_{MAP} = \hat{m}_{ms}$. We could have obtained this result directly by inspection, since we have shown that $f_{M|\mathbf{Y}}(m|\mathbf{y})$ is Gaussian. Consequently, the maximum of $f_{M|\mathbf{Y}}(m|\mathbf{y})$ occurs at its mean value.

Using the fact that the Gaussian density function is symmetric, and the $\hat{m}_{abs}$ is the median of the conditional density function $f_{M|\mathbf{Y}}(m|\mathbf{y})$, we conclude that

$$\begin{aligned}
\hat{m}_{abs} &= \hat{m}_{ms} = \hat{m}_{MAP} \\
&= \frac{1}{K+1}\sum_{k=1}^{K} Y_k
\end{aligned}$$

Example 4.6

Find $\hat{x}_{ms}$ the minimum mean square error and $\hat{x}_{map}$ the maximum *a posteriori* estimators of x from the observation:

$$Y = X + N$$

X and N are random variables with density functions:

$$f_X(x) = \frac{1}{2}\delta(x) + \frac{1}{2}\delta(x-1)$$

and

$$f_N(n) = \frac{1}{2}e^{-|n|} = \begin{cases} \frac{1}{2}e^n & , n \leq 0 \\ \frac{1}{2}e^{-n} & , n \geq 0 \end{cases}$$

Solution.

The estimate $\hat{x}_{map}$ maximizes the density function $f_{X|Y}(x|y)$. Since the probability density function:

$$f_{Y|X}(y|x) = \frac{1}{2} e^{-|n-x|}$$

the probability density function of Y is

$$
\begin{aligned}
f_Y(y) &= \int_{-\infty}^{\infty} f_{Y|X}(y|x) f_X(x) dx \\
&= \frac{1}{4} \int_{-\infty}^{\infty} e^{-|n-x|} [\delta(x) + \delta(x-1)] dx \\
&= \frac{1}{4} \{e^{-|n|} + e^{-|n-1|}\} = \begin{cases} \frac{1}{4}(e^n + e^{n-1}) & , n < 0 \\ \frac{1}{4}(e^{-n} + e^{n-1}) & , 0 \leq n < 1 \\ \frac{1}{4}(e^{-n} + e^{-n+1}) & , n \geq 1 \end{cases}
\end{aligned}
$$

The *a posteriori* density function is, from Equation (4.25), given by

$$
\begin{aligned}
f_{X|Y}(x|y) &= \frac{f_{Y|X}(y|x) f_X(x)}{f_Y(y)} \\
&= \frac{e^{-|n-x|} [\delta(x) + \delta(x-1)]}{e^{-|n|} + e^{-|n-1|}}
\end{aligned}
$$

$f_{X|Y}(x|y)$ is zero except when $x = 0$ and $x = 1$. The above expression is maximized when $|n - x|$ is minimized. Since x can take two values only, but must be close to n, we have

$$\hat{x}_{map} = \begin{cases} 1 & \text{for } n \geq \frac{1}{2} \\ 0 & \text{for } n < \frac{1}{2} \end{cases}$$

The mean-square error estimate is the mean of the *a posteriori* density function as given by Equation (4.15). Hence,

$$
\begin{aligned}
\hat{x}_{ms} &= \int_{-\infty}^{\infty} x f_{X|Y}(x|y) dx \\
&= \int_{-\infty}^{\infty} x \frac{e^{-|n-x|} [\delta(x) + \delta(x-1)]}{e^{-|n|} + e^{-|n-1|}} dx
\end{aligned}
$$

since $\int_{-\infty}^{\infty} \delta(t - t_0)g(t)dt = g(t_0)$, the mean-square estimate is

$$\hat{x}_{ms} = \frac{e^{-|n-1|}}{e^{-|n|} + e^{-|n-1|}}$$

We see that the $\hat{x}_{map}$ is not identical to $\hat{x}_{ms}$.

4.5 CRAMER – RAO INEQUALITY

From the MAP Equation of (4.24), if we set the density function of Θ to zero, for all θ, we obtain the likelihood Equation (4.3). That is, the ML estimate can be considered as a special case of the MAP estimate. In this case, to check whether the estimate is "good" we need to compute its bias, error variance, and determine its consistency. It may be very difficult to obtain an expression for the error variance. In this case the "goodness" of the estimator is studied in terms of a lower bound of the error variance. This bound is the *Cramer-Rao bound* and is given by the following theorem.

Theorem :   Let the vector $\mathbf{Y} = (Y_1, \cdots, Y_K)$ represent K observations, and $\hat{\theta}$ be the unbiased estimator of θ, then,

$$\text{var}[(\hat{\theta} - \theta)|\theta] \geq \frac{1}{E[\{\frac{\partial \ln f_{\mathbf{Y}|\Theta}(\mathbf{y}|\theta)}{\partial \theta}\}^2]} \tag{4.28}$$

where

$$E[\{\frac{\partial \ln f_{\mathbf{Y}|\Theta}(\mathbf{y}|\theta)}{\partial \theta}\}^2] = -E[\frac{\partial^2 \ln f_{\mathbf{Y}|\Theta}(\mathbf{y}|\theta)}{\partial \theta^2}] \tag{4.29}$$

Proof: For an unbiased estimator $\hat{\theta}$, we have

$$E[\hat{\theta}|\theta] = \theta \tag{4.30}$$

$$\Longrightarrow E[(\hat{\theta} - \theta)|\theta] = \int_{-\infty}^{\infty} (\hat{\theta} - \theta)f_{\mathbf{Y}|\Theta}(\mathbf{y}|\theta)d\mathbf{y} = 0 \tag{4.31}$$

Differentiating Equation (4.31) with respect to θ, we obtain

$$\int_{-\infty}^{\infty} (\hat{\theta} - \theta)\frac{\partial f_{\mathbf{Y}|\Theta}(\mathbf{y}|\theta)}{\partial \theta}d\mathbf{y} = \int_{-\infty}^{\infty} f_{\mathbf{Y}|\Theta}(\mathbf{y}|\theta)d\mathbf{y} \tag{4.32}$$

The second integral is one. Using the fact that

$$\frac{\partial \ln g(x)}{\partial x} = \frac{1}{g(x)}\frac{\partial g(x)}{\partial x} \tag{4.33}$$

where $g(x)$ is a function of x, we can express $\partial f_{\mathbf{Y}|\Theta}(\mathbf{y}|\theta)/\partial\theta$ as

$$\frac{\partial f_{\mathbf{Y}|\Theta}(\mathbf{y}|\theta)}{\partial\theta} = f_{\mathbf{Y}|\Theta}(\mathbf{y}|\theta)\frac{\partial \ln f_{\mathbf{Y}|\Theta}(\mathbf{y}|\theta)}{\partial\theta} \tag{4.34}$$

Substituting Equation (4.34) into Equation (4.32), we obtain

$$\int_{-\infty}^{\infty} (\hat{\theta} - \theta) f_{\mathbf{Y}|\Theta}(\mathbf{y}|\theta)\frac{\partial \ln f_{\mathbf{Y}|\Theta}(\mathbf{y}|\theta)}{\partial\theta} dy = 1 \tag{4.35}$$

Schwartz inequality states that

$$\{\int_{-\infty}^{\infty} x^2(t)dt\}\{\int_{-\infty}^{\infty} y^2(t)dt\} \geq \{\int_{-\infty}^{\infty} x(t)y(t)dt\}^2 \tag{4.36}$$

where $x(t)$ and $y(t)$ are two functions of t. Equality holds if and only if $y(t) = Kx(t)$; K constant. Rewriting Equation (4.35) in order to use the Schwartz inequality, we have

$$\int_{-\infty}^{\infty} [\frac{\partial \ln f_{\mathbf{Y}|\Theta}(\mathbf{y}|\theta)}{\partial\theta}\sqrt{f_{\mathbf{Y}|\Theta}(\mathbf{y}|\theta)}][(\hat{\theta} - \theta)\sqrt{f_{\mathbf{Y}|\Theta}(\mathbf{y}|\theta)}]dy = 1 \tag{4.37}$$

or

$$\{\int_{-\infty}^{\infty} (\hat{\theta} - \theta)^2 f_{\mathbf{Y}|\Theta}(\mathbf{y}|\theta)dy\}\{\int_{-\infty}^{\infty} (\frac{\partial \ln f_{\mathbf{Y}|\Theta}(\mathbf{y}|\theta)}{\partial\theta})^2 f_{\mathbf{Y}|\Theta}(\mathbf{y}|\theta)dy\} \geq 1 \tag{4.38}$$

The first integral between brackets is actually $\text{var}[(\hat{\theta}-\theta)|\theta]$. Hence, the inequality becomes

$$\text{var}[(\hat{\theta} - \theta)|\theta] \geq \frac{1}{E[\{\frac{\partial \ln f_{\mathbf{Y}|\Theta}(\mathbf{y}|\theta)}{\partial\theta}\}^2]} \tag{4.39}$$

which proves Equation (4.28).

We now prove Equation (4.29) which says that the Cramer–Rao bound can be expressed in a different form. We know that

$$\int_{-\infty}^{\infty} f_{\mathbf{Y}|\Theta}(\mathbf{y}|\theta)d\mathbf{y} = 1 \tag{4.40}$$

Differentiating both sides of the equation with respect to θ results in

$$\int_{-\infty}^{\infty} \frac{\partial f_{\mathbf{Y}|\Theta}(\mathbf{y}|\theta)}{\partial \theta} = 0 \tag{4.41}$$

rewriting Equation (4.41) and using Equation (4.33), we have

$$\int_{-\infty}^{\infty} \frac{\partial \ln f_{\mathbf{Y}|\Theta}(\mathbf{y}|\theta)}{\partial \theta} f_{\mathbf{Y}|\Theta}(\mathbf{y}|\theta) = 0 \tag{4.42}$$

Differentiating again with respect to θ, we obtain

$$\int_{-\infty}^{\infty} \frac{\partial^2 \ln f_{\mathbf{Y}|\Theta}(\mathbf{y}|\theta)}{\partial \theta^2} f_{\mathbf{Y}|\Theta}(\mathbf{y}|\theta)d\mathbf{y} + \int_{-\infty}^{\infty} \frac{\partial \ln f_{\mathbf{Y}|\Theta}(\mathbf{y}|\theta)}{\partial \theta} \frac{\partial f_{\mathbf{Y}|\Theta}(\mathbf{y}|\theta)}{\partial \theta} = 0 \tag{4.43}$$

Substituting Equation (4.42) for the second term of the second integral of Equation (4.43), and rearranging terms yields

$$E\left[\frac{\partial^2 \ln f_{\mathbf{Y}|\Theta}(\mathbf{y}|\theta)}{\partial \theta^2}\right] = -E\left[\left\{\frac{\partial \ln f_{\mathbf{Y}|\Theta}(\mathbf{y}|\theta)}{\partial \theta}\right\}^2\right] \tag{4.44}$$

which is the same as Equation (4.29), and the proof of the theorem is complete.

An important observation about Equation (4.38) is that equality holds, if and only if

$$\frac{\partial \ln f_{\mathbf{Y}|\Theta}(\mathbf{y}|\theta)}{\partial \theta} = K(\theta)[\hat{\theta} - \theta] \tag{4.45}$$

Any unbiased estimator that satisfies the equality in the Cramer-Rao inequality of Equation (4.28) is said to be an *efficient estimator*.

If an efficient estimate exists, it can easily be shown that it equals the ML estimate. The ML equation is given by

$$\frac{\partial \ln f_{\mathbf{Y}|\Theta}(\mathbf{y}|\theta)}{\partial \theta}\bigg|_{\theta=\hat{\theta}_{ml}} = 0 \tag{4.46}$$

Using Equation (4.45), provided that an efficient estimate exists, we have

$$\frac{\partial \ln f_{\mathbf{Y}|\Theta}(\mathbf{y}|\theta)}{\partial \theta}\bigg|_{\theta=\hat{\theta}_{ml}} = K(\theta)[\hat{\theta} - \theta]\big|_{\theta=\hat{\theta}_{ml}} \qquad (4.47)$$

which equals zero when $\hat{\theta} = \hat{\theta}_{ml}$.

Example 4.7

Consider K observations such that

$$Y_k = m + N_k \ , \ k = 1, 2, \cdots, K$$

where m is unknown and N_k's are statistically independent zero mean Gaussian random variables with unknown variance σ^2.
(a) Find the estimates $\hat{m}$ and $\hat{\sigma}^2$ for m and σ^2, respectively.
(b) Is $\hat{m}$ an efficient estimator ?
(c) Find the conditional variance of the error $\text{var}[(\hat{m} - m)|m]$.

Solution.
(a) Using the likelihood Equation of (4.5), we can determine $\hat{m}$ and $\hat{\sigma}^2$ simultaneously. The conditional density function of $\mathbf{Y}$ given m and σ^2 is

$$f_{\mathbf{Y}}(\mathbf{y}|m,\sigma^2) = \prod_{k=1}^{K} \frac{1}{\sqrt{2\pi}\sigma} e^{-\frac{(y_k-m)^2}{2\sigma^2}}$$

Taking the logarithm, we have

$$\ln f_{\mathbf{Y}}(\mathbf{y}|m,\sigma^2) = -\frac{K}{2} \ln(2\pi\sigma^2) - \sum_{k=1}^{K} \frac{(y_k - m)^2}{2\sigma^2}$$

We take the derivative of the above equation with respect to m and σ^2 to obtain two equations in two unknowns. That is,

$$\frac{\partial \ln f_{\mathbf{Y}}(\mathbf{y}|m,\sigma^2)}{\partial m} = 2 \sum_{k=1}^{K} \frac{y_k - m}{2\sigma^2} = 0$$

and

$$\frac{\partial \ln f_{\mathbf{Y}}(\mathbf{y}|m,\sigma^2)}{\partial \sigma^2} = -\frac{K}{2\sigma^2} + \sum_{k=1}^{K} \frac{(y_k - m)^2}{2\sigma^4} = 0$$

solving for $\hat{m}_{ml}$ and $\hat{\sigma}^2_{ml}$ simultaneously, we obtain

$$\hat{m}_{ml} = \frac{1}{K} \sum_{k=1}^{K} Y_k$$

and

$$\hat{\sigma}^2_{ml} = \frac{1}{K} \sum_{k=1}^{K} (Y_k - \frac{1}{K} \sum_{k=1}^{K} Y_k)^2$$

$$= \frac{1}{K} \sum_{k=1}^{K} (Y_k - \hat{m}_{ml})^2$$

(b) $\hat{m}_{ml}$ is an unbiased estimator, since

$$E[\hat{m}_{ml}] = \frac{1}{K} E[\sum_{k=1}^{K} Y_k] = m$$

To check if the estimator is efficient, we use Equation (4.45) to obtain

$$\frac{\partial \ln f_{\mathbf{Y}}(\mathbf{y}|m,\sigma^2)}{\partial m} = \sum_{k=1}^{K} \frac{y_k - m}{\sigma^2}$$

$$= \frac{K}{\sigma^2} [\frac{1}{K} \sum_{k=1}^{K} Y_k - m]$$

where $K(m) = K/\sigma^2$ and $\hat{m} = (1/K) \sum_{k=1}^{K} Y_k = \hat{m}_{ml}$. Hence, the estimator is efficient.

(c) To determine the conditional variance of the error, we use Equations (4.28) and (4.29). Taking the derivative of the likelihood equation with respect to m, we obtain

$$\frac{\partial^2 \ln f_{\mathbf{Y}}(\mathbf{y}|m,\sigma^2)}{\partial m^2} = -\frac{K}{\sigma^2}$$

Hence,

$$\text{var}[(\hat{m} - m)|m] = -\frac{1}{E[\frac{\partial^2 \ln f_{\mathbf{Y}}(\mathbf{y}|m,\sigma^2)}{\partial m^2}]}$$

$$= \frac{\sigma^2}{K}$$

4.6 GENERALIZED LIKELIHOOD RATIO TEST

In Example 3.21, we solved the hypothesis testing problem where the alternative hypothesis was composite. The parameter m under hypothesis H_1 was unknown, although it was known that m was either positive or negative. When m was positive only (negative only) a UMP test existed and the decision rule was

$$y \begin{array}{c} H_1 \\ > \\ < \\ H_0 \end{array} \frac{\sigma^2}{m} \ln\eta + \frac{m}{2} = \gamma_1 \qquad (4.48)$$

for m positive and

$$y \begin{array}{c} H_0 \\ > \\ < \\ H_1 \end{array} \frac{\sigma^2}{m} \ln\eta + \frac{m}{2} = \gamma_2 \qquad (4.49)$$

for m negative. Since the test designed for m positive was not the same as the test designed for m negative, we concluded that a UMP test did not exist for all possible values of m; positive and negative. This requires that different tests may be used. One approach is to use the concepts developed in this Chapter. That is, we use the required data to estimate θ, as though hypothesis H_1 is true. We also estimate θ, as though hypothesis H_0 is true. Then, we use these estimates in the likelihood ratio test as if they are the correct values. If the estimates used are the maximum likelihood estimates, the result is called the *generalized likelihood ratio test* and is given by

$$\Lambda_g(\mathbf{y}) = \frac{\overset{max}{\theta_1} f_{\mathbf{Y}|\Theta_1}(\mathbf{y}|\theta_1)}{\overset{max}{\theta_0} f_{\mathbf{Y}|\Theta_0}(\mathbf{y}|\theta_0)} \begin{array}{c} H_1 \\ > \\ < \\ H_0 \end{array} \eta \qquad (4.50)$$

θ_1 and θ_0 are the unknown parameters to be estimated under hypotheses H_1 and H_0, respectively.

Example 4.8

Consider the problem of Example 3.12, where m is an *unknown* parameter. Obtain the generalized likelihood ratio test and compare it to the optimum Neyman–Pearson test.

Solution.
Since the K observations are independent, the conditional density functions under both hypotheses H_1 and H_0 are

$$H_0 : f_{\mathbf{Y}|M,H_0}(\mathbf{y}|m, H_0) = \prod_{k=1}^{K} \frac{1}{\sqrt{2\pi}\sigma} \, e^{-\frac{y_k^2}{2\sigma^2}}$$

$$H_1 : f_{\mathbf{Y}|M,H_1}(\mathbf{y}|m, H_1) = \prod_{k=1}^{K} \frac{1}{\sqrt{2\pi}\sigma} \, e^{-\frac{(y_k-m)^2}{2\sigma^2}}$$

m is an unknown parameter. Since hypothesis H_0 does contain m (H_0 is simple), the estimation procedure is applicable to hypothesis H_1, only. From the likelihood equation given by Equation (4.3), the ML estimate of m under H_1 is given by

$$\frac{\partial \ln f_{\mathbf{Y}|M,H_1}(\mathbf{y}|m, H_1)}{\partial m} = 0$$

Substituting for $f_{\mathbf{Y}|M,H_1}(\mathbf{y}|m, H_1)$ in the above equation, we have

$$\frac{\partial}{\partial m}[-\sum_{k=1}^{K} \frac{(y_k - m)^2}{2\sigma^2}] = 0$$

or

$$\hat{m} = \frac{1}{M} \sum_{k=1}^{K} Y_k$$

The details are given in Example 4.1. The likelihood ratio test becomes

$$\Lambda_g(\mathbf{y}) = \frac{\prod_{k=1}^{K} \frac{1}{\sqrt{2\pi}\sigma} \, e^{-\frac{1}{2\sigma^2}(y_k - \hat{m})^2}}{\prod_{k=1}^{K} \frac{1}{\sqrt{2\pi}\sigma} \, e^{-\frac{y_k^2}{2\sigma^2}}} \underset{H_0}{\overset{H_1}{\gtrless}} \eta$$

Substituting for the obtained value of $\hat{m}$ in the above expression, and simplifying after taking the logarithm, the test becomes

$$\frac{1}{2\sigma^2 K}(\sum_{k=1}^{K} y_k)^2 \underset{H_0}{\overset{H_1}{\underset{<}{>}}} \ln\eta$$

Since $[1/(2\sigma^2 K)](\sum_{k=1}^{K} y_k)^2$ is nonnegative, the decision will always be H_1, if η is less than one ($\ln\eta$ negative), or η set equal to one. Consequently, η can always be chosen greater than or equal to one. Thus, an equivalent test is

$$(\frac{1}{\sqrt{K}} \sum_{k=1}^{K} y_k)^2 \underset{H_0}{\overset{H_1}{\underset{<}{>}}} 2\sigma^2 \ln\eta = \gamma_1^2$$

where $\gamma_1 \geq 0$. Equivalently, we can use the test:

$$|Z| = |\frac{1}{\sqrt{K}} \sum_{k=1}^{K} y_k| \underset{H_0}{\overset{H_1}{\underset{<}{>}}} \gamma_1$$

The decision regions are shown in Figure 4.3

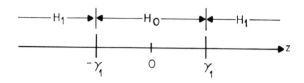

Figure 4.3 Decision regions of generalized likelihood ratio test.

Given the desired probability of false alarm, the value of γ_1 can be determined. Before we can get an expression for P_F, the probability of false alarm, we need to determine the density function of Z. Since

$$Z = \frac{1}{\sqrt{K}} \sum_{k=1}^{K} Y_k$$

the mean and variance of Y under hypothesis H_0 are zero and σ^2, respectively. All the observations are Gaussian, and statistically independent. Thus, the density function of $Z_1 = \sum_{k=1}^{K} Y_k$ is Gaussian with mean zero, and variance $K\sigma^2$. Consequently, Z is Gaussian with mean zero, and variance σ^2. That is,

$$f_{Z|H_0}(z|H_0) = \frac{1}{\sqrt{2\pi}\sigma} e^{-\frac{z^2}{2\sigma^2}}$$

The probability of false alarm, from Figure 4.4, is

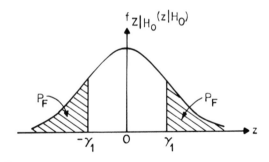

Figure 4.4 Density function of Z under H_0.

$$
\begin{aligned}
P_F &= P(\text{ decide } H_1|H_0 \text{ true }) \\
&= \int_{-\infty}^{-\gamma_1} \frac{1}{\sqrt{2\pi}\sigma} e^{-\frac{z^2}{2\sigma^2}} dz + \int_{\gamma_1}^{\infty} \frac{1}{\sqrt{2\pi}\sigma} e^{-\frac{z^2}{2\sigma^2}} dz \\
&= 2Q(\frac{\gamma_1}{\sigma})
\end{aligned}
$$

We observe that we are able to determine the value γ_1 from the derived probability of false alarm without any knowledge of m. However, the probability of detection cannot be determined without m, but can be evaluated with m as a parameter. Under hypothesis H_1, $Z_1 = \sum_{k=1}^{K} Y_k$ is Gaussian, with mean Km and variance $K\sigma^2$. Hence, the density function of Z is Gaussian with mean $\sqrt{K}m$ and variance σ^2. That is,

$$f_{Z|H_1}(z|H_1) = \frac{1}{\sqrt{2\pi}\sigma} e^{-\frac{(z-\sqrt{K}m)^2}{2\sigma^2}}$$

The probability of detection for a given value of m, from Figure 4.5, is

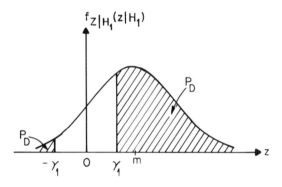

Figure 4.5 Density function of Z under H_1.

$$
\begin{aligned}
P_D &= P(\text{ decide } H_1|H_1 \text{ true }) \\
&= \int_{-\infty}^{-\gamma_1} \frac{1}{\sqrt{2\pi}\sigma} e^{-\frac{(z-\sqrt{K}m)^2}{2\sigma^2}} dz + \int_{\gamma_1}^{\infty} \frac{1}{\sqrt{2\pi}\sigma} e^{-\frac{(z-\sqrt{K}m)^2}{2\sigma^2}} dz \\
&= 1 - Q(\frac{-\gamma_1 - \sqrt{K}m}{\sigma}) + Q(\frac{\gamma_1 - \sqrt{K}m}{\sigma}) \\
&= Q(\frac{\gamma_1 + \sqrt{K}m}{\sigma}) + Q(\frac{\gamma_1 - \sqrt{K}m}{\sigma})
\end{aligned}
$$

In Figure 2.31 on page 95 of Van trees, it is shown that the generalized likelihood ratio test performs nearly as well as the Neyman–Pearson test.

4.7 SUMMARY

In this chapter, we have developed the concept of parameter estimation. We used the maximum likelihood estimation to estimate nonrandom parameters. We first obtained the likelihood function in terms of the parameters to be estimated. Then, we maximized the likelihood function to obtain the estimator, which resulted from solving the likelihood equation. Measuring criteria for the estimator, such as bias and consistency, were presented to determine the quality of the estimator.

When the parameter to be estimated was a random variable, we used Bayes estimation. In Bayes estimation, we minimized the risk, which is a function of the error between the estimate and the true value. Three cases were considered; the squared error, the absolute value error, and the uniform cost function. It was shown that the minimum mean-squared error represents the conditional mean of the parameter (associated with the observation random variable) to be estimated. The resulting minimum risk was just the conditional variance. In the absolute value error case, the estimate turned out to be the median of the conditional density function of the parameter to be estimated, given the observation random variable. For the uniform Bayes cost, the estimator was actually the solution to the MAP equation. In comparing the ML estimate and the MAP estimate, it was observed that the ML estimate was just a special case of the MAP estimate, and is obtained by setting to zero the density function of the parameter to be estimated in the MAP equation. In order to measure the "goodness" of the estimator, the Cramer–Rao bound was given as an alternate way to measure the error variance, since an expression for the error variance was difficult to obtain.

We concluded the chapter with a section about the generalized likelihood ratio test, which linked this chapter to the previous one. In the generalized likelihood ratio test, we used the maximum likelihood estimate of the unknown parameter in the composite hypothesis as its true value and then performed the likelihood ratio test. This was an alternative to cases where UMP tests did not exist.

PROBLEMS

4.1 Let $Y_1, Y_2, \cdots, Y_K$ be the observed random variables such that

$$Y_k = a + bx_k + Z_k \ , \ k = 1, 2, \cdots, K$$

The constants $x_k, k = 1, 2, \cdots, K$, are known while the constants a and b are not known. The random variables $Z_k, k = 1, 2, \cdots, K$, are statistically independent, each with zero mean and variance σ^2 known. Obtain the ML estimate of (a, b).

4.2 Let Y be a Gaussian random variable with mean zero and variance σ^2.
(a) Obtain the ML estimates of σ and σ^2.
(b) Are the estimates efficient ?

4.3 Let Y_1 and Y_2 be two statistically independent Gaussian random variables such that $E[Y_1] = m$, $E[Y_2] = 3m$ and $\text{var}(Y_1) = \text{var}(Y_2) = 1$; m is unknown.
(a) Obtain the ML estimates of m.
(b) If the estimator of m is of the form $a_1 Y_1 + a_2 Y_2$, determine a_1 and a_2 so that the estimator be unbiased ?

4.4 Let the observation Y satisfy the binomial law such that the density function of Y is

$$f_Y(y) = \binom{n}{k} p^n (1 - p)^{n-k}$$

(a) Find an unbiased estimate for p.
(b) Is the estimate consistent?

4.5 Obtain the ML estimates of the mean m and the variance σ^2 for the independent observations $Y_1, Y_2, \cdots, Y_K$, such that

$$f_{Y_k}(y_k) = \frac{1}{\sqrt{2\pi}\sigma} \, e^{-\frac{(y_k-m)^2}{2\sigma^2}} \, , \, k = 1, 2, \cdots, K$$

4.6 Let x be an unknown deterministic parameter that can have any value in the interval [-1,1]. Suppose we take two observations of x with independent samples of zero-mean Gaussian noise, and variance σ^2 superimposed on each of the observations.
(a) Obtain the ML estimate of x.
(b) Is $\hat{x}_{ml}$ unbiased ?

4.7 Let $Y_1, \cdots, Y_K$ be K independent observations each having the Poisson distribution given by

$$P_{Y_k}(n_k) = \begin{cases} e^{-\lambda} \frac{\lambda^{n_k}}{n_k!} & , n_k = 0, 1, 2, \cdots \text{ and } \lambda > 0 \\ 0 & , \text{otherwise} \end{cases}$$

$k = 1, 2, \cdots, K$ and the mean λ is unknown.
(a) Find the ML estimate of λ.
(b) Does the ML estimate of λ exist for all possible values of $Y_1, \cdots, Y_K$? If not, how can λ be changed so that the ML estimates exist?

4.8 Find $\hat{x}_{ms}$ the minimum mean square error and $\hat{x}_{map}$ the maximum a posteriori estimators of X from the observations:

$$Y = X + N$$

X and N are random variables with density functions

$$f_X(x) = \frac{1}{2}[\delta(x-1) + \delta(x+1)]$$

and

$$f_N(n) = \frac{1}{\sqrt{2\pi}\sigma} \, e^{-\frac{x^2}{2\sigma^2}}$$

4.9 The conditional density function of the observed random variable given a random parameter X is given by

$$f_{Y|X}(y|x) = \begin{cases} x\, e^{-xy} & ,y \geq 0 \text{ and } x > 0 \\ 0 & ,y < 0 \end{cases}$$

The *a priori* probability density function of X is

$$f_X(x) = \begin{cases} \frac{\alpha^r}{\Gamma(r)} x^{r-1}\, e^{-\alpha x} & ,x \geq 0 \\ 0 & ,x < 0 \end{cases}$$

where α is a parameter, r is a positive integer, and $\Gamma(r)$ is the Gamma function

(a) Obtain the *a priori* mean and variance of X.

(b) For Y given,

(i) obtain the minimum mean square error estimate of X.

(ii) What is the variance of this estimate ?

(c) Suppose we take K independent observations of Y_k, $k = 1, 2, \cdots, K$, such that

$$f_{Y_k|X}(y_k|x) = \begin{cases} x\, e^{-xy_k} & ,y_k \geq 0 \text{ and } x > 0 \\ 0 & ,y_k < 0 \end{cases}$$

(i) Determine the minimum mean square error estimate of X.

(ii) What is the variance of this estimate?

(d) Verify if the MAP estimate equals the minimum mean square error estimate.

4.10 Consider the problem where the observation is given by

$$Y = \ln X + N$$

where X is the parameter to be estimated. X is uniformly distributed over the interval $[0,1]$, and N has an exponential distribution given by

$$f_N(n) = \begin{cases} e^{-n} & ,n \geq 0 \\ 0 & \text{otherwise} \end{cases}$$

Obtain
(a) the mean square estimate, $\hat{x}_{ms}$
(b) the MAP estimate, $\hat{x}_{map}$
(c) $\hat{x}_{abs}$

4.11 The observation Y is given by

$$Y = X + N$$

where X and N are two random variables. N is normal with mean one, and variance σ^2, and X is uniformly distributed over the interval $[0, 2]$. Determine the MAP estimate of the parameter X.

REFERENCES

1. Dudewicz, E.J., *Introduction to Statistics and Probability*, Holt, Rinehart and Winston, New York, 1976.

2. Melsa, J.L., and D.L. Cohn, *Decision and Estimation Theory*, McGraw-Hill, New York, 1978.

3. Mohanty, N., *Signal Processing: Signals, Filtering, and Detection*, Van Nostrand Reinhold, New York, 1987.

4. Nahi, N.E., *Estimation Theory and Applications*, John Wiley and Sons, New York, 1969.

5. Sage, A.P., and J.L. Melsa, *Estimation Theory with Applications to Communications and Control*, McGraw-Hill, New York, 1971.

6. Shanmugan, K.S., and A.M. Breipohl, *Random Signals: Detection, Estimation and Data Analysis*, John Wiley and Sons, New York, 1979.

7. Sorenson, H.W., *Parameter Estimation: Principles and Problems*, Marcel Dekker, New York, 1980.

8. Srinath, M.D., and P.K. Rajasekaran, *An Introduction to Statistical Signal Processing with Applications*, John Wiley and Sons, New York, 1979.

9. Stark, H., and J.W. Woods, *Probability, Random Processes, and Estimation Theory for Engineers*, Prentice-Hall, Englewood Cliffs, NJ, 1986.

10. Urkowitz, H., *Signal Theory and Random Processes*, Artech House, Norwood, MA, 1983.

11. Van Trees, H.L., *Detection, Estimation, and Modulation Theory*, Part I, John Wiley and Sons, New York, 1968.

12. Whalen, A.D., *Detection of Signals in Noise*, Academic Press, New York, 1971.

Chapter 5

Filtering

5.1 INTRODUCTION

In the previous chapter, we developed techniques for estimating random and nonrandom parameters. We also studied measures to determine the "goodness" of the estimates. In many applications, the goal was to estimate a signal waveform from a noisy version of the signal, in an "optimal" manner.

In this chapter, we assume that the received signal is corrupted by an additive noise. We would like to extract the desired signal from the received one based on the *Linear Minimum Mean-Square Error* criterion. The received process signal, $Y(t)$, is observed over some interval of time $t\epsilon[t_i, t_f]$. t_i denotes initial time and t_f denotes final time. The problem is to determine $\hat{Y}(t)$; a linear estimate of $Y(t)$. When t is outside the interval, we talk about *prediction*. If $t < t_i$, $\hat{Y}(t)$ is a *backward predictor*. If $t > t_f$, $\hat{Y}(t)$ is a *forward predictor*. When $t\epsilon[t_i, t_f]$, the problem is referred to as *smoothing*. The process of extracting the information carrying signal $S(t)$ from the observed signal $Y(t)$, $Y(t) = S(t) + N(t)$ and $N(t)$ a noise process, is called *filtering*. In Section 5.2, we define linear transformation and present some related theorems. The orthogonality principle theorem will be discussed. We also show how the orthogonality principle is used in different problems.

In Section 5.3, we discuss the problem of filtering by deriving the impulse response of the system for both realizable and unrealizable filters. We conclude this chapter with a brief section on Kalman filtering.

5.2 LINEAR TRANSFORMATION; ORTHOGONALITY PRINCIPLE

The estimate to be determined $\hat{Y}(t)$ is a *linear* transformation of the received signal $Y(t)$. In this section, we present some useful properties about linear transformations and discuss an important theorem known as " the orthogonality principle" before deriving the *estimation rule*. The estimate may be written as

$$\hat{Y}(t) = L[Y(t)] \tag{5.1}$$

where the operator $L[\cdot]$ denotes linear transformation. The estimation rule is based on the minimum mean-square error. Hence, defining the error as

$$\varepsilon(t) = Y(t) - \hat{Y}(t) \tag{5.2}$$

we would like to derive the estimation rule $L[\cdot]$ so that the mean-square error

$$E[|\varepsilon(t)|^2] = E[|Y(t) - \hat{Y}(t)|^2] \tag{5.3}$$

is minimized.

By definition, a transformation $L[\cdot]$ is linear, provided that

$$L[a_1 Y_1(t) + a_2 Y_2(t)] = a_1 L[Y_1(t)] + a_2 L[Y_2(t)] \tag{5.4}$$

for all constants a_1 and a_2, and processes $Y_1(t)$ and $Y_2(t)$. The difference transformation is also linear. That is, if $L_1[\cdot]$ and $L_2[\cdot]$ are two linear transformations such that

$$L_1[a_1 Y_1(t) + a_2 Y_2(t)] = a_1 L_1[Y_1(t)] + a_2 L_1[Y_2(t)] \tag{5.5}$$

and

$$L_2[a_1 Y_1(t) + a_2 Y_2(t)] = a_1 L_2[Y_1(t)] + a_2 L_2[Y_2(t)] \tag{5.6}$$

then, the difference transformation:

$$L[\cdot] = L_2[\cdot] - L_1[\cdot] \tag{5.7}$$

is linear. The proof is straight forward by direct substitutions of Equations (5.5) and (5.6) in Equation (5.7). Also, for a linear transformation, it can be shown that

$$E\{L[\cdot]\} = L\{E[\cdot]\} \tag{5.8}$$

where the operator $E[\cdot]$ denotes expectation.

If $Z(t)$ is a process orthogonal to $Y(\xi)$ for all ξ in the interval $[t_i, t_f]$, then *any linear transformation* on $Y(\xi)$ is also orthogonal to $Z(t)$ in the interval $\xi\epsilon[t_i, t_f]$. To prove this statement, we start from the definition that $Y(\xi)$ is orthogonal to $Z(t)$ in the given interval; that is,

$$E[Y(\xi)Z^*(t)] = 0 \text{ for } \xi\epsilon[t_i, t_f] \tag{5.9}$$

Let $L[Y(\xi)]$ be a linear transformation of $Y(\xi)$. Since linear operations and expectations are interchangeable as given in Equation (5.8), we have

$$
\begin{aligned}
E\{L[Y(\xi)]Z^*(t)\} &= E\{L[Y(\xi)Z^*(t)]\} \\
&= L\{E[Y(\xi)Z^*(t)]\} \\
&= 0 \qquad\qquad \text{for } \xi\epsilon[t_i, t_f]
\end{aligned}
\tag{5.10}
$$

which proves that the linear transformation of $Y(\xi)$ is orthogonal to $Z(t)$.

Theorem: Orthogonality Principle

The linear transformation $L[\cdot]$ is the minimum mean-square error estimate, if and only if the error $\varepsilon(t)$ is orthogonal to $Y(\xi)$ for $\xi\epsilon[t_i, t_f]$.

Proof: Let all processes, $Y(t) = S(t) + N(t)$, be real and stationary. Consider the linear transformation $L_1[\cdot], L_1[Y(\xi)] = \hat{S}(t)$ for all ξ, such that the mean-square error $E[\varepsilon_1^2(t)], \varepsilon_1(t) = S(t) - \hat{S}(t)$, is minimum. That is

$$L_1[Y(\xi)] = \hat{S}(t) \tag{5.11}$$

is the optimum estimator and

$$
\begin{aligned}
e_m &= E[\varepsilon_1^2(t)] \\
&= E[\{S(t) - L_1[Y(\xi)]\}^2] = 0
\end{aligned}
\tag{5.12}
$$

Consider the linear transformation $L_2[\cdot]$, $L_2[Y(\xi)] = \hat{S}(t)$ for all ξ, such that the error $\varepsilon_2(t)$, $\varepsilon_2(t) = S(t) - \hat{S}(t)$, is orthogonal to the data $Y(\xi)$ for all ξ. That is,

$$
E[\varepsilon_2(t)Y(\xi)] = E[\{S(t) - \hat{S}(t)\}Y(\xi)] = 0
\tag{5.13}
$$

The error $\varepsilon_1(t)$ can be expressed in terms of $\varepsilon_2(t)$ as

$$
\begin{aligned}
\varepsilon_1(t) &= S(t) - L_1[Y(\xi)] \\
&= S(t) + \{L_2[Y(\xi)] - L_2[Y(\xi)]\} - L_1[Y(\xi)] \\
&= \varepsilon_2(t) + L_2[Y(\xi)] - L_1[Y(\xi)] \\
&= \varepsilon_2(t) + L[Y(\xi)]
\end{aligned}
\tag{5.14}
$$

where the difference transformation $L[\cdot] = L_2[\cdot] - L_1[\cdot]$ is linear as given by Equation (5.7). Substituing Equation (5.14) into Equation (5.12), the linear mean-square error using the optimum estimator becomes

$$
\begin{aligned}
e_m &= E[\{S(t) - L_1[Y(\xi)]\}^2] \\
&= E[\{\varepsilon_2(t) + L[Y(\xi)]\}^2] \\
&= E[\varepsilon_2^2(t)] + 2E[\varepsilon_2(t)L[Y(\xi)]] + E[\{L[Y(\xi)]\}^2]
\end{aligned}
\tag{5.15}
$$

Since $\varepsilon_2(t)$ is orthogonal to the data, $\varepsilon_2(t)$ is also orthogonal to $L[Y(\xi)]$ as shown in Equation (5.9). Thus,

$$
E\{\varepsilon_2(t)L[Y(\xi)]\} = 0
\tag{5.16}
$$

and the minimum mean-square error reduces to

$$
e_m = E[\varepsilon_2^2(t)] + E[\{L[Y(\xi)]\}^2]
\tag{5.17}
$$

where $E[\varepsilon_2^2(t)]$ in the mean-square error with $L_2[\cdot]$ as the estimator. Therefore,

$$
\begin{aligned}
e_m &= E[\varepsilon_2^2(t)] + E[\{L[Y(\xi)]\}^2] \\
&= E[\varepsilon_2^2(t)]
\end{aligned}
\tag{5.18}
$$

if and only if the nonnegative quantity $E[\{L[Y(\xi)]\}^2]$ is zero. That is,

$$L[\cdot] = L_2[\cdot] - L_1[\cdot] = 0 \tag{5.19}$$

Hence, this proves our theorem, which says that the linear transformation for which the error is orthogonal to data results in the minimum mean-square error linear estimator and vice-versa.

We now derive a simple expression for e_m, the minimum mean-square error, using the fact that the error is orthogonal to the data which is given by

$$E[\{S(t) - \hat{S}(t)\}Y(\xi)] = 0 \tag{5.20}$$

Substituting Equation (5.9) into Equation (5.20), the above expression becomes

$$E[\{S(t) - L[Y(\xi)]\}L[Y(\xi)]] = 0 \tag{5.21}$$

Consequently, the minimum mean-square error reduces to

$$e_m = E[\{S(t) - L[Y(\xi)]\}S(t)] \tag{5.22}$$

The Yule-Walker Equations

We now consider the problem where we are given K random variables, $Y_1, Y_2, \cdots, Y_K$, and we need to determine the linear minimum mean-square error estimator for the random variable S. Since the estimator is linear, the estimate $\hat{S}$ is given by

$$\hat{S} = a_1 Y_1 + a_2 Y_2 + \cdots + a_K Y_K \tag{5.23}$$

From the orthogonality principle, the mean-square error is minimum, if and only if the constants $a_1, a_2, \cdots, a_K$ are chosen so that the error is orthogonal to the data. That is,

$$E[(S - a_1 Y_1 - a_2 Y_2 - \cdots - a_K Y_K)Y_k] = 0 \text{ for } k = 1, 2, \cdots, K \tag{5.24a}$$

or

$$E[SY_k] - a_1 E[Y_1 Y_k] - a_2 E[Y_2 Y_k] - \cdots - a_k E[Y_K Y_k] = 0 \tag{5.24b}$$

for $k = 1, 2, \cdots, K$. Defining

$$E[SY_k] = R_{0k} \qquad (5.25a)$$

and

$$E[Y_j Y_k] = R_{jk} \qquad (5.25b)$$

we obtain from Equation (5.22) the following set of K equations in K unknowns, known as the *Yule-Walker equations*:

$$
\begin{aligned}
a_1 R_{11} + a_2 R_{12} + \cdots + a_K R_{1K} &= R_{01} \\
a_1 R_{21} + a_2 R_{22} + \cdots + a_K R_{2K} &= R_{02} \\
&\vdots \\
a_1 R_{K1} + a_2 R_{K2} + \cdots + a_K R_{KK} &= R_{0K}
\end{aligned} \qquad (5.26)
$$

The solution yields the constants $a_1, a_2, \cdots, a_K$. In matrix form, Equation (5.26) can be written as

$$\mathbf{R}\mathbf{a} = \mathbf{R}_0 \qquad (5.27a)$$

where

$$\mathbf{R} = \begin{bmatrix} R_{11} & R_{12} & \cdots & R_{1K} \\ R_{21} & R_{22} & \cdots & R_{2K} \\ \vdots & \vdots & \vdots & \vdots \\ R_{K1} & R_{K2} & \cdots & R_{KK} \end{bmatrix} \qquad (5.27b)$$

$$\mathbf{a} = \begin{bmatrix} a_1 \\ a_2 \\ \vdots \\ a_K \end{bmatrix}, \quad \mathbf{R}_0 = \begin{bmatrix} R_{01} \\ R_{02} \\ \vdots \\ R_{0K} \end{bmatrix} \qquad (5.27c)$$

Since $R_{jk} = R_{kj}$, $j, k = 1, 2, \cdots, K$, in the Yule-Walker equations, the coefficients are obtained from Equation (5.27a) to be

$$\mathbf{a} = \mathbf{R}^{-1}\mathbf{R}_0 \qquad (5.28)$$

Note that the data correlation matrix is given by

$$\mathbf{R} = E[\mathbf{Y}\mathbf{Y}^T] \qquad (5.29a)$$

where $\mathbf{Y}^T$ denotes the transpose of $\mathbf{Y}$ and

$$\mathbf{Y} = \begin{bmatrix} Y_1 \\ Y_2 \\ \vdots \\ Y_K \end{bmatrix} \qquad (5.29b)$$

The estimator is simply the inner product of $\mathbf{a}$ and $\mathbf{Y}$; that is,

$$\hat{S} = \mathbf{a}^T\mathbf{Y} = \mathbf{Y}^T\mathbf{a} \qquad (5.30)$$

Geometrically, $\hat{S}$ is the projection of S onto the surface spanned by $Y_1, Y_2, \cdots, Y_K$. This is illustrated in Figure 5.1 for $K = 2$.

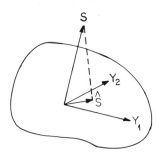

Figure 5.1 Projection of S onto plane spanned by Y_1 and Y_2.

The mean-square vector is

$$\begin{aligned} E[\hat{S}^2] &= E[\hat{S}\hat{S}^T] \\ &= E[\mathbf{a}^T\mathbf{Y}\mathbf{Y}^T\mathbf{a}] \\ &= \mathbf{a}^T\mathbf{R}\mathbf{a} \end{aligned} \qquad (5.31)$$

Similarly, we can determine the minimum mean-square error to be

$$\begin{aligned} e_m &= E[(S - \hat{S})S] \\ &= E[(S - a_1Y_1 - a_2Y_2 - \cdots - a_KY_K)S] \\ &= R_{00} - a_1R_{01} - a_2R_{02} - \cdots - a_KR_{0K} \end{aligned} \qquad (5.32)$$

where $R_{00} = E[S^2]$. In matrix form, the minimum mean-square error is expressed as

$$e_m = R_{00} - \mathbf{a}^T \mathbf{R}_0 \tag{5.33}$$

substituting for Equation (5.28) in (5.33), we have

$$
\begin{aligned}
e_m &= R_{00} - (\mathbf{R}^{-1}\mathbf{R}_0)^T \mathbf{R}_0 \\
&= R_{00} - \mathbf{R}_0^T (\mathbf{R}^{-1})^T \mathbf{R}_0
\end{aligned} \tag{5.34}
$$

Example 5.1

Let the observation process be $Y(t) = S(t) + N(t)$ where $S(t)$ and $N(t)$ are zero mean wide-sense stationary processes. Obtain an estimate of $S(t)$ in terms of the present value of $Y(t)$ and determine the minimum mean-square error.

Solution.

In this case, the problem is to estimate the constant a such that the estimate $\hat{S}(t)$ is given by

$$\hat{S}(t) = aY(t)$$

The linear minimum mean-square error estimator results in requiring the error $\varepsilon(t) = S(t) - \hat{S}(t)$ be orthogonal to the observed data $Y(t)$. That is,

$$
\begin{aligned}
E[\{S(t) - \hat{S}(t)\}Y(t)] &= E[\{S(t) - aY(t)\}Y(t)] \\
&= E[S(t)Y(t)] - aE[Y^2(t)] \\
&= R_{SY}(0) - aR_{YY}(0) = 0
\end{aligned}
$$

Solving for a, we obtain

$$a = \frac{R_{SY}(0)}{R_{YY}(0)}$$

The minimum mean-square error is given by

$$
\begin{aligned}
e_m &= E[\{S(t) - \hat{S}(t)\}S(t)] \\
&= E[\{S(t) - aY(t)\}S(t)] \\
&= E[S^2(t)] - aE[Y(t)S(t)] \\
&= R_{SS}(0) - aR_{YS}(0)
\end{aligned}
$$

substituting for the value of a and noting that $R_{SY}(0) = R_{YS}(0)$, the minimum mean-square error becomes

$$
\begin{aligned}
e_m &= R_{SS}(0) - \frac{R_{SY}^2(0)}{R_{YY}(0)} \\
&= \frac{R_{SS}(0)R_{YY}(0) - R_{SY}^2(0)}{R_{YY}(0)}
\end{aligned}
$$

If in addition, the signal process and the noise process are statistically independent, then,

$$
\begin{aligned}
R_{SY}(0) &= E[S(t)Y(t)] \\
&= E[S(t)\{S(t) + N(t)\}] \\
&= R_{SS}(0)
\end{aligned}
$$

since $E[S(t)N(t)] = E[S(t)]E[N(t)] = 0$. Also,

$$
\begin{aligned}
R_{YY}(0) &= E[\{S(t) + N(t)\}\{S(t) + N(t)\}] \\
&= R_{SS}(0) + R_{NN}(0)
\end{aligned}
$$

Therefore,

$$
a = \frac{R_{SS}(0)}{R_{SS}(0) + R_{NN}(0)}
$$

and

$$
\begin{aligned}
e_m &= \frac{R_{SS}(0)R_{NN}(0)}{R_{SS}(0) + R_{NN}(0)} \\
&= aR_{NN}(0)
\end{aligned}
$$

It should be noted that if the processes were not zero mean, the estimator must solve for the constants a and b, such that

$$
\hat{S}(t) = aY(t) + b
$$

Example 5.2

Estimate $Y(t)$ in the time interval $t\epsilon[0,T]$ given $Y(0)$ and $Y(T)$. Determine the minimum mean-square error.

Solution.

The problem of estimating a signal at any instant in an interval of time, given the values of the signal at the end of the interval, is known as *interpolation*. Using a linear estimator, the estimate $\hat{Y}(t)$ may be written as

$$\hat{Y}(t) = aY(0) + bY(T)$$

where a and b are the constants to be determined. Since we require that the error be orthogonal to the data, we have

$$E[\{Y(t) - aY(0) - bY(T)\}Y(0)] = 0$$

and

$$E[\{Y(t) - aY(0) - bY(T)\}Y(T)] = 0$$

It follows that

$$R_{YY}(t) = aR_{YY}(0) + bR_{YY}(T)$$

$$R_{YY}(T - t) = aR_{YY}(T) + bR_{YY}(0)$$

We have two equations in two unknowns. Solving for a and b, we obtain

$$a = \frac{R_{YY}(0)R_{YY}(t) - R_{YY}(T)R_{YY}(T - t)}{R_{YY}^2(0) - R_{YY}^2(T)}$$

and

$$b = \frac{R_{YY}(0)R_{YY}(T - t) - R_{YY}(t)R_{YY}(T)}{R_{YY}^2(0) - R_{YY}^2(T)}$$

The minimum mean-square error is

$$
\begin{aligned}
e_m &= E[\{Y(t) - aY(0) - bY(T)\}Y(t)] \\
&= R_{YY}(0) - aR_{YY}(t) - bR_{YY}(t - T)
\end{aligned}
$$

where a and b are as given above.

5.3 WIENER FILTERS

We now consider the case where the data is not a finite number of random variables as in Examples 5.1 and 5.2, but rather a random process, $X(\xi)$, observed over an interval of time $\xi \epsilon [t_i, t_f]$. The goal is to estimate another process, $Y(t)$, by a *linear function* of $X(\xi)$ such that

$$\hat{Y}(t) = \int_{t_i}^{t_f} h(\xi)X(\xi)d\xi \tag{5.35}$$

The weighting function $h(\xi)$ is to be determined based on the minimum mean-square error criterion. The orthogonality principle requires that the error $\varepsilon(t) = Y(t) - \hat{Y}(t)$ be orthogonal to the observed data $X(\xi)$, during the interval $\xi \epsilon [t_i, t_f]$. Hence,

$$E[\{Y(t) - \int_{t_i}^{t_f} h(\xi, \lambda)X(\xi)d\xi\}X(\lambda)] = 0 \ , \ \lambda \epsilon [t_i, t_f] \tag{5.36}$$

Assuming that the processes are stationary, in which case the filter used is time-invariant, Equation (5.36) becomes

$$R_{YX}(t - \lambda) = \int_{t_i}^{t_f} R_{XX}(\xi - \lambda)h(\xi)d\xi \ , \ \lambda \epsilon [t_i, t_f] \tag{5.37}$$

Solving the integral Equation of (5.37), results in the desired weighting function $h(\xi)$.

The minimum mean-square error is

$$e_m = E[\{Y(t) - \int_{t_i}^{t_f} h(\xi)X(\xi)\}Y(t)] \tag{5.38}$$

or

$$e_m = R_{YY}(0) - \int_{t_i}^{t_f} h(\xi)R_{YX}(t - \xi)d\xi \tag{5.39}$$

5.3.1 The Optimum Unrealizable Filter

We now let $Y(t) = S(t) + N(t)$, where $S(t)$ is the desired signal to be estimated from the data $Y(t)$ for all time t. That is, we wish to obtain the optimum linear time-invariant filter such that the mean-square error estimate of $S(t)$ is minimum. We assume all processes to be stationary, real, and zero mean. The system is required to be time-invariant and consequently, the desired estimate may be expressed as

$$\hat{S}(t) = \int_{-\infty}^{\infty} h(t - \xi)Y(\xi)d\xi \tag{5.40}$$

Note that in this case, the filter is not constrained to be realizable, since we do not require a causal system. The impulse response $h(t - \xi)$ may be zero for $t < \xi$. This *filtering* problem can be represented by the block diagram of Figure 5.2.

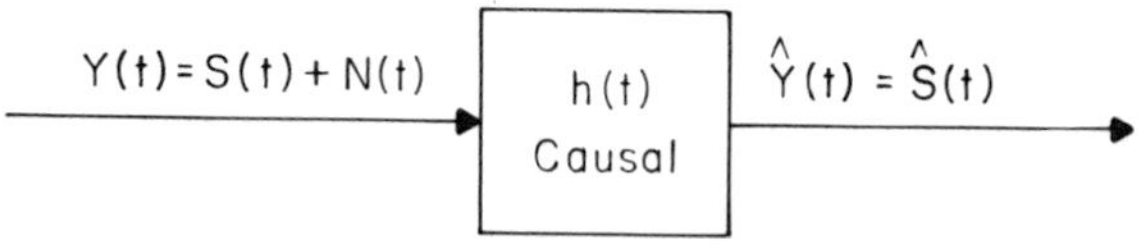

Figure 5.2 Filtering $S(t)$.

The orthogonality principle requires that

$$E[\{S(t) - \int_{-\infty}^{\infty} h(\xi)Y(t - \xi)d\xi\}Y(\lambda)] = 0 \text{ for all } \lambda \tag{5.41}$$

That is,

$$R_{SY}(t - \lambda) = \int_{-\infty}^{\infty} R_{YY}(t - \xi - \lambda)h(\xi)d\xi \tag{5.42}$$

Let $t - \lambda = \tau$, then Equation (5.42) becomes

$$\begin{aligned} R_{SY}(\tau) &= \int_{-\infty}^{\infty} R_{YY}(\tau - \xi)h(\xi)d\xi \\ &= R_{YY}(\tau) * h(\tau) \end{aligned} \tag{5.43}$$

where $*$ denotes convolution. It is easier to solve the above equation in the frequency domain. Taking the Fourier transform of Equation (5.43), we obtain

$$S_{SY}(f) = S_{YY}(f)H(f) \tag{5.44}$$

or

$$H(f) = \frac{S_{SY}(f)}{S_{YY}(f)} \tag{5.45}$$

The minimum mean-square error is given by

$$
\begin{aligned}
e_m &= E[\{S(t) - \int_{-\infty}^{\infty} Y(t - \xi)h(\xi)d\xi\}S(t)] \\
&= R_{SS}(0) - \int_{-\infty}^{\infty} R_{SY}(\xi)h(\xi)d\xi \tag{5.46a}
\end{aligned}
$$

where

$$\int_{-\infty}^{\infty} R_{SY}(\xi)h(\xi)d\xi = \{R_{SY}(-\tau) * h(\tau)\}|_{\tau=0} \tag{5.46b}$$

Expressing the error of Equation (5.46) in terms of the power spectral density, we obtain

$$e_m = \int_{-\infty}^{\infty} [S_{SS}(f) - S_{SY}(-f)H(f)]df \tag{5.47}$$

Substituting Equation (5.45) in Equation (5.46), the error becomes

$$e_m = \int_{-\infty}^{\infty} [S_{SS}(f) - \frac{S_{SY}(-f)S_{SY}(f)}{S_{YY}(f)}]df \tag{5.48}$$

If $S(t)$ and $N(t)$ are statistically independent, then

$$E[S(t)N(t)] = E[S(t)]E[N(t)] = 0 \tag{5.49}$$

since they are zero-mean. It follows that

$$S_{YY}(f) = S_{SS}(f) + S_{NN}(f) \tag{5.50}$$

and

$$S_{SY}(f) = S_{SS}(f) \tag{5.51}$$

Therefore, the transfer function $H(f)$ is

$$H(f) = \frac{S_{SY}(f)}{S_{YY}(f)} = \frac{S_{SS}(f)}{S_{SS}(f) + S_{NN}(f)} \tag{5.52}$$

The resulting minimum mean-square error in this case is

$$
\begin{aligned}
e_m &= \int_{-\infty}^{\infty} [S_{SS}(f) - \frac{S_{SS}^2(f)}{S_{SS}(f) + S_{NN}(f)}] df \\
&= \int_{-\infty}^{\infty} \frac{S_{SS}(f) S_{NN}(f)}{S_{SS}(f) + S_{NN}(f)} df \tag{5.53}
\end{aligned}
$$

We note that if the power spectral densities $S_{SS}(f)$ and $S_{NN}(f)$ do not overlap, then $S_{SS}(f)$ ($S_{NN}(f)$) is zero, when $S_{NN}(f)$ ($S_{SS}(f)$) is different than zero. In this case, the transfer function $H(f)$ becomes

$$H(f) = \begin{cases} 1 & \text{for } f \text{ such that } S_{SS}(f) \neq 0 \\ 0 & \text{for } f \text{ such that } S_{NN}(f) \neq 0 \end{cases} \tag{5.54}$$

Also, for these power spectral densities of $S_{SS}(f)$ and $S_{NN}(f)$, which are nonoverlapping, the product of $S_{SS}(f)$ and $S_{NN}(f)$ is zero, and thus, the minimum mean-square error is zero.

Example 5.3

Let the observation process for all time t be $Y(t) = S(t) + N(t)$. (a) Obtain the linear mean-square error of $S'(t)$, the derivative of the signal $S(t)$. (b) Determine the impulse response $h(t)$ given that

$$R_{SS} = e^{-\alpha \tau^2} \quad , \quad R_{SN}(\tau) = 0 \quad \text{and} \quad R_{NN}(\tau) = kS(\tau)$$

Solution.

(a) The linear mean-square error estimate of $S'(t)$ is given by

$$\hat{S}'(t) = \int_{-\infty}^{\infty} Y(t - \xi) h(\xi) d\xi$$

since the error is orthogonal to the data, we have

$$E[\{S'(t) - \int_{-\infty}^{\infty} Y(t - \alpha)h(\alpha)d\alpha\}Y(\xi)] = 0 \text{ for all } \xi$$

or

$$R_{S'Y}(t - \xi) = \int_{-\infty}^{\infty} R_{YY}(t - \alpha - \xi)h(\alpha)d\alpha$$

Let $\tau = t - \xi$; then

$$R_{S'Y}(\tau) = \int_{-\infty}^{\infty} R_{YY}(\tau - \alpha)h(\alpha)d\alpha \text{ for all } \tau$$

Taking the Fourier transform of the above expression, we obtain

$$jfS_{SY}(f) = S_{YY}(f)H(f)$$

or

$$H(f) = jf\frac{S_{SY}(f)}{S_{YY}(f)}$$

Hence, the optimum filter for estimating $S'(t)$ is a cascade of two systems; the first system with transfer function $S_{SY}(f)/S_{YY}(f)$ while the second system is a differentiator, as shown in Figure 5.3.

Figure 5.3 Filter for estimating $S'(t)$.

(b) Using Equations (5.50) and (5.51), we have $R_{YY}(\tau) = R_{SS}(\tau) + R_{NN}(\tau)$ and $R_{SY}(\tau) = R_{SS}(\tau)$ since $R_{SN}(\tau) = 0$. Consequently, the transfer function becomes

$$H(f) = jf\frac{S_{SS}(f)}{S_{SS}(f) + S_{NN}(f)}$$

we need to determine $S_{SS}(f)$ from $R_{SS}(\tau)$. The Fourier transform of the autocorrelation $R_{SS}(\tau)$ is

$$S_{SS}(f) = \int_{-\infty}^{\infty} e^{-\alpha\tau^2} e^{-j2\pi f\tau} d\tau$$

$$= \sqrt{\frac{\pi}{\alpha}}\, e^{-\frac{\pi^2 f^2}{\alpha}}$$

where we have used the fact that

$$\int_{-\infty}^{\infty} e^{-(ax^2 + 2bx + c)} dx = \sqrt{\frac{\pi}{a}}\, e^{\frac{b^2 - ac}{a}} \quad \text{for } a > 0$$

The transfer function becomes

$$H(f) = jf \frac{\sqrt{\frac{\pi}{\alpha}}\, e^{-\frac{\pi^2 f^2}{\alpha}}}{\sqrt{\frac{\pi}{\alpha}}\, e^{-\frac{\pi^2 f^2}{\alpha}} + k}$$

$$= jf \frac{\frac{1}{k}\sqrt{\frac{\pi}{\alpha}}\, e^{-\frac{\pi^2 f^2}{\alpha}}}{1 + \frac{1}{k}\sqrt{\frac{\pi}{\alpha}}\, e^{-\frac{\pi^2 f^2}{\alpha}}}$$

$$= jf G(f)$$

where

$$G(f) = \frac{\frac{1}{k}\sqrt{\frac{\pi}{\alpha}}\, e^{-\frac{\pi^2 f^2}{\alpha}}}{1 + \frac{1}{k}\sqrt{\frac{\pi}{\alpha}}\, e^{-\frac{\pi^2 f^2}{\alpha}}}$$

Thus, $h(t) = dg(t)/dt$ where $g(t)$ is the inverse Fourier transform of $G(f)$. It can be shown that the impulse response $h(t)$ is given by

$$h(t) = \frac{2\alpha^{\frac{3}{2}}}{\sqrt{\pi}} t \sum_{n=1}^{\infty} (-1)^n \frac{1}{n^{\frac{3}{2}}} \left(\frac{1}{k}\sqrt{\frac{\pi}{\alpha}}\right)^n e^{-\frac{t^2}{\alpha n}}$$

The above series converges, provided that $(1/k)\sqrt{\pi/\alpha} < 1$.

Example 5.4

(a) Determine the optimum unrealizable filter $h(t)$ of the observation process $Y(t) = S(t) + N(t)$. $S(t)$ and $N(t)$ are uncorrelated, with autocorrelations:

$$R_{SS}(\tau) = e^{-\alpha|\tau|} \quad \text{and} \quad R_{NN}(\tau) = \frac{N_0}{2}\delta(\tau)$$

(b) Calculate the minimum mean-square error.

Solution.
From Equation (5.52), the transfer function $H(f)$ is

$$H(f) = \frac{S_{SS}(f)}{S_{SS}(f) + S_{NN}(f)}$$

where $S_{SS}(f)$ and $S_{NN}(f)$ are the Fourier transforms of $R_{SS}(\tau)$, and $R_{NN}(\tau)$, respectively. Hence,

$$
\begin{aligned}
H(f) &= \frac{2\alpha/(\alpha^2 + 4\pi^2 f^2)}{[2\alpha/(\alpha^2 + 4\pi^2 f^2)] + (N_0/2)} \\
&= \frac{a}{b^2 + 4\pi^2 f^2}
\end{aligned}
$$

where $a = 4\alpha/N_0$ and $b^2 = \alpha^2 + (4\alpha/N_0)$. It follows that

$$
\begin{aligned}
h(t) &= \int_{-\infty}^{\infty} \frac{a}{b^2 + 4\pi^2 f^2} e^{j2\pi ft} df \\
&= \frac{a}{2b} e^{-b|t|} = \frac{2\alpha}{N_0\sqrt{\alpha^2 + \frac{4\alpha}{N_0}}} e^{-\sqrt{\alpha^2 + \frac{4\alpha}{N_0}}|t|}
\end{aligned}
$$

From the plot of $h(t)$ shown in the Figure 5.4, we observe that the impulse response $h(t)$ is noncausal, and thus nonrealizable.

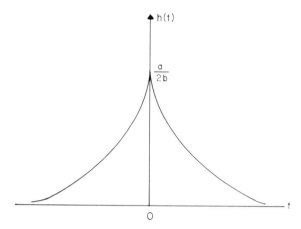

Figure 5.4 Filter $h(t)$ for Example 5.4.

(b) The minimum mean-square error is given in Equation (5.53). Substituting for the expression of $S_{SS}(f)$ and $S_{NN}(f)$, and solving the integral, we obtain the minimum mean-square error to be

$$e_m = \frac{\alpha}{b} = \frac{\alpha}{\sqrt{\alpha^2 + \frac{4\alpha}{N_0}}}$$

5.3.2 The Optimum Realizable Filter

In the previous section, we solved for the optimum unrealizable filter to extract the desired signal from the observation process $Y(t) = S(t) + N(t)$. We now consider the same problem with the constraint that the filter $h(t)$ is *realizable*; that is, $h(t) = 0$ for $t < 0$. The system representing the problem is shown in Figure 5.5.

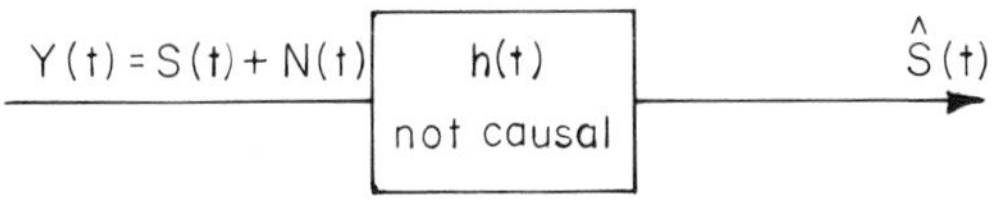

Figure 5.5 Optimum realizable filter.

We assume that the signal process $Y(t)$ is known only up to the present moment t. Therefore, the estimate of $S(t)$ is

$$\hat{S}(t) = \int_{-\infty}^{t} h(t - \xi)Y(\xi)d\xi$$
$$= \int_{0}^{\infty} Y(t - \xi)h(\xi)d\xi \tag{5.55}$$

The linear mean-square estimation requires that $h(t)$ be chosen so that the mean-square error is minimum. From the orthogonality principle, we have

$$E[\{S(t) - \int_{0}^{\infty} Y(t - \xi)h(\xi)d\xi\}Y(t')] = 0 \text{ for all } t' \leq t \tag{5.56}$$

The impulse response of the optimum realizable filter satisfies the integral equation:

$$R_{SY}(t - t') = \int_{0}^{\infty} R_{YY}(t - \xi - t')h(\xi)d\xi \tag{5.57}$$

Let $t - t' = \tau$, then the integral equation becomes

$$R_{SY}(\tau) = \int_{0}^{\infty} R_{YY}(\tau - \xi)h(\xi)d\xi \text{ for all } \tau \geq 0 \tag{5.58}$$

Equation (5.58) is called the *Wiener-Hopf Integral Equation*. Furthermore, the mean-square error of the estimate reduces to

$$e_m = E[\{S(t) - \int_{0}^{\infty} Y(t - \xi)h(\xi)d\xi\}S(t)]$$
$$= R_{SS}(0) - \int_{0}^{\infty} R_{SY}(\xi)h(\xi)d\xi \tag{5.59}$$

The solution to the Wiener-Hopf integral equation is not as easy as in the case of the unrealizabe filter, since the integral is valid for τ positive only. Equation (5.58) can be written as

$$R_{SY}(\tau) = \int_{-\infty}^{\infty} R_{YY}(\tau - \xi)h(\xi)d\xi \quad \tau \geq 0 \tag{5.60}$$

It can be shown that the choice of the above integral for τ negative is not arbitrary, when the additional constraint $h(\xi)$ causal is imposed. Moreover,

$$\int_{-\infty}^{\infty} R_{YY}(\tau - \xi)h(\xi)d\xi \neq R_{SY}(\tau) \quad \tau < 0 \tag{5.61}$$

Consequently, we cannot obtain the impulse response of the optimum realizable filter, by simply using the frequency domain approach, as we did with the unrealizable filter.

To obtain an integral equation valid of all τ, $-\infty < \tau < \infty$, we combine Equations (5.60) and (5.61), which results in

$$\int_{-\infty}^{\infty} R_{YY}(\tau - \xi)h(\xi)d\xi - R_{SY}(\tau) = a(\tau) \text{ for all } \tau \tag{5.62}$$

where $a(\tau)$ is zero for τ positive and

$$a(\tau) = \int_{-\infty}^{\infty} A(f) \, e^{+j2\pi f\tau}df \quad \tau < 0 \tag{5.63}$$

Assume that the power spectral density $S_{YY}(f)$ is a rational function, and can be factored into

$$S_{YY}(f) = S_{YY}^{+}(f)S_{YY}^{-}(f) \tag{5.64}$$

$S_{YY}^{+}(f)$ and its conjugate $S_{YY}^{-}(f)$ are called the *spectral factorizations* of $S_{YY}(f)$. $S_{YY}^{+}(f)$ has all its poles and zeros in the left half-plane (LHP) of the S-plane ($S = j\omega, \omega = 2\pi f$), whereas $S_{YY}^{-}(f)$ has all its poles and zeros in the right half-plane (RHP). Taking the Fourier transform of Equation (5.62), we have

$$S_{YY}(f)H(f) - S_{SY}(f) = A(f) \tag{5.65}$$

or

$$S_{YY}^+(f)S_{YY}^-(f)H(f) - S_{SY}(f) = A(f) \qquad (5.66)$$

Dividing Equation (5.66) by $S_{YY}^-(f)$, we obtain

$$S_{YY}^+(f)H(f) - \frac{S_{SY}(f)}{S_{YY}^-(f)} = \frac{A(f)}{S_{YY}^-(f)} \qquad (5.67)$$

Note that since $S_{YY}^+(f)$ and $H(f)$ have all their poles in the LHP, the product $S_{YY}^+(f)H(f)$ has all its poles in the LHP and consequently, the corresponding time function is zero for τ negative. Also, since $S_{YY}^-(f)$ has its zeros in the RHP, and $A(f)$ has all its poles in the RHP, the quotient $A(f)/S_{YY}^-(f)$ has all its poles in the RHP. Consequently, the corresponding time function is zero for $\tau > 0$. The ratio $S_{SY}(f)/S_{YY}^-(f)$ has poles in the LHP and the RHP. Thus, its corresponding time function is valid for all τ. Splitting the poles and zeros, the ratio $S_{SY}(f)/S_{YY}^-(f)$ may be expressed as

$$\frac{S_{SY}(f)}{S_{YY}^-(f)} = [\frac{S_{SY}(f)}{S_{YY}^-(f)}]^+ + [\frac{S_{SY}(f)}{S_{YY}^-(f)}]^- \qquad (5.68)$$

$[S_{SY}(f)/S_{YY}^-(f)]^+$ has all its poles and zeros in the LHP, whereas $[S_{SY}(f)/S_{YY}^-(f)]^-$ has all its poles and zeros in the RHP. Substituting Equation (5.68) into (5.67), we obtain

$$\underbrace{S_{YY}^+(f)H(f)}_{LHP} - \underbrace{[\frac{S_{SY}(f)}{S_{YY}^-(f)}]^+}_{LHP} - \underbrace{[\frac{S_{SY}(f)}{S_{YY}^-(f)}]^-}_{RHP} = \underbrace{\frac{A(f)}{S_{YY}^-(f)}}_{RHP} \qquad (5.69)$$

Define

$$B^+(f) = [\frac{S_{SY}(f)}{S_{YY}^-(f)}]^+ \qquad (5.70a)$$

and

$$B^-(f) = [\frac{S_{SY}(f)}{S_{YY}^-(f)}]^- \qquad (5.70b)$$

Equating the terms marked by LHP (all poles in LHP) in Equation (5.69), we have

$$S_{YY}^+(f)H(f) - B^+(f) = 0 \qquad (5.71)$$

or, the transfer function $H(f)$ is

$$H(f) = \frac{B^+(f)}{S_{YY}^+(f)} \tag{5.72}$$

All poles of $H(f)$ are in LHP and, consequently, the filter response $h(t)$ is zero for t negative and thus are realizable. Therefore, the optimum filter is

$$h(t) = \int_{-\infty}^{\infty} \frac{B^+(f)}{S_{YY}^+(f)} e^{j2\pi ft} df \tag{5.73}$$

The corresponding minimum mean-square error is obtained by substituting for Equation (5.72) into (5.59) and taking the Fourier transform which results in

$$e_m = \int_{-\infty}^{\infty} [S_{SS}(f) - S_{SY}(-f) \frac{B^+(f)}{S_{YY}^+(f)}] df \tag{5.74}$$

Example 5.5

Consider the problem where the signal $S(t)$ and the noise $N(t)$ are uncorrelated and with autocorrelation functions

$$R_{SS}(\tau) = e^{-\alpha|\tau|} \quad \text{and} \quad R_{NN}(\tau) = \frac{N_0}{2}\delta(\tau)$$

For simplicity assume $\alpha = N_0/2 = 1$.
(a) Obtain the optimum realizable filter.
(b) Calculate the minimum mean-square error.

Solution.
(a) The optimum realizable filter is given by Equation (5.73) where $B^+(f)$ and $S_{YY}^+(f)$ are defined in Equations (5.70a) and (5.64), respectively. Since $S(t)$ and $N(t)$ are uncorrelated, the power spectral density of $Y(t)$ is as given by Equation (5.50); i.e.,

$$S_{YY}(f) = S_{SS}(f) + S_{NN}(f)$$

where

$$S_{SS}(f) = \frac{1}{1 + 4\pi^2 f^2} \quad \text{and} \quad S_{NN}(f) = 1$$

Substituting for the expressions of $S_{SS}(f)$ and $S_{NN}(f)$, we obtain

$$S_{YY}(f) = \frac{1}{1 + 4\pi^2 f^2} + 1$$

$$= \frac{2 + 4\pi^2 f^2}{1 + 4\pi^2 f^2} = \frac{S^2 - 2}{S^2 - 1}$$

where $S = j2\pi f$. In order to differentiate between S denoting the power spectral density, and $S = j2\pi f$ the Laplace transform, we denote the Laplace transform by V; that is, $V = j2\pi f$. Consequently,

$$S_{YY}(V) = \frac{V^2 - 2}{V^2 - 1}$$

$$= \frac{(V + \sqrt{2})(V - \sqrt{2})}{(V + 1)(V - 1)} = S_{YY}^+(V)S_{YY}^-(V)$$

where

$$S_{YY}^+(V) = \frac{V + \sqrt{2}}{V + 1} \text{ and } S_{YY}^-(V) = \frac{V - \sqrt{2}}{V - 1}$$

we need to determine the cross-spectral density $S_{SY}(f)$. From Equation (5.51), we have

$$S_{SY}(f) = S_{SS}(f) = \frac{1}{1 + 4\pi^2 f^2}$$

or, in the Laplace domain $(V = j2\pi f)$

$$S_{SS}(V) = \frac{-1}{V^2 - 1} = \frac{-1}{(V + 1)(V - 1)}$$

Hence, from Equation (5.68),

$$\frac{S_{SY}(V)}{S_{YY}^-(V)} = B^+(V) + B^-(V)$$

$$= \frac{-1}{(V + 1)(V - \sqrt{2})} = \frac{1/(1 + \sqrt{2})}{V + 1} + \frac{-1/(1 + \sqrt{2})}{V - \sqrt{2}}$$

Now that we have $B^+(V)$ and $S_{YY}^+(V)$, we substitute in Equation (5.72) to obtain

$$H(V) = \frac{B^+(V)}{S_{YY}^+(V)} = \frac{1/(1 + \sqrt{2})}{V + 1} \cdot \frac{V + 1}{V + \sqrt{2}}$$

$$= \frac{1/1 + \sqrt{2})}{V + \sqrt{2}}$$

or

$$H(f) = \frac{1}{1 + \sqrt{2}} \cdot \frac{1}{j2\pi f + \sqrt{2}}$$

Taking the inverse Laplace transform of $H(V)$, we obtain the optimum realizable filter $h(t)$ to be

$$h(t) = \frac{1}{1 + \sqrt{2}} e^{-\sqrt{2}t} U(t)$$

where $U(t)$ is the unit step function.

(b) The minimum mean-square error is given by Equation (5.59) to be

$$e_m = R_{SS}(0) - \int_0^\infty R_{SY}(\xi) h(\xi) d\xi$$

From Part (a) we found that

$$S_{SS}(V) = S_{SY}(V) = \frac{-1}{V^2 - 1}$$

Taking the inverse Laplace transform, we have

$$R_{SS}(\tau) = R_{SY}(\tau) = \frac{1}{2} e^{-|\tau|}$$

Substituting for the expression of $R_{SS}(0) = \frac{1}{2}$ and $R_{SY}(\tau)$ into e_m, we obtain

$$e_m = \frac{1}{2} - \frac{1}{2(1 + \sqrt{2})^2}$$

Example 5.6

Let $Y(t) = S(t) + N(t)$, where the signals $S(t)$ and $N(t)$ are statistically independent, with zero mean and autocorrelation functions:

$$R_{SS}(\tau) = e^{-|\tau|} \qquad \text{and} \qquad R_{NN}(\tau) = \delta(\tau) + 2e^{-|\tau|}$$

(a) Find the optimum unrealizable filter.
(b) Find the optimum realizable filter.

Solution.

(a) The optimum unrealizable filter is given by Equation (5.45) to be

$$H(f) = \frac{S_{SY}(f)}{S_{YY}(f)}$$

where $S_{SY}(f) = S_{SS}(f)$ and $S_{YY}(f) = S_{SS}(f) + S_{NN}(f)$, since the signal and noise are uncorrelated, and zero-mean. Hence,

$$S_{SS}(f) = \frac{1}{1 + 4\pi^2 f^2} = S_{SY}(f)$$

$$S_{NN}(f) = 1 + \frac{2}{1 + 4\pi^2 f^2}$$

and

$$S_{YY}(f) = \frac{4 + 4\pi^2 f^2}{1 + 4\pi^2 f^2}$$

Substituting into the expression of $H(f)$, we obtain

$$H(f) = \frac{1}{4 + 4\pi^2 f^2}$$

Consequently,

$$h(t) = \frac{1}{4} e^{-2|t|} = \begin{cases} \frac{1}{4} e^{-2t} & ,t \geq 0 \\ \frac{1}{4} e^{2t} & ,t < 0 \end{cases}$$

(b) The optimum realizable filter is given by Equation (5.72). First, we factor $S_{YY}(f)$ into $S_{YY}^+(f)$, and $S_{YY}^-(f)$. Let $V = j2\pi f$ denote Laplace transform. Then,

$$S_{YY}(V) = \frac{V^2 - 4}{V^2 - 1} = \frac{(V - 2)(V + 2)}{(V - 1)(V + 1)}$$

where

$$S_{YY}^+(V) = \frac{V + 2}{V + 1} \quad \text{and} \quad S_{YY}^-(V) = \frac{V - 2}{V - 1}$$

Also,

$$S_{SY}(V) = S_{SS}(V) = \frac{-1}{V^2 - 1} = \frac{-1}{(V + 1)(V - 1)}$$

Consequently,

$$\frac{S_{SY}(V)}{S_{YY}^-(V)} = \frac{-1}{(V+1)(V-2)} = \frac{1/3}{V+1} + \frac{-1/3}{V-2}$$

where $B^+(V) = 1/[3(V+1)]$ and $B^-(V) = -1/[3(V-2)]$. The transfer function of the realizable filter is

$$H(V) = \frac{B^+(V)}{S_{YY}^+(V)} = \frac{1/3}{V+2}$$

Taking the inverse Laplace transform of $H(V)$, we obtain the Wiener filter to be

$$h(t) = \begin{cases} \frac{1}{3} e^{-2t} & ,t \geq 0 \\ 0 & ,t < 0 \end{cases}$$

5.3.3 Discrete Wiener Filter

We now consider the filtering problem where the observed signal is a discrete random sequence, and the goal is to estimate another random sequence. The incoming sequence composed of the signal sequence $S(n)$ $n = 0, 1, 2, \cdots$ and the additive noise sequence $N(n)$, $n = 0, 1, 2, \cdots$ enters a linear discrete-time filter with impulse response denoted by the sequence $h(n)$, $n = 0, 1, 2, \cdots$, as shown in Figure 5.6.

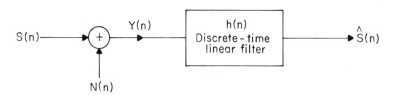

Figure 5.6 Filtering the sequence $S(n)$.

We assume that the sequences $S(n)$ and $N(n)$ are uncorrelated zero-mean random variables. We wish to find $\hat{S}(n)$ the minimum linear mean-square error estimator. The estimate $\hat{S}(n)$ may be expressed as the correlation sum of the sequences $Y(n)$, and $h(n)$ given by

$$\hat{S}(n) = \sum_{k=-\infty}^{\infty} h(k)Y(n-k) \tag{5.75}$$

or

$$\hat{S}(n) = \sum_{k=0}^{\infty} h(k)Y(n-k) \tag{5.76}$$

Equation (5.75) indicates that all data $Y(n)$ is available for all n. The sequence is not finite, and thus the filter is *not realizable*. Equation (5.76) indicates that only present and past values of Y are used in estimating $S(n)$. Thus, we have a finite sequence and the filter is *causal* or *realizable*.

Unrealizable filter

In this case, the estimator is given by Equation (5.75). The criterion used to determine the filter sequence $h(n)$ is the mean-square error and is given by

$$E[\{S(n) - \hat{S}(n)\}^2] \tag{5.77}$$

The mean-square error is minimized by applying the orthogonality principle; that is, the error is orthogonal to the data. Hence,

$$E[\{S(n) - \sum_{k=-\infty}^{\infty} h(k)Y(n-k)\}Y(n-m)] \text{ for all } m \tag{5.78}$$

or

$$R_{SY}(m) = \sum_{k=-\infty}^{\infty} h(k)R_{YY}(m-k) \text{ for all } m \tag{5.79a}$$

where

$$R_{SY}(m) = E[S(n)Y(n-m)] \tag{5.79b}$$

and

$$R_{YY}(n-m) = E[Y(n)Y(m)] \tag{5.79c}$$

We define the Fourier transform of a discrete sequence $f(k)$, $k = 0, \pm 1, \pm 2, \cdots$, as

$$F(e^{j\omega}) = \sum_{k=-\infty}^{\infty} f(k)e^{-kj\omega} \tag{5.80}$$

where $\omega = 2\pi f$. Making the change of variable $z = e^{j2\pi f}$, Equation (5.80) becomes

$$F(z) = \sum_{k=-\infty}^{\infty} f(k)z^{-k} \tag{5.81}$$

Note that the trajectory of $z = e^{j\omega}$ is the unit circle on the z-plane. $F(e^{j\omega})$ is periodic with period 2π and hence, the spectrum is usually plotted for $\omega \epsilon [-\pi, \pi]$ or $f \epsilon [-\frac{1}{2}, \frac{1}{2}]$. Taking the Fourier transform of Equation (5.79), we obtain

$$S_{SY}(z) = H(z)S_{YY}(z) \tag{5.82}$$

or

$$H(z) = \frac{S_{SY}(z)}{S_{YY}(z)} \tag{5.83}$$

The resulting mean-square error is

$$e_m = R_{SS}(0) - \sum_{k=-\infty}^{\infty} h(k)R_{YS}(-k) \tag{5.84}$$

Realizable Filter

In this case the filter is causal, and the estimator is given by Equation (5.76); that is,

$$\hat{S}(n) = \sum_{k=0}^{\infty} h(k)Y(n - k) \tag{5.85}$$

Applying the orthogonality principle, we obtain the discrete version of the Wiener-Hopf equation given by

$$R_{SY}(m) = \sum_{k=0}^{\infty} h(k)R_{YY}(m - k) \tag{5.86}$$

where $R_{SY}(m)$ and $R_{YY}(m-k)$ are as defined in Equations (5.79b) and (5.79c), respectively. Let the spectral density $S_{YY}(z)$ be

$$S_{YY}(z) = \sum_{k=-\infty}^{\infty} R_{YY}(k)z^{-k} \qquad (5.87)$$

Following a similar procedure as in the continuous case, we first factor $S_{YY}(z)$ into

$$S_{YY}(z) = S_{YY}^{+}(z)S_{YY}^{-}(z) \qquad (5.88)$$

such that the poles and zeros of $S_{YY}(z)$ inside the circle $|z| < 1$ are assigned to $S_{YY}^{+}(z)$, whereas the poles and zeros in the $|z| > 1$ are assigned to $S_{YY}^{-}(z)$. Consequently, $S_{YY}^{+}(z)$ is analytic inside the unit circle, and $S_{YY}^{-}(z)$ is analytic in $|z| > 1$. Dividing Equation (5.88) by $S_{YY}^{-}(z)$, and applying partial fraction expansion, we obtain

$$\frac{S_{SY}(z)}{S_{YY}^{-}(z)} = [\frac{S_{SY}(z)}{S_{YY}^{-}(z)}]^{+} + [\frac{S_{SY}(z)}{S_{YY}^{-}(z)}]^{-} \qquad (5.89)$$

where $[\cdot]^{+}$ denotes poles and zeros inside $|z| < 1$, and $[\cdot]^{-}$ denotes poles and zeros in $|z| > 1$. Let

$$B^{+}(z) = [\frac{S_{SY}(z)}{S_{YY}^{-}(z)}]^{+} \qquad (5.90a)$$

and

$$B^{-}(z) = [\frac{S_{SY}(z)}{S_{YY}^{-}(z)}]^{-} \qquad (5.90b)$$

the optimum causal filter is

$$\begin{aligned} H(z) &= \frac{B^{+}(z)}{S_{YY}^{-}(z)} \\ &= \frac{1}{S_{YY}^{+}(z)}[\frac{S_{SY}(z)}{S_{YY}^{-}(z)}]^{+} \end{aligned} \qquad (5.91)$$

We see that the optimum discrete realizable filter, is a cascade of two filters, as shown in Figure 5.7.

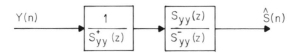

Figure 5.7 Wiener filter.

The mean-square error is

$$e_m = R_{SS}(0) - \sum_{k=0}^{\infty} h(k) R_{SY}(k) \tag{5.92}$$

Example 5.7

Consider the problem where the received sequence is $Y(n) = S(n) + N(n)$. The signal sequence $S(n)$ is stationary and zero-mean with power spectrum

$$S_{SS}(e^{j\omega}) = \frac{2}{5 - 4\cos\omega}$$

The noise sequence $N(n)$ is independent of the signal sequence $S(n)$, and has power spectrum

$$S_{NN}(e^{j\omega}) = 1$$

(a) Obtain the realizable filter.
(b) Find the unrealizable filter.

Solution.

(a) Since the signal and the noise sequences are independent, then,

$$S_{SY}(e^{j\omega}) = S_{SS}(e^{j\omega})$$

Making the change of variable $z = e^{j\omega}$, we have

$$S_{NN}(z) = 1$$

and

$$S_{SY}(z) = S_{SS}(z) = \frac{2z}{-2z^2 + 5z - 2}$$

Thus,

$$\begin{aligned}
S_{YY}(z) &= S_{SS}(z) + S_{NN}(z) \\
&= \frac{2z^2 - 7z + 2}{2z^2 - 5z + 2} \\
&= \frac{(z - 3.186)(z - 0.314)}{(z - 2)(z - 0.5)}
\end{aligned}$$

Hence,

$$S_{YY}^+(z) = \frac{z - 0.314}{z - 0.5}$$

and

$$S_{YY}^-(z) = \frac{z - 3.186}{z - 2}$$

Dividing $S_{SY}(z)$ by $S_{YY}^-(z)$, we obtain

$$\begin{aligned}
\frac{S_{SY}(z)}{S_{YY}^-(z)} &= \frac{-z}{(z - 3.186)(z - 0.5)} \\
&= \frac{-1.186}{z - 3.186} + \frac{0.186}{z - 0.5}
\end{aligned}$$

where

$$B^+(z) = \frac{0.186}{z - 0.5}$$

Using Equation (5.91), the optimum realizable filter is

$$\begin{aligned}
H(z) &= \frac{B^+(z)}{S_{YY}^+(z)} = \frac{0.186}{z - 0.5} \cdot \frac{z - 0.5}{z - 0.314} \\
&= \frac{0.186}{z - 0.314}
\end{aligned}$$

or

$$h(n) = 0.186(0.314)^n \ , \ n = 0, 1, 2, \cdots$$

(b) The optimum unrealizable filter is given by Equation (5.83) to be

$$H(z) \;=\; \frac{S_{SY}(z)}{S_{YY}(z)} = \frac{-z}{z^2 - 3.5z + 1}$$

$$\;=\; \frac{-z}{(z - 3.186)(z - 0.314)}$$

Note that the pole at $z = 3.186$ outside the unit circle makes this filter unstable, and thus unrealizable in real time.

5.4 KALMAN FILTER

In this section, we give a brief description of the optimum Kalman filter. We consider the state model approach. In this case, filtering means estimating the state vector at the present time, based upon past observed data. Prediction is estimating the state vector at a future time. Since it can be shown that the filtered estimate of the state vector is related to the one-step prediction of the state, we first develop the concept of prediction, and then derive the equations for the filtered state.

Let the state model be of the form:

$$\mathbf{S}(n) = \mathbf{\Phi}(n)\mathbf{S}(n - 1) + \mathbf{W}(n) \qquad (5.93)$$

where $\mathbf{S}(n)$ is the $(m \times 1)$ state vector, $\mathbf{\Phi}(n)$ is an $(m \times m)$ known state transition matrix, and $\mathbf{W}(n)$ is an $(m \times 1)$ noise vector. We assume that the vector random sequence $\mathbf{S}(n)$ is zero-mean Gaussian, and the noise vector process $\mathbf{W}(n)$ is also zero-mean and white with autocorrelation:

$$E[\mathbf{W}(n)\mathbf{W}^T(k)] = \begin{cases} \mathbf{Q}(n) & ,n = k \\ 0 & ,n \neq k \end{cases} \qquad (5.94)$$

Let $\mathbf{Y}(n)$ be the $(p \times 1)$ observation vector consisting of a Gaussian random sequence. The observation can be modeled as

$$\mathbf{Y}(n) = \mathbf{H}(n)\mathbf{S}(n) + \mathbf{V}(n) \qquad (5.95)$$

where $\mathbf{H}(n)$ is a $(p \times m)$ measurement matrix relating the state vector to the observation vector, and $\mathbf{V}(n)$ is a known $(p \times 1)$ measurement error. $\mathbf{V}(n)$ is a Gaussian zero-mean white noise sequence with auto-correlation:

$$E[\mathbf{V}(n)\mathbf{V}^T(k)] = \begin{cases} \mathbf{R}(n) & n = k \\ 0 & n \neq k \end{cases} \qquad (5.96)$$

In order to obtain the Kalman filter state, $\hat{s}(n)$, we first solve for $\hat{\mathbf{S}}(n + 1)$, the one step linear predictor, using the concept of innovations.

5.4.1 Innovations

In this subsection, we first present the concept of innovations for random variables, and give some important properties. The result, which will then be generalized to random vectors, will be used to solve for the Kalman filter. Let $Y(1), Y(2), \cdots, Y(n)$ be a sequence of zero-mean Gaussian random variables. The innovations process $INV(n)$ represents the new information, which is not carried from the observed data $Y(1), Y(2), \cdots, Y(n-1)$, to obtain the predicted estimate $\hat{Y}(n)$ of the observed random variable. Specifically, let $\hat{S}(n-1)$ be the linear minimum mean-square estimate of the random variable $S(n-1)$, based on the observation data $Y(1), Y(2), \cdots, Y(n-1)$. Suppose that we take an additional observation $Y(n)$, and desire to obtain $\hat{S}(n)$ the estimate of $S(n)$. In order to avoid redoing the computations from the beginning as for $\hat{S}(n-1)$, it is more efficient to use the previous estimate $\hat{S}(n-1)$ based on the $(n-1)$ observation random variables $Y(1), Y(2), \cdots, Y(n-1)$ and compute $\hat{S}(n)$ *recursively* based on the n random variables; $Y(1), Y(2), \cdots, Y(n-1)$ and the additional new observation variable $Y(n)$. We define

$$INV(n) = Y(n) - \hat{Y}(n|Y(1), \cdots, Y(n-1)) , \ n = 1, 2, \cdots \qquad (5.97)$$

where $INV(n)$ denotes the innovations process and $\hat{Y}(n|Y(1), \cdots, Y(n-1))$ is the estimate of $Y(n)$ based on the $(n-1)$ observations, $Y(1), \cdots, Y(n-1)$. We see from Equation (5.97) that because $INV(n)$ represents a new information *measure* in the observation variable $Y(n)$, it is referred to as "innovation". It can be shown that the estimate $\hat{S}(n)$ based on the n observations, $Y(1), \cdots, Y(n)$, is related to the estimate $\hat{S}(n-1)$ based on the $(n-1)$ observations, $Y(1), \cdots, Y(n-1)$, by the following recursive rule:

$$\hat{S}(n) = \hat{S}(n-1) + b_n INV(n) \tag{5.98}$$

where the constant b_n is given by

$$b_n = \frac{E[S(n)INV(n)]}{E[\{INV(n)\}^2]} \tag{5.99}$$

We now mention two important properties of $INV(n)$ without giving any formal proof.

(i) The innovations $INV(n)$ is orthogonal to the past observation variables, $Y(1), \cdots, Y(n-1)$; that is,

$$E[INV(n) \, Y(k)] = 0 \ , \ k = 1, \cdots, n-1 \tag{5.100}$$

(ii) The innovations $INV(k)$, $k = 1, 2, \cdots, n$ are orthogonal to each other; that is,

$$E[INV(n)INV(k)] = 0 \ k \neq n \tag{5.101}$$

Generalizing the result given in Equations $(5.97), (5.100)$, and (5.101), to random vectors, we obtain

$$\mathbf{INV}(n) = \mathbf{Y}(n) - \hat{\mathbf{Y}}(n|\mathbf{Y}(1), \cdots, \mathbf{Y}(n-1)) \ n = 1, 2, \cdots \tag{5.102}$$

$$E[\mathbf{INV}(n)\mathbf{Y}^T(k)] = \mathbf{0} \quad k = 1, \cdots, n-1 \tag{5.103}$$

and

$$E[\mathbf{INV}(n)\mathbf{INV}^T(k)] = \mathbf{0} \quad k \neq n \tag{5.104}$$

5.4.2 Prediction and Filtering

The optimum linear mean-square error one-step predictor is

$$\hat{\mathbf{S}}(n+1) = E[\mathbf{S}(n+1)|\mathbf{Y}(1), \cdots, \mathbf{Y}(n)] \qquad (5.105)$$

Using Equation (5.102), and the fact that there is a one-to-one correspondence between the set of observation vectors and the set representing the innovations process, we can write that

$$\hat{\mathbf{S}}(n+1) = \sum_{k=1}^{n} \mathbf{A}(n, k)\mathbf{INV}(k) \qquad (5.106)$$

where $\mathbf{A}(n, k)$ is an $(m \times p)$ matrix to be determined.

In accordance with the orthogonality principle, the error is orthogonal to the innovations process. Hence,

$$E[\{\mathbf{S}(n+1) - \hat{\mathbf{S}}(n+1)\}\mathbf{INV}(k)] = \mathbf{0} \quad k = 1, \cdots, n \qquad (5.107)$$

Substituting Equation (5.106) in Equation (5.107) and simplifying, we obtain

$$\begin{aligned} E[\mathbf{S}(n+1)\mathbf{INV}^T(l)] &= \mathbf{A}(n, l)E[\mathbf{INV}(l)\mathbf{INV}^T(l)] \\ &= \mathbf{A}(n, l)\mathbf{\Sigma}(l) \end{aligned} \qquad (5.108)$$

where $\mathbf{\Sigma}(l)$ is the correlation matrix of the innovations process. Solving for $\mathbf{A}(n, l)$, and substituting in Equation (5.106), the predictor state becomes

$$\hat{\mathbf{S}}(n+1) = \sum_{k=1}^{\infty} E[\mathbf{S}(n+1)\mathbf{INV}^T(k)]\mathbf{\Sigma}^{-1}(k)\mathbf{INV}(k) \qquad (5.109)$$

Upgrading the state Equation (5.93) to $(n+1)$, and substituting into (5.109), we have

$$E[\mathbf{S}(n+1)\mathbf{INV}^T(k)] = \mathbf{\Phi}(n+1)E[\mathbf{S}(n)\mathbf{INV}^T(k)] , \ k = 0, 1, \cdots, n \qquad (5.110)$$

where we have used the fact that

$$E[\mathbf{Y}(k)\mathbf{W}^T(n)] = \mathbf{0} \qquad (5.111)$$

and the fact that the innovations depend on the observation vectors. Substituting Equation (5.110) into (5.109), and after some manipulations, the predictor state becomes

$$\hat{\mathbf{S}}(n+1) = \mathbf{\Phi}(n+1)\hat{\mathbf{S}}(n) + \mathbf{G}(n)\mathbf{INV}(n) \qquad (5.112)$$

where $\mathbf{G}(n)$ is an $(m \times p)$ matrix called the *predictor gain matrix* and defined as

$$\mathbf{G}(n) = \mathbf{\Phi}(n+1)E[\mathbf{S}(n)\mathbf{INV}^T(n)]\mathbf{\Sigma}^{-1}(n) \qquad (5.113)$$

The above expressions given in Equations (5.112) and (5.113), can be simplified further for computational purposes. If we define

$$\boldsymbol{\varepsilon}(n) = \mathbf{S}(n) - \hat{\mathbf{S}}(n-1) \qquad (5.114)$$

and

$$\mathbf{K}(n) = E[\boldsymbol{\varepsilon}(n)\boldsymbol{\varepsilon}^T(n)] \qquad (5.115)$$

where $\boldsymbol{\varepsilon}(n)$ is called the *predicted state-error vector*, and $\mathbf{K}(n)$ the *predicted state-error correlation matrix*. Then, it can be shown that [10,11]

$$\mathbf{G}(n) = \mathbf{\Phi}(n+1)\mathbf{K}(n)\mathbf{H}^T(n)\mathbf{\Sigma}^{-1}(n) \qquad (5.116)$$

It can also be shown that $\mathbf{K}(n)$ can be updated recursively as

$$\begin{aligned} \mathbf{K}(n+1) = {} & \{\mathbf{\Phi}(n+1) - \mathbf{G}(n)\mathbf{H}(n)\}\mathbf{K}(n)\{\mathbf{\Phi}(n+1) - \mathbf{G}(n)\mathbf{H}(n)\}^T \\ & + \mathbf{Q}(n) + \mathbf{G}(n)\mathbf{R}(n)\mathbf{G}^T(n) \qquad (5.117) \end{aligned}$$

and that the filter state is

$$\hat{\mathbf{S}}(n) = \mathbf{\Phi}(n)\hat{\mathbf{S}}(n-1) + \mathbf{P}(n)\mathbf{\Sigma}(n)$$

where $\mathbf{P}(n)$ is a $(m \times m)$ matrix called the *filter gain matrix*, and given by

$$\mathbf{P}(n) = \mathbf{\Phi}(n)\mathbf{G}(n) \qquad (5.118)$$

Equation (5.117) can be decomposed into a pair of coupled equations to constitute the *Ricatti difference equations*.

Relationship between Kalman and Wiener Filters

The Kalman filter can also be derived for continuous time. If all signal processes considered are stationary, the measurement noise is white, and uncorrelated with the signal, and the observation interval is semi-infinite, the Kalman filter reduces to the Wiener filter. That is, both the Kalman filter and the Wiener filter lead to the same result in estimating a stationary process.

In discrete time, the Kalman filter, which is an optimum recursive filter based on the concept of innovations, has the ability to consider nonstationary processes; whereas the Wiener filter, which is an optimum nonrecursive filter, does not.

5.5 SUMMARY

In this chapter, we have covered the concept of filtering. We first presented the orthogonality principle theorem, the definition of linear transformations and related theorems. Realizable and unrealizable Wiener filters for continuous-time were presented in Section 5.3. To obtain the linear mean-square error realizable filter we needed to solve the Wiener-Hopf integral equation. An approach using Laplace transform to solve for the Wiener-Hopf equation was shown. Then, we extended the concept of Wiener filters to discrete-time. We concluded this chapter with a brief section about Kalman filtering.

PROBLEMS

5.1 Let the observation process be $Y(t) = S(t) + N(t)$. The signal process $S(t)$, and the zero-mean white noise process $N(t)$ are uncorrelated with power spectral densities:

$$S_{SS}(f) = \frac{2\alpha}{\alpha^2 + 4\pi^2 f^2} \qquad \text{and} \qquad S_{NN}(f) = \frac{N_0}{2}$$

(a) Obtain the optimum unrealizable linear filter for estimating the delayed signal $S(t - t_0)$.
(b) Compute the minimum mean-square error.

5.2 Let the observation process be $Y(t) = S(t) + N(t)$. The signal process $S(t)$, and the zero-mean noise process $N(t)$ are uncorrelated with autocorrelations:

$$R_{SS}(\tau) = e^{-0.5|\tau|} \qquad \text{and} \qquad R_{NN}(\tau) = \delta(\tau)$$

(a) Find the optimum unrealizable filter.
(b) Obtain the optimum realizable filter.
(c) Compute the minimum mean-square error for both filters and compare the results.

5.3 Let the observation process be $Y(t) = S(t) + N_1(t)$. The signal process $S(t)$, and the zero-mean noise process $N_1(t)$ are uncorrelated. The autocorrelation of $N_1(t)$ is

$$R_{N_1 N_1}(\tau) = e^{-|\tau|}$$

Assume that the signal $S(t)$ is given by the expression $S'(t) + S(t) = N_2(t)$ for t positive. $S'(t)$ denotes derivative of $S(t)$ with respect to t. $N_2(t)$ is a white Gaussian noise with power spectral density 2. Determine the Wiener filter if the processes $N_1(t)$ and $N_2(t)$ are independent.

5.4 Let the observation process be $Y(t) = S(t) + N(t)$, for $-\infty < t \leq \xi$. The signal process $S(t)$ and the noise process $N(t)$ are uncorrelated with power spectral densities:

$$S_{SS}(f) = \frac{1}{1 + 4\pi^2 f} \qquad \text{and} \qquad S_{NN}(f) = \frac{1}{2}$$

Obtain the optimum linear filter to estimate $S'(t)$; $S'(t)$ is the derivative of the signal $S(t)$ with respect to t.

5.5 Let the observation process be $Y(t) = S(t) + N(t)$. The signal process $S(t)$ and the zero-mean noise process $N(t)$ are uncorrelated with autocorrelations:

$$R_{SS}(\tau) = \frac{5}{3} e^{-\frac{1}{2}|\tau|} \qquad \text{and} \qquad R_{NN}(\tau) = \frac{7}{6} e^{-|\tau|}$$

Obtain the optimum linear filter to estimate $S(t + \alpha)$, $\alpha > 0$.

5.6 Let $Y(n) = S(n) + N(n)$ be the received sequence. The signal sequence $S(n)$ and the white noise sequence $N(n)$ are independent, and zero-mean with autocorrelations:

$$R_{SS}(n) = \frac{1/2^{|n|}}{1 - \frac{1}{4}} \qquad \text{and} \qquad R_{NN}(n) = \begin{cases} 2 & , n = 0 \\ 0 & , n \neq 0 \end{cases}$$

(a) Obtain the optimum realizable filter.
(b) Compute the mean-square error.

5.7 Let $Y(n) = S(n) + N(n)$ represent the received sequence. The signal sequence $S(n)$ and the noise sequence $N(n)$ are zero-mean and independent with autocorrelations:

$$R_{SS}(n) = \frac{1}{2^{|n|}} \qquad \text{and} \qquad R_{NN}(n) = \begin{cases} 1 & , n = 0 \\ 0 & , n \neq 0 \end{cases}$$

(a) Obtain the optimum realizable filter.
(b) Compute the mean-square error.

REFERENCES

1. Anderson, B.D.O., and J.B. Moore, *Optimal Filtering,* Prentice-Hall, Englewood Cliffs, NJ, 1979.

2. Brogan, W.L., *Modern Control Theory,* Quantum, New York, 1974.

3. Candy, J.V., *Signal Processing: The Modern Approach,* McGraw-Hill, New York, 1988.

4. Chen, C.T., *One-Dimensional Digital Signal Processing,* Marcel Dekker, New York, 1979.

5. Cowan, C.F.N., and P.M. Grant, *Adaptive Filters,* Prentice-Hall, Englewood Cliffs, NJ, 1985.

6. Cox, D.R., and H.D. Miller, *The Theory of Stochastic Processes,* John Wiley and Sons, New York, 1965.

7. Davenport, Jr., W.B., and W.L. Root, *An Introduction to the Theory of Random Signals and Noise,* McGraw-Hill, New York, 1958.

8. Gelb, A., *Applied Optimal Estimation,* M.I.T., Cambridge, MA., 1974.

9. Goodwin, G.C., and K.S. Sin, *Adaptive Filtering, Prediction and Control,* Prentice-Hall, Englewood Cliffs, NJ, 1984.

10. Haykin, S., *Adaptive Filter Theory,* Prentice-Hall, Englewood Cliffs, NJ, 1986.

11. Haykin, S., *Modern Filters,* Macmillan, New York, 1989.

12. Kailath, T., *Lectures on Wiener and Kalman Filtering,* Springer-Verlag, New York, 1981.

13. Kay, S.M., *Modern Spectral Estimation: Theory and Application,* Prentice-Hall, Englewood Cliffs, NJ, 1988.

14. Lee, Y.W., *Statistical Theory of Communication*, John Wiley and Sons, New York, 1960.

15. Maybeck, P.S., *Stochastic Models, Estimation, and Control Volume 1*, Academic Press, New York, 1979.

16. Melsa, J.L., and D.L. Cohn, *Decision and Estimation Theory*, McGraw-Hill, New York, 1978.

17. Mohanty, N., *Signal Processing: Signals, Filtering, and Detection*, Van Nostrand Reinhold, New York, 1987.

18. Morrison, N., *Introduction to Sequential Smoothing and Prediction*, McGraw-Hill, New York, 1969.

19. Nahi, N.E., *Estimation Theory and Applications*, John Wiley and Sons, New York, 1969.

20. Papoulis, A., *Probability, Random Variables, and Stochastic Processes*, McGraw-Hill, New York, 1986.

21. Sage, A.P., and J.L. Melsa, *Estimation Theory with Applications to Communications and Control*, McGraw-Hill, New York, 1971.

22. Sage, A.P., and C.C. White, III, *Optimum Systems Control*, Prentice-Hall, Englewood Cliffs, NJ, 1977.

23. Shanmugan, K.S., and A.M. Breipohl, *Random Signals: Detection, Estimation and Data Analysis*, John Wiley and Sons, New York, 1988.

24. Sorenson, H.W., *Parameter Estimation: Principles and Problems*, Marcel Dekker, New York, 1980.

25. Srinath, M.D., and P.K. Rajaskaran, *An Introduction to Statistical Signal Processing with Applications*, John Wiley and Sons, New York, 1979.

Chapter 6

Representation of Signals

6.1 INTRODUCTION

In this chapter we study some mathematical principles which will be very useful to us in the last two chapters. First, we define the meaning of orthogonal functions, which are used to represent deterministic signals in a series expansion known as the generalized Fourier series. We use the Gram-Schmidt procedure to transform a set of M linear dependent or independent functions into a set of K, $K \leq M$, orthogonal functions. We also discuss geometric representation of signals in the signal space, which can be used to determine decision regions in M-ary detection of signals in noise as will be seen later. In Section 6.3, we review the concepts of eigenvalues, eigenvectors, and eigenfunctions and show how they are obtained. This will help us in solving the general Gaussian problem. Then, integral equations are studied. The relation between integral equations and their corresponding linear differential equations are established through a Green's function or kernel. In solving integral equations we present an approach by which we obtain the eigenfunctions and eigenvalues from the linear differential equation, and then by back substitution into the integral equation. In Section 6.5, we discuss the series representation of random processes by orthogonal functions known as Karhunen-Loeve expansion. Specifically, we consider processes with rational power spectral densities, the Wiener process, and the white Gaussian noise process. We show how the Wiener process is obtained from the random walk, and how the white Gaussian noise process is obtained from the Wiener process. A possible Karhunen-Loeve series expansion of the white Gaussian noise process is also presented.

6.2 ORTHOGONAL FUNCTIONS

From vector analysis, we say that two vectors $\mathbf{X}$ and $\mathbf{Y}$ are orthogonal (perpendicular) if their dot or inner product is zero. That is

$$\mathbf{X} \cdot \mathbf{Y} = 0 \tag{6.1}$$

Let $\mathbf{X}$ and $\mathbf{Y}$ be two vectors in R^K, such that $\mathbf{X} = [x_1, \cdots, x_K]^T$ and $\mathbf{Y} = [y_1, \cdots, y_K]^T$, then

$$\mathbf{X} \cdot \mathbf{Y} = x_1 y_1 + x_2 y_2 + \cdots + x_K y_K \tag{6.2}$$

The distance $d(x, y)$ between the points x and y is given by

$$d(x, y) = \sqrt{(y_1 - x_1)^2 + \cdots + (y_K - x_K)^2} \tag{6.3}$$

The length or the norm of the vector $\mathbf{X}$, denoted $|\mathbf{X}|$, is defined by

$$|\mathbf{X}| = \sqrt{\mathbf{X} \cdot \mathbf{X}} = \sqrt{x_1^2 + \cdots + x_K^2} \tag{6.4}$$

If the length $|\mathbf{X}| = 1$, we say that $\mathbf{X}$ is a *normalized* vector. Geometrically, Equation (6.1) says that the angle θ between the vectors $\mathbf{X}$ and $\mathbf{Y}$ is 90°. For arbitrary angle θ between the two vectors $\mathbf{X}$ and $\mathbf{Y}$, θ is defined by

$$\cos\theta = \frac{\mathbf{X} \cdot \mathbf{Y}}{|\mathbf{X}|\ |\mathbf{Y}|} \tag{6.5}$$

We now generalize the above concepts to continuous functions of time. Let $\{s_k(t)\}$, $k = 1, 2, \cdots$ be a set of deterministic functions with finite energies defined over the interval $t\epsilon[t_i, t_f]$. Let $\mathcal{E}_k$ denote energy of $s_k(t)$. Then,

$$\mathcal{E}_k = \int_{t_i}^{t_f} |s_k(t)|^2 dt < \infty \tag{6.6}$$

The norm of $s_k(t)$, $k = 1, 2, \cdots$, can be written as

$$|s_k(t)| = \{\int_{t_i}^{t_f} s_k^2(t)dt\}^{\frac{1}{2}} \tag{6.7}$$

Geometrically, Equation (6.7) represents the square root of the area under the curve $s_k^2(t)$. The "distance" between the two signals $s_m(t)$, and $s_n(t)$ is

$$|s_m(t) - s_n(t)| = \{\int_{t_i}^{t_f} [s_m(t) - s_n(t)]^2 dt\}^{\frac{1}{2}} \tag{6.8}$$

We say that the set of functions (signals), $\{s_k(t)\}$, $k = 1, 2, \cdots$, are orthogonal, when

$$\int_{t_i}^{t_f} s_m(t)s_n(t)dt = 0 \quad m \neq n \tag{6.9}$$

A set of functions $\{\phi_k(t)\}$, $k = 1, 2, \cdots$, are *orthonormal* if

$$\int_{t_i}^{t_f} \phi_m(t)\phi_n(t)dt = \delta_{mn} \tag{6.10}$$

where δ_{mn} is the Kronecker's delta. Equation (6.10) can be rewritten as

$$\int_{t_i}^{t_f} \phi_m(t)\phi_n(t) = \begin{cases} 0 & \text{if } m \neq n \\ 1 & \text{if } m = n \end{cases} \tag{6.11}$$

Note that the set of functions $\{\phi_k(t)\}$, $k = 1, 2, \cdots$, are normalized.

6.2.1 Generalized Fourier Series

Let $s(t)$ be a deterministic signal with finite energy $\mathcal{E}$, and observed over the interval $t\epsilon[t_i, t_f]$. Given an orthonormal set of functions $\{\phi_k(t)\}$, $k = 1, 2, \cdots$, for the specified time $t\epsilon[t_i, t_f]$, it may be possible to represent the signal $s(t)$ as a linear combination of the functions $\phi_k(t)$, $k = 1, 2, \cdots$, as

$$
\begin{aligned}
s(t) &= s_1\phi_1(t) + s_2\phi_2(t) + \cdots + s_k\phi_k(t) + \cdots \\
&= \sum_{k=1}^{\infty} s_k\phi_k(t)
\end{aligned} \tag{6.12}
$$

Assuming the series of Equation (6.12) converges to $s(t)$, then

$$
\int_{t_i}^{t_f} s(t)\phi_k(t)dt = s_k \tag{6.13}
$$

where we have used the fact that $\int_{t_i}^{t_f} \phi_m(t)\phi_n(t)dt = \delta_{mn}$. In this case, the coefficients s_k, $k = 1, 2, \cdots$, are called the *generalized Fourier coefficients*. The series in Equation (6.12) with the coefficients as given by Equation (6.13) is called the *generalized Fourier series*.

If there exists a set of orthonormal functions $\{\phi_k(t)\}$, $k = 1, 2, \cdots, K$, such that the signal $s(t)$ may be expressed as

$$
s(t) = \sum_{k=1}^{K} s_k\phi_k(t) \tag{6.14}
$$

where s_k's are as given by Equation (6.13) then, the set of orthonormal functions $\{\phi_k(t)\}$, $k = 1, 2, \cdots, K$, is said to be *complete*.

Consider the finite sum $s_K(t)$ such that

$$
s_K(t) = \sum_{k=1}^{K} s_k\phi_k(t) \tag{6.15}
$$

where $s_K(t)$ is an approximate to the signal $s(t)$ observed over the interval $t \epsilon [t_i, t_f]$. In general, it is practical to only use a finite number of terms K. The goal is to select the coefficients s_k such that the mean-square error is minimum. We define the error $\varepsilon_K(t)$ as

$$\varepsilon_K(t) = s(t) - s_K(t) \tag{6.16}$$

and its corresponding energy as

$$\mathcal{E}_{\varepsilon_K} = \int_{t_i}^{t_f} \varepsilon_K^2(t) dt \tag{6.17}$$

The mean-square error is

$$< \varepsilon_K^2(t) > = \frac{1}{t_f - t_i} \int_{t_i}^{t_f} \varepsilon_K^2(t) dt \tag{6.18}$$

where $< \cdot >$ denotes time average. We observe from Equations (6.17) and (6.18), that minimizing the mean-square error is equivalent to minimizing the energy. Hence,

$$\mathcal{E}_{\varepsilon_K} = \int_{t_i}^{t_f} [s(t) - \sum_{k=1}^{K} s_k \phi_k(t)]^2 dt \tag{6.19}$$

Differentiating Equation (6.19) with respect to s_k, we obtain

$$\frac{d\mathcal{E}_{\varepsilon_K}}{ds_k} = -2 \int_{t_i}^{t_f} [s(t) - \sum_{k=1}^{K} s_k \phi_k(t)] \phi_j(t) dt$$

$$= -2 \int_{t_i}^{t_f} s(t) \phi_j(t) dt + 2 \sum_{k=1}^{K} s_k \int_{t_i}^{t_f} \phi_k(t) \phi_j(t) dt \tag{6.20}$$

Setting Equation (6.20) equal to zero and using Equation (6.10), the coefficients of s_k are given by

$$s_k = \int_{t_i}^{t_f} s(t) \phi_k(t) dt \tag{6.21}$$

Note that the second derivative $d^2\mathcal{E}_{\varepsilon_k}/ds_k^2 = 2$ is positive and thus, the coefficients s_k, $k = 1, 2, \cdots, K$, minimize the energy or the mean-square error. The set $\{\phi_k(t)\}$ forms a complete orthonormal set in the interval $[t_i, t_f]$. That is,

$$\lim_{K\to\infty} \int_{t_i}^{t_f} [s(t) - s_K(t)]^2 dt = 0 \tag{6.22}$$

or

$$\underset{K\to\infty}{\text{l.i.m.}} \; s_K(t) = s(t) \tag{6.23}$$

Equation (6.23) is read as *the limit in the mean of $s_K(t)$ as $K \to \infty$ equals $s(t)$*, or $s_K(t)$ *converges in the mean to $s(t)$ as $K \to \infty$.* Substituting the result of Equation (6.21) in (6.19) and solving for $\mathcal{E}_{\varepsilon_K}$, we obtain

$$
\begin{aligned}
\mathcal{E}_{\varepsilon_K} &= \int_{t_i}^{t_f} s^2(t)dt - \sum_{k=1}^{K} s_k^2 \\
&= \mathcal{E} - \sum_{k=1}^{K} s_k^2
\end{aligned}
\tag{6.24}
$$

We observe that $\mathcal{E}_{\varepsilon_K}$ is minimum when the set of orthonormal signals $\{\phi_k\}$ is complete. That is,

$$\mathcal{E} = \int_{t_i}^{t_f} s^2(t)dt = \sum_{k=1}^{\infty} s_k^2 \tag{6.25}$$

s_k^2 may be interpreted as the energy of the signal in the kth component. Equation (6.25) is referred to as *Parseval's identity for orthonormal series of functions.*

Let the signal observation time $t \epsilon [0, T]$, then the coefficients s_k, $k = 1, 2, \cdots, K$, may be determined by a correlation operation as shown in Figure 6.1.

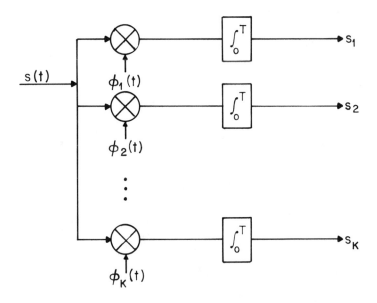

Figure 6.1 Correlation operation for generating the set of coefficients $\{s_K\}$.

An equivalent operation is filtering. The signal $s(t)$ is passed through a set of linear filters (matched filters) with impulse response $h_k(\tau) = \phi_k(T - \tau)$. Observe the outputs of the filters at time $t = T$. This is shown in Figure 6.2.

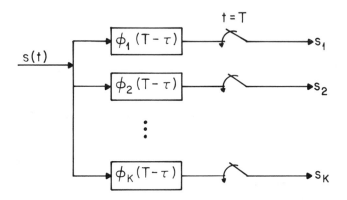

Figure 6.2 Filtering operation for generating the set of coefficients $\{s_K\}$.

Let the output of kth channel be $y_k(T)$. The output of the kth filter is

$$
\begin{aligned}
y_k(t) &= \int_0^T s(\tau) h_k(t - \tau) d\tau \\
&= \int_0^T s(\tau) \phi_k(T - t + \tau) d\tau
\end{aligned}
\tag{6.26}
$$

Sampling $y_k(t)$ at $t = T$, we obtain

$$
\begin{aligned}
y_k(T) &= \int_0^T s(\tau) \phi_k(T - T + \tau) d\tau \\
&= \int_0^T s(\tau) \phi_k(\tau) d\tau = s_k
\end{aligned}
\tag{6.27}
$$

6.2.2 Gram-Schmidt Orthogonalization Procedure

Given a set of M signals $s_k(t)$, $k = 1, 2, \cdots, M$, we would like to represent these signals as a linear combination of K *orthonormal basis functions*, $K \leq M$. The signals $s_1(t), s_2(t), \cdots, s_M(t)$ are real-valued, and each is of duration T. From Equation (6.14), we may represent these energy signals in the form:

$$s_k(t) = \sum_{j=1}^{K} s_{kj} \phi_j(t) \qquad \begin{array}{l} 0 \leq t \leq T \\ k = 1, 2, \cdots, M \end{array} \qquad (6.28)$$

where the coefficients s_{kj}, $j = 1, 2, \cdots, K$, of the signal $s_k(t)$ are defined by

$$s_{kj} = \int_0^T s_k(t) \phi_j(t) dt \qquad \begin{array}{l} k = 1, 2, \cdots, M \\ j = 1, 2, \cdots, K \end{array} \qquad (6.29)$$

The orthonormal functions $\phi_j(t)$, $j = 1, 2, \cdots, K$ are as defined in (6.10); that is,

$$\int_0^T \phi_k(t) \phi_j(t) dt = \delta_{kj} \qquad (6.30)$$

The orthogonalization procedure is as follows.

(i) Normalize the first signal $s_1(t)$ to obtain $\phi_1(t)$. That is,

$$\phi_1(t) = \frac{s_1}{\sqrt{\int_0^T s_1^2(t) dt}} = \frac{s_1}{\sqrt{\mathcal{E}_1}} \qquad (6.31)$$

where $\mathcal{E}_1$ is the energy of $s_1(t)$. Thus,

$$\begin{aligned} s_1(t) &= \sqrt{\mathcal{E}_1} \phi_1(t) \\ &= s_{11} \phi_1(t) \end{aligned} \qquad (6.32)$$

where the coefficient $s_{11} = \sqrt{\mathcal{E}_1}$.

(ii) Using the signal $s_2(t)$, we compute the projection of $\phi_1(t)$ onto $s_2(t)$, which is

$$s_{21} = \int_0^T s_2(t) \phi_1(t) dt \qquad (6.33)$$

We then substract $s_{21}\phi_1(t)$ from $s_2(t)$ to yield

$$f_2(t) = s_2(t) - s_{21}\phi_1(t) \tag{6.34}$$

which is orthogonal to $\phi_1(t)$ over the interval $0 \leq t \leq T$; $\phi_2(t)$ is obtained by normalizing $f_2(t)$; that is,

$$
\begin{aligned}
\phi_2(t) &= \frac{f_2(t)}{\sqrt{\int_0^T f_2^2(t)dt}} \\
&= \frac{s_2(t) - s_{21}\phi_1(t)}{\sqrt{\mathcal{E}_2 - s_{21}^2}}
\end{aligned}
\tag{6.35}
$$

where $\mathcal{E}_2$ is the energy of the signal $s_2(t)$. Note that from Equation (6.35)

$$\int_0^T \phi_2^2(t)dt = 1 \tag{6.36}$$

and

$$\int_0^T \phi_2(t)\phi_1(t) = 0 \tag{6.37}$$

that is, $\phi_1(t)$ and $\phi_2(t)$ are orthonormal.

(iii) Continuing in this manner, we can determine all $K(K \leq M)$ orthonormal functions to be

$$\phi_k(t) = \frac{f_k(t)}{\sqrt{\int_0^T f_k^2(t)dt}} \tag{6.38}$$

where

$$f_k(t) = s_k - \sum_{j=1}^{k-1} s_{kj}\phi_j(t) \tag{6.39}$$

and the coefficients s_{kj}, $j = 1, 2, \cdots, k-1$, defined by

$$s_{kj} = \int_0^T s_k(t)\phi_j(t)dt \tag{6.40}$$

If all M signals $s_1(t), s_2(t), \cdots, s_M(t)$ are independent, that is, no signal is a linear combination of the other, then the dimensionality K, of the signal space, is equal to M.

Modified Gram-Schmidt

The proposed Gram-Schmidt procedure defined in (6.38), (6.39), and (6.40) is referred to as the *Classical Gram-Schmidt* (CGS) procedure. The concept of subtracting away the components in the direction of $\phi_1(t), \phi_2(t), \cdots, \phi_{k-1}(t)$, is sometimes numerically unstable. A slight modification in the algorithm makes it stable and efficient. This modification yields the *Modified Gram-Schmidt* (MGS) procedure. For simplicity, we show only the first two steps. We compute the projection of $s_k(t)$ onto $\phi_1(t), \phi_2(t), \cdots, \phi_{k-1}(t)$. We start with $s_{k1}\phi_1(t)$ and subtract it *immediately*. That is, we are left with a new function $s_k^1(t)$ such that

$$s_k^1(t) = s_k(t) - s_{k1}\phi_1(t) \tag{6.41}$$

and s_{k1} ia as defined in (6.40). Then, we project $s_k^1(t)$ instead of the original signal $s_k(t)$ onto $\phi_2(t)$ and subtract that projection. That is,

$$s_k^2(t) = s_k^1(t) - s_{21}\phi_2(t) \tag{6.42}$$

where

$$s_{21}^1 = \int_0^T s_k^1(t)\phi_2(t)dt \tag{6.43}$$

and the power 2 on $s_k(t)$ denotes a *superscript*. Observe that this is identical in principle to the classical Gram-Schmidt procedure, which projects $s_k(t)$ onto both $\phi_1(t)$ and $\phi_2(t)$ to yield $f_k(t)$. Substituting Equations (6.41) and (6.43) into Equation (6.42), we obtain

$$
\begin{aligned}
s_k^2(t) &= [s_k(t) - s_{k1}\phi_1(t)] - \{\phi_2(t)[s_k(t) - s_{k1}\phi_1(t)]\phi_2(t)\} \\
&= s_k(t) - s_{k1}\phi_1(t) - s_{k2}\phi_2(t) \\
&\equiv f_k(t)
\end{aligned} \tag{6.44}
$$

since $\int_0^T \phi_1(t)\phi_2(t)dt = 0$.

6.2.3 Geometric Representation

In order to have a geometric interpretation of the signals, we write the M signals by their corresponding *vectors* of coefficients. That is, the M *signal vectors* are

$$
\mathbf{s}_k = \begin{bmatrix} s_{k1} \\ s_{k2} \\ \vdots \\ s_{KK} \end{bmatrix} \qquad k = 1, 2, \cdots, M \qquad (6.45)
$$

The vectors $\mathbf{s}_k$, $k = 1, 2, \cdots, M$, may be visualized as M points in an K-*dimensional Euclidean space*. The K mutually perpendicular axes are labeled $\phi_1, \phi_2, \cdots, \phi_K$. This K-dimensional Euclidean space is referred to as the *signal space*.

Using Equation (6.4), we say that the inner product of the vector $\mathbf{s}_k$ with itself, which is the norm of $\mathbf{s}_k$, is

$$
\begin{aligned}
|\mathbf{s}_k|^2 &= (\mathbf{s}_k, \mathbf{s}_k) \\
&= \sum_{j=1}^{K} s_{kj}^2 \qquad (6.46)
\end{aligned}
$$

Since the K orthonormal functions form a complete set, Equation (6.46) also represents the energy of the signal $s_k(t)$ as shown in the previous subsection. Thus,

$$
\mathcal{E}_k = \sum_{j=1}^{K} s_{kj}^2 \qquad (6.47)
$$

Also, from Equations (6.3), (6.45), and (6.47), the Euclidean distance between the points represented by the signal vectors $\mathbf{s}_k$ and $\mathbf{s}_j$ can be written as

$$|\mathbf{s}_k - \mathbf{s}_j|^2 = \sum_{i=1}^{K}(s_{ki}^2 - s_{ji}^2)$$

$$= \int_0^T [s_k(t) - s_j(t)]^2 dt \qquad (6.48)$$

The correlation coefficient between the signals $s_k(t)$ and $s_j(t)$ is defined by

$$
\begin{aligned}
s_{kj} &= \frac{\int_0^T s_k(t)s_j(t)dt}{\sqrt{\mathcal{E}_k \mathcal{E}_j}} \\
&= \frac{\int_0^T \{\sum_{i=1}^K s_{ki}\phi_i(t)\}\{\sum_{i=1}^K s_{ji}\phi_i(t)\}dt}{\sqrt{\mathcal{E}_k \mathcal{E}_j}} \\
&= \frac{\sum_{i=1}^K s_{ki}s_{ji}}{\sqrt{\mathcal{E}_k \mathcal{E}_j}} \\
&= \frac{\mathbf{s}_k^T \mathbf{s}_j}{|\mathbf{s}_k||\mathbf{s}_j|} \qquad (6.49)
\end{aligned}
$$

where $\mathbf{s}_k$ is given in Equation (6.45) and $\mathbf{s}_j$ is

$$
\mathbf{s}_j = \begin{bmatrix} s_{j1} \\ s_{j2} \\ \vdots \\ s_K \end{bmatrix} \qquad (6.50)
$$

Example 6.1

Consider the signals $s_1(t), s_2(t), s_3(t)$, and $s_4(t)$, as shown in Figure 6.3. Use the Gram-Schmidt procedure to determine orthonormal basis functions for $s_k(t)$, $k = 1, 2, 3, 4$.

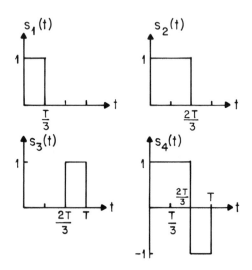

Figure 6.3 Set of signals $\{s_k(t)\}$.

Solution.

From Equation (6.31), the first function $\phi_1(t)$ is

$$\phi_1(t) = \frac{s_1(t)}{\sqrt{\mathcal{E}_1}}$$

$$= \begin{cases} \sqrt{\frac{3}{T}} & ,0 \le t \le \frac{T}{3} \\ 0 & , \text{ otherwise} \end{cases}$$

where $\mathcal{E}_1 = \int_0^{\frac{T}{3}} (1)^2 dt = T/3$. To find $\phi_2(t)$, we first use Equation (6.33) to determine s_{21}; i.e.,

$$s_{21} = \int_0^T s_2(t)\phi_1(t)dt$$

$$= \sqrt{\frac{T}{3}}$$

From Equation (6.34), $f_2(t)$ is given by

$$
\begin{aligned}
f_2(t) &= s_2(t) - s_{21}\phi_1(t) \\
&= \begin{cases} 1 & , \frac{T}{3} \le t \le \frac{2T}{3} \\ 0 & , \text{ otherwise} \end{cases}
\end{aligned}
$$

Normalizing $f_2(t)$, we have

$$
\begin{aligned}
\phi_2(t) &= \frac{f_2(t)}{\sqrt{\int_0^T f_2^2(t)dt}} \\
&= \begin{cases} \sqrt{\frac{3}{T}} & , \frac{T}{3} \le t \le \frac{2T}{3} \\ 0 & , \text{ otherwise} \end{cases}
\end{aligned}
$$

We use Equations (6.39) and (6.40) to find the coefficients s_{31} and s_{32}; that is,

$$
\begin{aligned}
s_{31} &= \int_0^T s_3(t)\phi_1(t)dt \\
&= 0
\end{aligned}
$$

$$
\begin{aligned}
s_{32} &= \int_0^T s_3(t)\phi_2(t)dt \\
&= 0
\end{aligned}
$$

Thus, $f_3(t) = s_3(t)$ and the normalized signal $\phi_3(t)$ is

$$
\begin{aligned}
\phi_3(t) &= \frac{s_3(t)}{\sqrt{\mathcal{E}_3}} \\
&= \begin{cases} \sqrt{\frac{3}{T}} & , \frac{2T}{3} \le t \le T \\ 0 & , \text{ otherwise} \end{cases}
\end{aligned}
$$

We observe that $s_4(t) = s_2(t) - s_3(t)$ is a linear combination of $s_2(t)$ and $s_3(t)$. The complete set of orthonormal functions is $\phi_1(t), \phi_2(t)$ and $\phi_3(t)$; that is, the dimensionality is $K = 3$. The basis functions are shown in Figure 6.4.

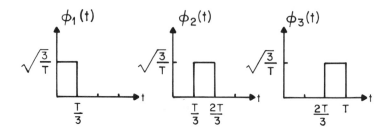

Figure 6.4 Orthonormal basis functions $\{\phi_k(t)\}$.

Example 6.2

(a) Find a set of orthonormal basis functions that can be used to represent the signal shown in Figure 6.5.
(b) Find the vector corresponding to each signal for the orthonormal basis set found in (a) and sketch the location of each signal in the signal space.

Solution.

(a) In this example, we are not going to do a formal mathematical derivation as we did in the previous one, but instead we solve it by inspection. We see that the given waveforms can be decomposed into two basis functions $\phi_1(t)$ and $\phi_2(t)$ as shown in Figure 6.6.

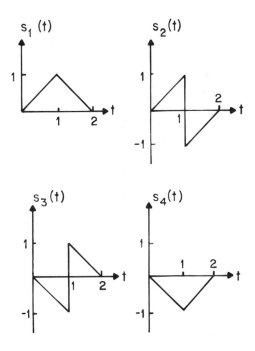

Figure 6.5 Signal set for Example 6.2.

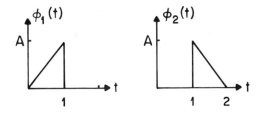

Figure 6.6 Basis functions for Example 6.2.

Since $\phi_1(t)$ and $\phi_2(t)$ must have unit energy, we have

$$\mathcal{E} = \int_0^1 (At)^2 dt = 1$$

or

$$A = \sqrt{3}$$

(b) The signal vectors are

$$s_1 = [\frac{1}{\sqrt{3}}, \frac{1}{\sqrt{3}}] \ , \ s_3 = [\frac{-1}{\sqrt{3}}, \frac{1}{\sqrt{3}}]$$

$$s_2 = [\frac{1}{\sqrt{3}}, \frac{-1}{\sqrt{3}}] \ , \ s_4 = [\frac{-1}{\sqrt{3}}, \frac{-1}{\sqrt{3}}]$$

Thus, the signal space is as shown in Figure 6.7.

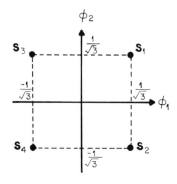

Figure 6.7 Signal space for Example 6.2.

Example 6.3

Consider the three possible functions

$$\phi_k(t) = E \cos\frac{2k\pi}{T}t, \quad \begin{array}{l} k = 1,2,3 \\ 0 \le t \le T \end{array}$$

(a) Do the ϕ_k's constitute an orthonormal set?
(b) What geometric figure do the s_i's form in the signal space?

Solution.
(a) To check for orthogonality, we do

$$
\begin{aligned}
(\phi_k, \phi_j) &= \int_0^T \phi_k(t)\phi_j(t) \\
&= E^2 \int_0^T \cos\frac{2k\pi}{T}t \cdot \cos\frac{2j\pi}{T} \, t dt \\
&= \frac{E^2}{2}[\int_0^T \cos\frac{2\pi}{T}t(k-j)dt + \int_0^T \cos\frac{2\pi}{T}t(k+j)dt] \\
&= \frac{E^2}{2}\{\left[\tfrac{T}{2\pi(k-j)} \sin\tfrac{2\pi t}{T}(k+j) \right]_0^T + \left[\tfrac{T}{2\pi(k+j)} \sin\tfrac{2\pi t}{T}(k+j) \right]_0^T\} \\
&= 0 \text{ , for } k \ne j
\end{aligned}
$$

If $k = j$, we have $(\phi_k, \phi_k) = E^2 \int_0^T \cos[(2k\pi)/T]tdt = E^2T/2$. Hence, the ϕ's constitute an orthonormal set.

(b) The signal vectors for the set of signals $\{s_k(t)\}$ is

$$s_1 = [1,0,0] \text{ , } s_3[0,0,1]$$

$$s_2 = [0,1,0] \text{ , } s_4[1,1,1]$$

Hence, they are as shown in Figure 6.8.

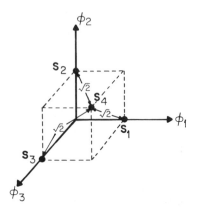

Figure 6.8 Signal space for Example 6.3.

6.3 EIGENVALUES AND EIGENVECTORS

In this section, we define eigenvalues and eigenvectors. We present methods of determining eigenvalues and eigenvectors, and some related properties. Eigenvalues and eigenvectors are extremely useful in many applications of signal processing and modern control theory. In the context of this book, eigenvalues and eigenvectors will be used in representing stochastic processes, which will be covered in later sections of this chapter, and solving the general Gaussian problem which will be covered in the next chapter.

We define a *linear transformation* or *linear operator* or *linear mapping* **T** from vector space $\mathcal{X}$, called the domain, to a vector space $\mathcal{Y}$, called the range (or codomain) as a correspondence that assigns to every vector **X** in $\mathcal{X}$ a vector **T(X)** in $\mathcal{Y}$, such that

$$\mathbf{T}(\alpha \mathbf{X} + \beta \mathbf{Y}) = \alpha \mathbf{T}(\mathbf{X}) + \beta \mathbf{T}(\mathbf{Y}) \tag{6.51}$$

where α and β are constants, and $\mathbf{X}$ and $\mathbf{Y}$ are vectors in $\mathcal{X}$.

It can be shown that any equation involving a linear operator on a finite dimensional space can be converted into an equivalent matrix operator. If the transform $\mathbf{T} : \mathcal{V} \to \mathcal{V}$ maps elements in $\mathcal{V}$ into other elements in $\mathcal{V}$, we can define $\mathbf{T}$ by a matrix $\mathbf{A}$.

Using the above concept of the linear transformation, we are now ready to define the concept of eigenvalues and eigenvectors. An *eigenvalue* (or *characteristic value*) of a linear operator $\mathbf{T}$ on a vector space $\mathcal{X}$ is a scalar λ such that

$$\mathbf{AX} = \lambda\mathbf{X} \qquad (6.52)$$

for a nonzero vector $\mathbf{X}$ in $\mathcal{V}$. Every nonzero vector $\mathbf{X}$ satisfying the relation $\mathbf{AX} = \lambda\mathbf{X}$ is called an *eigenvector* of $\mathbf{A}$ associated with the eigenvalue λ. The matrix representation of Equation (6.52) is

$$(\mathbf{A} - \mathbf{I}\lambda)\mathbf{X} = \mathbf{0} \qquad (6.53)$$

where $\mathbf{I}$ is the identity matrix.

If the operator $\mathbf{T}$ acts on a function space, then the eigenvectors associated with the eigenvalues are called *eigenfunctions*.

6.3.1 Eigenvalues

If $\mathbf{A}$ is a $n \times n$ matrix, a necessary condition for the n homogeneous Equations in (6.53) to yield nonzero solutions is that the rank of the matrix $(\mathbf{A} - \mathbf{I}\lambda)$ be less than n. That is, the determinant

$$|\mathbf{A} - \mathbf{I}\lambda| = 0 \qquad (6.54)$$

Equation (6.54) is called the *characteristic equation of the matrix* $\mathbf{A}$ (or of the operator $\mathbf{T}$ representd by $\mathbf{A}$). Expanding Equation (6.54), we define the *characteristic polynomial of* $\mathbf{A}$ as

$$c(\lambda) \overset{\Delta}{=} |\lambda \mathbf{I} - \mathbf{A}| = (-1)^n |\mathbf{A} - \mathbf{I}\lambda|$$
$$= \lambda^n + c_{n-1}\lambda^{n-1} + c_{n-2}\lambda^{n-2} + \cdots + c_1\lambda + c_0 \qquad (6.55)$$

where $c(\lambda)$ is an nth degree polynormial in λ. Solving for λ from the characteristic equation results in n roots $(\lambda_1, \lambda_2, \cdots, \lambda_n)$ if all roots are distinct. Consequently, $c(\lambda)$ can be written as

$$c(\lambda) = (\lambda - \lambda_1)(\lambda - \lambda_2) \cdots (\lambda - \lambda_n) \qquad (6.56)$$

However, if the roots are not distinct so that λ_1 has multiplicity m_1, λ_2 has multiplicity m_2, and so on. Then,

$$c(\lambda) = (\lambda - \lambda_1)^{m_1} (\lambda - \lambda_2)^{m_2} \cdots (\lambda - \lambda_p)^{m_p} \qquad (6.57)$$

where $m_1 + m_2 + \cdots + m_p = n$.

6.3.2 Eigenvectors

Once the eigenvalues are determined from the characteristic equation, we substitute for λ in Equations (6.52), or (6.53), and solve for the corresponding vector $\mathbf{X}$. However, in determining λ, two possible cases arise. (i) All eigenvalues are distinct, and (ii) some eigenvalues have multiplicity greater than one.

(i) *Case 1*: When all eigenvalues are distinct, the eigenvectors are solved for directly from Equation (6.52) or (6.53). If $\mathbf{X}_k$ is an eigenvector corresponding to the eigenvalue λ_k, then $\alpha \mathbf{X}_k$ is also an eigenvector for any nonzero scalar α. Since all eigenvalues and their corresponding eigenvectors satisfy the equation:

$$\mathbf{AX}_1 = \lambda_1 \mathbf{X}_1$$

$$\mathbf{AX}_2 = \lambda_2 \mathbf{X}_2$$

$$\vdots$$

$$\mathbf{AX}_n = \lambda_n \mathbf{X}_n \qquad (6.58)$$

we can write that

$$\mathbf{AM} = \mathbf{M\Lambda} \qquad (6.59)$$

where the $n \times n$ matrix $\mathbf{M}$ is called the *modal matrix* and defined by

$$\mathbf{M} = [\mathbf{X}_1 | \mathbf{X}_2 | \cdots | \mathbf{X}_n] \qquad (6.60)$$

The rank of the matrix $\mathbf{M}$ is n since the eigenvectors are linearly independent. $\mathbf{\Lambda}$ is a diagonal matrix defined by

$$\mathbf{\Lambda} = \begin{bmatrix} \lambda_1 & 0 & 0 & \cdots & 0 \\ 0 & \lambda_2 & 0 & \cdots & 0 \\ \vdots & \vdots & \vdots & \vdots & \vdots \\ 0 & 0 & 0 & \cdots & \lambda_n \end{bmatrix} \qquad (6.61)$$

also denoted $\mathbf{\Lambda} = \mathrm{diag}\,[\lambda_1, \lambda_2, \cdots, \lambda_n]$. Solving for $\mathbf{\Lambda}$ from Equation (6.59), we have

$$\mathbf{\Lambda} = \mathbf{M}^{-1}\mathbf{AM} \qquad (6.62)$$

where $\mathbf{M}^{-1}$ is the inverse matrix $\mathbf{M}$. Equation (6.62) is known as *similarity transformation*. If the eigenvectors are orthogonal, then $\mathbf{M}^{-1} = \mathbf{M}^T$, where T denotes transpose, and the matrix $\mathbf{A}$ is diagonalized by the *orthogonal transformation*:

$$\mathbf{\Lambda} = \mathbf{M}^T\mathbf{AM} \qquad (6.63)$$

Example 6.4

(a) Find the eigenvalues and the eigenvectors of the matrix $\mathbf{A}$.

$$\mathbf{A} = \begin{bmatrix} -3 & 0 & 0 \\ -5 & 2 & 0 \\ -5 & 1 & 1 \end{bmatrix}$$

(b) Find the characteristic polynomial of $\mathbf{A}$.
(c) Diagonalize $\mathbf{A}$ by similarity transformation.

Solution.
(a) The characteristic equation is $|\mathbf{A} - \mathbf{I}\lambda| = 0 \implies$

$$\begin{vmatrix} -3-\lambda & 0 & 0 \\ -5 & 2 & 0 \\ -5 & 1 & 1-\lambda \end{vmatrix} = (\lambda+3)(\lambda-2)(\lambda-1) = 0$$

Thus, the eigenvalues are $\lambda_1 = -3, \lambda_2 = 1$ and $\lambda_3 = 2$ all distinct. The eigenvectors $\mathbf{X}_1, \mathbf{X}_2$ and $\mathbf{X}_3$ are obtained by solving the equations $\mathbf{A}\mathbf{X}_1 = \lambda_1\mathbf{X}_1, \mathbf{A}\mathbf{X}_2 = \lambda_2\mathbf{X}_2$ and $\mathbf{A}\mathbf{X}_3 = \lambda_3\mathbf{X}_3$. For $\lambda = -3$, we have

$$\begin{bmatrix} -3 & 0 & 0 \\ -5 & 2 & 0 \\ -5 & 1 & 1 \end{bmatrix} \begin{bmatrix} a \\ b \\ c \end{bmatrix} = -3 \begin{bmatrix} a \\ b \\ c \end{bmatrix}$$

where $\mathbf{X}^T = [\,a\ b\ c\,]$. This results 3 equations in 3 unknowns; i.e.,

$$\begin{array}{rlll} -3a & & & = -3a \\ -5a & +2b & & = -3b \\ -5a & +b & +c & = -3c \end{array}$$

solving the equations, we obtain $a = b = c$. Thus,

$$\mathbf{X}_1 = \alpha \begin{bmatrix} 1 \\ 1 \\ 1 \end{bmatrix}$$

is the eigenvector and α is any constant.

Similarly, we solve for $\mathbf{X}_2$ and $\mathbf{X}_3$ to obtain

$$\mathbf{X}_2 = \begin{bmatrix} 0 \\ 0 \\ 1 \end{bmatrix} \text{ and } \mathbf{X}_3 = \begin{bmatrix} 0 \\ 1 \\ 1 \end{bmatrix}$$

(b) The characteristic polynomial of $\mathbf{A}$ is $c(\lambda) = |\lambda\mathbf{I} - \mathbf{A}| \implies$

$$c(\lambda) = \begin{vmatrix} \lambda + 3 & 0 & 0 \\ -5 & \lambda - 2 & 0 \\ -5 & 1 & \lambda - 1 \end{vmatrix} = \lambda^3 - 7\lambda + 6$$

(c) Using the similarity transformation, $\Lambda = \mathbf{M}^{-1}\mathbf{A}\mathbf{M}$, we have

$$\mathbf{M} = \begin{bmatrix} 1 & 0 & 0 \\ 1 & 0 & 1 \\ 1 & 1 & 1 \end{bmatrix} \implies \mathbf{M}^{-1} = \begin{bmatrix} 1 & 0 & 0 \\ 0 & -1 & 1 \\ -1 & 1 & 0 \end{bmatrix}$$

and

$$\Lambda = \begin{bmatrix} 1 & 0 & 0 \\ 0 & -1 & 1 \\ -1 & 1 & 0 \end{bmatrix} \begin{bmatrix} -3 & 0 & 0 \\ -5 & 2 & 0 \\ -5 & 1 & 1 \end{bmatrix} \begin{bmatrix} 1 & 0 & 0 \\ 1 & 0 & 1 \\ 1 & 1 & 1 \end{bmatrix} = \begin{bmatrix} -3 & 0 & 0 \\ 0 & 2 & 0 \\ 0 & 0 & 1 \end{bmatrix}$$

(i) *Case 2*: When all eigenvalues are not distinct, the corresponding eigenvectors may or may not be linearly independent. If m_i is the order of an eigenvalue, called *algebraic multiplicity*, the corresponding number of independent vectors q_i, $q_i \le m_i$, is called *geometric multiplicity* or *degeneracy*. q_i is given by

$$q_i = n - \text{ rank } (\mathbf{A} - \mathbf{I}\lambda_i) \tag{6.64}$$

If $q_i = m_i$, all eigenvectors associated with λ_i are independent and can be solved for as in Case 1.

If $q_i = 1; (m_i > 1)$, there is one eigenvector associated with λ_i. The other $(m_i - 1)$ vectors are called *generalized eigenvectors*. A generalized eigenvector of rank k is a nonzero vector for which

$$(\mathbf{A} - \lambda_i\mathbf{I})^k\mathbf{X}_k = \mathbf{0} \tag{6.65a}$$

and

$$(\mathbf{A} - \lambda_i\mathbf{I})^{k-1}\mathbf{X}_{k-1} \neq \mathbf{0} \qquad (6.65b)$$

The eigenvector $\mathbf{X}_1$ is found as before; that is,

$$(\mathbf{A} - \mathbf{I}\lambda_i)\mathbf{X}_1 = \mathbf{0} \qquad (6.66)$$

whereas, the rest $(m_i - 1)$ generalized eigenvectors are found by

$$\begin{aligned}
(\mathbf{A} - \mathbf{I}\lambda_i)\mathbf{X}_2 &= \mathbf{X}_1 \\
(\mathbf{A} - \mathbf{I}\lambda_i)\mathbf{X}_3 &= \mathbf{X}_2 \\
&\vdots \\
(\mathbf{A} - \mathbf{I}\lambda_i)\mathbf{X}_j &= \mathbf{X}_{j-1} \\
&\vdots \\
(\mathbf{A} - \mathbf{I}\lambda_i)\mathbf{X}_{m_i} &= \mathbf{X}_{m_{i-1}}
\end{aligned} \qquad (6.67)$$

If the modal matrix $\mathbf{M}$ is formed as before, the generalized m_{i-1} eigenvectors are included, and the similarity transformation becomes

$$\mathbf{AM} = \mathbf{MJ} \qquad (6.68a)$$

or

$$\mathbf{J} = \mathbf{M}^{-1}\mathbf{AM} \qquad (6.68b)$$

$\mathbf{J}$ is an $n \times n$ diagonal matrix, called the *Jordan form*, such that

$$\mathbf{J} = \text{diag}\,[\mathbf{J}_1, \mathbf{J}_2, \cdots, \mathbf{J}_p] \qquad (6.69a)$$

and

$$\mathbf{J}_i = \begin{bmatrix}
\lambda_j & 1 & 0 & \cdots & 0 & 0 \\
0 & \lambda_j & 1 & \cdots & 0 & 0 \\
\vdots & \vdots & \vdots & \vdots & \vdots & \vdots \\
0 & 0 & 0 & \cdots & \lambda_j & 1 \\
0 & 0 & 0 & \cdots & \cdots & \lambda_j
\end{bmatrix} \qquad (6.69b)$$

Equation (6.69b) says that each submatrix $\mathbf{J}_i$ has the same eigenvalue along its main diagonal, ones for all elements in the diagonal above the main diagonal, and the rest of the elements are zero.

If $1 \leq q_i \leq m_i$, there may be more than one Jordan block for each eigenvector. Assume that we have two eigenvalues λ_1 of order 5 and λ_3 of order 1, and $q_1 = 1$. Then, we have two eigenvectors X_1 and X_2 and three generalized eigenvectors for λ_1; and one eigenvector X_3 for λ_3. The generalized eigenvectors may be associated with X_1, or with X_2, or with both X_1 and X_2. That is, we may have two Jordan blocks of the form

$$J_1 = \begin{bmatrix} \lambda_1 & 1 & 0 \\ 0 & \lambda_1 & 1 \\ 0 & 0 & \lambda_1 \end{bmatrix} , J_2 = \begin{bmatrix} \lambda_1 & 1 \\ 0 & \lambda_1 \end{bmatrix} \qquad (6.70a)$$

or

$$J_1 = \begin{bmatrix} \lambda_1 & 1 & 0 & 0 \\ 0 & \lambda_1 & 1 & 0 \\ 0 & 0 & \lambda_1 & 1 \\ 0 & 0 & 0 & \lambda_1 \end{bmatrix} , J_2 = [\lambda_1] \qquad (6.70b)$$

or vice versa. The approach to determine the Jordan blocks will be shown by an example. Assume that we have the case of Equation $(6.70a)$ then, the generalized eigenvalues and eigenvectors are determined by

$$
\begin{aligned}
(A - I\lambda_1)X_{13} &= X_{12} \\
(A - I\lambda_1)X_{12} &= X_1 \\
(A - I\lambda_1)X_1 &= 0 \\
(A - I\lambda_1)X_2 &= X_{22} \\
(A - I\lambda_1)X_2 &= 0 \\
(A - I\lambda_1)X_3 &= 0
\end{aligned}
\qquad (6.71)
$$

The modal matrix M is

$$M = [X_1|X_{12}|X_{13}|X_2|X_{22}|X_3] \qquad (6.72)$$

The similarity transformation is as given by Equation (6.68) where J is

$$J = \left[\begin{array}{ccc|cc|c} \lambda_1 & 1 & 0 & 0 & 0 & 0 \\ 0 & \lambda_1 & 1 & 0 & 0 & 0 \\ 0 & 0 & \lambda_1 & 0 & 0 & 0 \\ \hline 0 & 0 & 0 & \lambda_1 & 1 & 0 \\ 0 & 0 & 0 & 0 & \lambda_1 & 0 \\ 0 & 0 & 0 & 0 & 0 & \lambda_1 \end{array} \right] \qquad (6.73)$$

Example 6.5

(a) Find the eigenvalues and the eigenvectors of the matrix $\mathbf{A}$.

$$\mathbf{A} = \begin{bmatrix} 3 & 0 & 0 & 1 \\ 0 & 2 & 0 & 0 \\ 1 & 1 & 3 & 1 \\ -1 & 0 & 0 & 1 \end{bmatrix}$$

(b) Find the Jordan form by the transformation $\mathbf{J} = \mathbf{M}^{-1}\mathbf{A}\mathbf{M}$

Solution.

(a) The characteristic equation is given by $|\mathbf{A} - \mathbf{I}\lambda| = 0$; i.e.,

$$\begin{vmatrix} 3-\lambda & 0 & 0 & 1 \\ 0 & 2-\lambda & 0 & 0 \\ 1 & 1 & 3-\lambda & 1 \\ -1 & 0 & 0 & 1-\lambda \end{vmatrix} = (2-\lambda)^3(3-\lambda) = 0$$

Hence, two eigenvalues $\lambda_1 = 2$ with algebraic multiplicity $m_1 = 3$, and $\lambda_2 = 3$, with $m_2 = 1$. We need to determine the number of independent eigenvectors and generalized eigenvectors associated with λ_1. The rank of $|\mathbf{A} - \mathbf{I}\lambda|_{\lambda=2} = 2 = r$. Thus, $q_1 = n - r = 4 - 2 = 2$; i.e., we have two eigenvectors. Since $m_1 = 3$, there is only $m_1 - q = 1$ generalized eigenvector. Solving for $\mathbf{X}_1$ using the 4 equations $\mathbf{A}\mathbf{X}_1 = 2\mathbf{X}_1$, where $\mathbf{X}_1 = [\, a\ b\ c\ d]^T$, we obtain $a = -d$ and $b = -c$. Since we have two eigenvectors corresponding to $\lambda = 2$, we let $(a = 1, b = 0)$ to obtain $\mathbf{X}_1 = [\, 1\ 0\ 0\ -1]^T$, and $(a = 0, b = 1)$ to obtain $\mathbf{X}_2 = [\, 0\ 1\ -1\ 0]^T$. The generalized eigenvector $\mathbf{X}_{12}$ is given by $(\mathbf{A} - 2\mathbf{I})\mathbf{X}_{12} = \mathbf{X}_1$ to yield $\mathbf{X}_{12} = [\, 0\ 0\ -1\ -1]$. Similarly, we solve for $\mathbf{X}_3$ by using $\mathbf{A}\mathbf{X}_3 = 3\mathbf{X}_3$ to obtain $\mathbf{X}_3 = [\, 0\ 0\ 1\ 0\,]$.

(b) We form the modal matrix $\mathbf{M}$ as

$$\mathbf{M} = \begin{bmatrix} \mathbf{X}_1 | \mathbf{X}_{12} | \mathbf{X}_2 | \mathbf{X}_3 \end{bmatrix}$$

$$= \begin{bmatrix} 1 & 0 & 0 & 0 \\ 0 & 0 & 1 & 0 \\ 0 & -1 & -1 & 1 \\ -1 & 1 & 0 & 0 \end{bmatrix}$$

Perfoming the operation $\mathbf{J} = \mathbf{M}^{-1}\mathbf{A}\mathbf{M}$ results in

$$\mathbf{J} = \begin{bmatrix} 2 & 1 & 0 & 0 \\ 0 & 2 & 0 & 0 \\ 0 & 0 & 2 & 0 \\ 0 & 0 & 0 & 3 \end{bmatrix}$$

as expected. The inverse of $\mathbf{M}$ is

$$\mathbf{M}^{-1} = \begin{bmatrix} 1 & 0 & 0 & 0 \\ 1 & 0 & 0 & 1 \\ 0 & 1 & 0 & 0 \\ 1 & 1 & 1 & 1 \end{bmatrix}$$

6.4 INTEGRAL EQUATIONS

6.4.1 Green's Functions

Consider the nonhomogeneous linear equation defined over the interval $t \epsilon [0, T]$ and given by

$$\frac{d^2\phi(t)}{dt^2} + f(t) = 0 \quad , \quad \phi(0) = \alpha, \phi(T) = \beta \qquad (6.74)$$

The *forcing function*, $f(t)$, and the *boundary values*, $\phi(0)$ and $\phi(T)$, are known. It can be shown that the solution to Equation (6.74) is of the form:

$$\phi(t) = \int_0^T g(u, t) f(u) du + (1 - t)\alpha + \beta \qquad (6.75)$$

where $g(u,t)$, known as *Green's function*, is a function of the real variables u and t defined on the square $0 \leq u,t \leq T$. $g(u,t)$ is a weighting function for $d^2\phi/dt^2$ and if the boundary values are zero ($\phi(0) = \phi(T) = 0$), Equation (6.75) becomes

$$\phi(t) = \int_0^T g(u,t)f(u)du \tag{6.76}$$

The function $g(u,t)$ exists and is unique. It is explicitly given by

$$g(u,t) = \begin{cases} \frac{t(T-u)}{T} & ,0 \leq t < u \\ \frac{u(T-t)}{T} & ,u < t \leq T \end{cases} \tag{6.77}$$

If $T = 1$, $g(u,t)$ reduces to

$$g(u,t) = \begin{cases} t(1-u) & ,0 \leq t < u \\ u(1-t) & ,u < t \leq T \end{cases} \tag{6.78}$$

The Green's function associated with Equation (6.74) for $t\epsilon(0,1)$ and $\phi(0) = \phi(1) = 0$ satisfies

$$-\frac{d^2 g(u,t)}{dt^2} = \delta(t-u) \qquad , \begin{array}{c} 0 < u,t < 1 \\ g(0,u) = g(1,t) = 0 \end{array} \tag{6.79}$$

A different approach for solving the given nonhomogeneous linear differential equation:

$$\frac{d^2\phi(t)}{dt^2} + f(t) = 0 \;, \quad \begin{array}{c} 0 < t < 1 \\ \phi(0) = \phi(1) = 0 \end{array} \tag{6.80}$$

is to use eigenfunction expansion. The associated *eigenproblem* is given by

$$\frac{d^2\phi(t)}{dt^2} + \lambda\phi(t) = 0 \;, \quad \begin{array}{c} 0 < t < 1 \\ \phi(0) = \phi(1) = 0 \end{array} \tag{6.81}$$

The nontrivial solutions ($\phi(t) \neq 0$) to the boundary value problem of Equation (6.81) are called *eigenfunctions*. The corresponding values of λ are known as *eigenvalues*. It should be noted that an eigenfunction corresponds to a definite eigenvalue, but an eigenvalue may be associated with more than one independent eigenfunction. Also, any constant multiple of an eigenfunction is an eigenfunction corresponding to the same eigenvalue λ. In addition, if $\phi_1(t)$ and $\phi_2(t)$ are two eigenfunctions corresponding to the same λ, then $c_1\phi_1(t) + c_2\phi_2(t)$, where c_1 and c_2 are constants, is also an eigenfunction corresponding to that λ.

Solving the differential Equation in (6.81) and imposing the boundary conditions, we obtain the nontrivial solutions:

$$\phi_k(t) = \sin k\pi t , \; k = 1, 2, \cdots \qquad (6.82)$$

corresponding to the eigenvalues:

$$\lambda_k = k^2\pi^2 , \; k = 1, 2, \cdots \qquad (6.83)$$

The eigenfunctions corresponding to different eigenvalues are orthogonal; that is,

$$\int_0^1 \sin(m\pi t) \sin(n\pi t)dt = 0 \text{ for } m \neq n \qquad (6.84)$$

Note that the differential Equation (6.81) is a problem of the type of (6.80), with the forcing function $f(t) = \lambda\phi(t)$. Since the "solution" of (6.80) is given by Equation (6.76), it becomes

$$\phi(t) = \lambda \int_0^1 g(u,t)\phi(u)du , \; 0 < t < 1 \qquad (6.85)$$

The function $\phi(t)$ appears in both sides of Equation (6.85), that is, we really have not solved for $\phi(t)$, but we have shown that nonhomogeneous linear Equation (6.80) is equivalent to the *integral equation* (6.85). The boundary conditions are incorporated in the integral equation through its Green's function, known as *kernel $k(u,t)$*. Hence, we write

$$\phi(t) = \lambda \int_0^1 k(u,t)\phi(u)du \; 0 < t < 1 \qquad (6.86)$$

6.4.2 Fredholm Integral Equations

We now consider the linear integral equation of the form:

$$K\phi(t) - \lambda\phi(t) = f(t) \tag{6.87}$$

That is,

$$\int_0^T k(u,t)\phi(u)du - \lambda\phi(t) = f(t) \qquad ,0 \le t \le T \tag{6.88}$$

Given the kernel $k(u,t)$, the nonhomogeneous term $f(t)$ and the eigenvalue λ, we would like to determine the unknown function $\phi(t)$ and study its dependence on the function $f(t)$ and the eigenvalue λ.

The integral Equation (6.88) is known as *Fredholm equation of the second kind* when $\lambda \neq 0$, and as *Fredholm equation of the first kind* when $\lambda = 0$. When the function $f(t)$ is zero, we have the *eigenvalue problem*

$$K\phi(t) = \lambda\phi(t) \tag{6.89}$$

or

$$\int_0^T k(u,t)\phi(u)du = \lambda\phi(t) \qquad ,0 \le t \le T \tag{6.90}$$

To determine the eigenvalues for which Equation (6.89) has no trivial solution, we solve the eigenvalue problem

$$(\mathbf{K} - \mathbf{I}\lambda)\mathbf{\Phi} = \mathbf{0} \tag{6.91}$$

where $\mathbf{K}$ is a symmetric nonnegative definite matrix representing the transformation operator.

Example 6.6

Consider a differential system of the form

$$\frac{d\phi(t)}{dt} + \phi(t) = f(t) \quad , 0 \leq t \leq 1$$

$$\phi(0) + \phi(1) = \alpha$$

Determine the Green's function $g(u, t)$.

Solution.

A solution to the homogeneous equation

$$\frac{d\phi(t)}{dt} + \phi(t) = 0$$

is $\phi_h(t) = c\, e^{-t}$. We guess a solution for $g(u, t)$ as

$$g(u, t) = \begin{cases} c_1\, e^{-t} & , 0 \leq t \leq u \\ c_2\, e^{-t} & , u \leq t \leq 1 \end{cases}$$

As in (6.79), the Green's function satisfies $dg(u, t)/dt = \delta(t - u)$, and hence at $t = u$, we have

$$-c_1\, e^{-u} + c_2\, e^{-u} = 1$$

From the boundary condition, we have

$$c_1 + c_2\, e^{-1} = 0$$

Solving for c_1 and c_2, we obtain

$$c_1 = -\frac{e^{u-1}}{1 + e^{-1}} \; , c_2 = \frac{e^{-u}}{1 + e^{-1}}$$

Consequently, the Green's function is

$$g(u, t) = \begin{cases} -\frac{e^{-u-t-1}}{1+e^{-1}} & , 0 \leq t \leq u \\ \\ \frac{e^{u-t}}{1+e^{-1}} & , u \leq t \leq 1 \end{cases}$$

Example 6.7

Consider the homogeneous eigenvalue problem $K\phi - \lambda\phi = 0$. Let the associated eigenproblem be given by the following homogeneous differential equation:

$$\frac{d^2\phi(t)}{dt^2} + \lambda\phi(t) = 0 \quad , \quad \begin{array}{l} 0 < t < 1 \\ \phi'(0) = \phi'(1) = 0 \end{array}$$

(a) Determine the kernel and write the corresponding integral equation.
(b) Find the eigenvalues and eigenfunctions.

Solution.
   (a) Let $\phi''(t)$ denotes $d^2\phi(t)/dt^2$ and $\phi'(t)$ denotes $d\phi(t)/dt$. Integrating the differential equation with respect to t, we have

$$\phi'(t) - \phi'(0) + \lambda \int_0^t \phi(u)du = 0$$

Integrating again, results in

$$\phi(t) - \phi(0) - t\phi_{\prime}(0) + \lambda \int_0^t (t - u)\phi(u)du = 0$$

Applying the boundary condition, $\phi'(0) = 0$ yields

$$\phi(t) - \phi(0) + \lambda \int_0^t (t - u)\phi(u)du = 0$$

To determine $\phi(0)$ in the above equation, we apply the boundary condition $\phi(1) = 0$ at $t = 1$; that is,

$$\phi(1) - \phi(0) + \lambda \int_0^1 (1 - u)\phi(u)du = 0$$

or

$$\phi(0) = \lambda \int_0^1 (1 - u)\phi(u)du$$

Substituting for $\phi(0)$ into $\phi(t)$, we have

$$
\begin{aligned}
\phi(t) &= \lambda \int_0^1 (1-u)\phi(u)du - \lambda \int_0^t (t-u)\phi(u)du \\
&= \lambda \{ \int_0^t (1-u)\phi(u)du + \int_t^1 (1-u)\phi(u)du \} - \lambda \int_0^t (t-u)\phi(u)du \\
&= \lambda \int_0^t (1-t)\phi(u)du + \lambda \int_0^1 (1-u)\phi(u)du \\
&= \lambda \int_0^1 k(u,t)\phi(u)du
\end{aligned}
$$

where, the kernel $k(u,t)$ is

$$
k(u,t) = \begin{cases} 1-t & ,0 \le t \le u \\ 1-u & ,t \le u \le 1 \end{cases}
$$

(b) From the homogeneous differential equation $\phi''(t) + \lambda\phi'(t) = 0$, we have the general solution:

$$
\phi(t) = A \cos\sqrt{\lambda}t + B \sin\sqrt{\lambda}t
$$

the derivative of $\phi(t)$ is

$$
\phi'(t) = -\sqrt{\lambda}A \sin\sqrt{\lambda}t + \sqrt{\lambda}B \cos\sqrt{\lambda}t
$$

Applying the boundary condition $\phi'(0) = 0 \Longrightarrow$

$$
\phi'(0) = -\sqrt{\lambda}B = 0
$$

since $\lambda \ne 0 \Longrightarrow B = 0$ and $\phi(t)$ is

$$
\phi(t) = A \cos\sqrt{\lambda}t
$$

Applying the other boundary condition $\phi(1) = 0$ yields

$$
\phi(1) = A \cos\sqrt{\lambda} = 0 \Longrightarrow \sqrt{\lambda} = (2k-1)\frac{\pi}{2} \quad , k = 1,2,\cdots
$$

Hence, the eigenvalues are

$$
\lambda_k = \frac{(2k-1)^2\pi^2}{4} \quad , k = 1,2,\cdots
$$

The eigenfunctions are

$$\phi(t) = A \cos(2k - 1)\frac{\pi}{2}t \ , \ 0 \le t \le 1$$

but $\int_0^1 \phi^2(t)dt = 1 \implies A = \sqrt{2}$. Therefore, the nonzero eigenvalues and the normalized eigenfunctions are

$$\lambda_k = \frac{(2k - 1)^2\pi^2}{4} \quad , \ k = 1, 2, \cdots$$

and

$$\phi_k(t) = \sqrt{2} \cos(2k - 1)\frac{\pi}{2}t \quad , \ k = 1, 2, \cdots$$

6.5 REPRESENTATION OF RANDOM PROCESSES

In Section 6.2, we represented deterministic finite energy signals in terms of an orthogonal series expansion. We now extend this concept to random processes.

Let $X(t)$ be a random process to be represented by a complete set of orthonormal functions $\{\phi_k(t)\}$ specified over the interval $[0, T]$. That is, we write

$$X(t) = \lim_{k \to \infty} \sum_{k=1}^{K} X_i \phi_k(t) \tag{6.92}$$

where the random variable X_k is given by

$$X_k = \int_0^T X(t)\phi_k(t)dt \tag{6.93}$$

The above ordinary limit is not practical, since it requires that all sample functions of the random process satisfy Equation (6.92), which is not possible. Instead, we use a little more relaxed type of convergence which is the *mean-square convergence*. That is, we require

$$\lim_{K \to \infty} E\{[X(t) - \sum_{k=1}^{K} X_k \phi_k(t)]^2\} = 0 \tag{6.94}$$

equivalently, we say

$$X(t) = \lim_{K \to \infty} \sum_{k=1}^{K} X_k \phi_k(t) \tag{6.95}$$

Since in general it is easier to solve problems in which the random variables are uncorrelated, we would like to select the set $\{\phi_k(t)\}$ such that the coefficients X_k, $k = 1, 2, \cdots, K$, are uncorrelated. If $m_k = E[X_k] = 0$, then the coefficients are uncorrelated, provided that

$$E[X_k X_j] = \lambda_k \delta_{kj} \tag{6.96}$$

Substituting Equation (6.93) into Equation (6.96), we obtain

$$\begin{aligned} E[X_k X_j] &= E[\int_0^T X(t)\phi_k(t)dt \int_0^T X(u)\phi_j(u)du \\ &= \int_0^T \phi_k(t)dt \int_0^T K_{XX}(t,u)\phi_j(u)du = \lambda_k \delta_{kj} \end{aligned} \tag{6.97}$$

where $K_{XX}(t, u) = E[X(t)X(u)]$ is the autocovariance function of $X(t)$. Equation (6.97) is satisfied if

$$\int_0^T K_{XX}(t,u)\phi_j(u)du = \lambda_j \phi_j(t) \tag{6.98}$$

Equation (6.98) is the homogeneous linear integral equation as defined in the previous section, and the autocovariance function represents the kernel. The kernel $K_{XX}(t, u)$ can always be expanded in the series:

$$K_{XX}(t,u) = \sum_{k=1}^{\infty} \lambda_k \phi_k(t)\phi_k(u) \, , \, 0 \le t, u \le T \tag{6.99}$$

where the convergence is uniform for $0 \le t, u \le T$. This is known as *Mercer's theorem*. Also, the mean energy of the random process $X(t)$

is the infinite sum of the eigenvalues. That is

$$E[\int_0^T X^2(t)dt] = E[\int_0^T \sum_{j=1}^{\infty} \sum_{k=1}^{\infty} X_j X_k \phi_j(t)\phi_k(t)dt]$$

$$= \sum_{j=1}^{\infty} \sum_{k=1}^{\infty} E[X_j X_k] \int_0^T \phi_j(t)\phi_k(t)dt$$

$$= \sum_{k=1}^{\infty} E[X_k^2] \qquad (6.100)$$

Using Equation (6.96), Equation (6.100) reduces to

$$E[\int_0^T X^2(t)dt] = \int_0^T K_{XX}(t,t)dt = \sum_{k=1}^{\infty} \lambda_k \qquad (6.101)$$

We now show the mean-square convergence for the series representation of the process $X(t)$. We define the error $\varepsilon_K(t)$ as

$$\varepsilon_K(t) = E\{[X(t) - \sum_{k=1}^{K} X_k \phi_k(t)]^2\} \qquad (6.102)$$

Expanding the expression in (6.102) and using the fact that $E[X_j X_k] = \lambda_k \delta_{jk}$ and $\int_0^T K_{XX}(t,u)\phi_k(u)du = \lambda_k \phi_k(t)$, results in

$$\varepsilon_K(t) = K_{XX}(t,t) - \sum_{k=1}^{K} \lambda_k \phi_k^2(t) \qquad (6.103)$$

From Mercer's theorem, we have

$$K_{XX}(t,u) = \sum_{k=1}^{\infty} \lambda_k \phi_k(t)\phi_k(u) \qquad (6.104a)$$

or

$$K_{XX}(t,t) = \sum_{k=1}^{\infty} \lambda_k \phi_k^2(t) \qquad (6.104b)$$

Hence,

$$\lim_{K \to \infty} \varepsilon_K(t) = 0 \qquad (6.105)$$

and the mean-square convergence is satisfied. This series expansion of $X(t)$ is also known as *Karhunen-Loeve* expansion.

6.5.1 Rational Power Spectral Densities

Let $X(t)$ be a zero-mean wide-sense stationary process with a rational power spectrum density of the form

$$S_{XX}(\omega) = \frac{N(\omega^2)}{D(\omega^2)} \tag{6.106}$$

where $\omega = 2\pi f$. $S_{XX}(\omega)$ is an even function of ω and forms a Fourier transform pair with the autocorrelation $R_{XX}(\tau)$, which is equal to the autocovariance function, since $E[X(t)] = 0$. Thus,

$$R_{XX}(\tau) = E[X^2(t)] = K_{XX}(\tau) \tag{6.107}$$

and

$$E[X^2(t)] = \frac{1}{2\pi} \int_{-\infty}^{\infty} S_{XX}(\omega) d\omega \tag{6.108}$$

Let $N(\omega^2)$ be a polynomial of degree q in ω^2, and $D(\omega^2)$ a polynomial of degree r in ω^2, where $q < r$ since the mean-square value $E[X^2(t)]$ is assumed finite.

For a rational function of the form given in Equation (6.106), the solution to the integral equation can always be obtained from the corresponding linear differential equation with constant coefficients. The integral equation is

$$\lambda \phi(t) = \int_0^T K_{XX}(t-u)\phi(u)du \qquad , 0 \le t \le T \tag{6.109}$$

Since the ϕ's are zero outside the interval $[0, T]$, Equation (6.109) can be rewritten as

$$\lambda \phi(t) = \int_{-\infty}^{\infty} K_{XX}(t-u)\phi(u)du \qquad , -\infty < t < \infty \tag{6.110}$$

Taking the Fourier transform, we have

$$\begin{aligned} \lambda \Phi(j\omega) &= S_{XX}(\omega)\Phi(j\omega) \\ &= \frac{N(\omega^2)}{D(\omega^2)}\Phi(j\omega) \end{aligned} \tag{6.111}$$

or

$$[\lambda D(\omega^2) - N(\omega^2)]\Phi(j\omega) = 0 \qquad (6.112)$$

Let $V = j\omega$, then $V^2 = -\omega^2$. Substituting in Equation (6.112), we have

$$[\lambda D(-V^2) - N(-V^2)]\Phi(V) = 0 \qquad (6.113)$$

where V can be implemented as an operator and thus, Equation (6.113) can be transformed into a homogeneous linear differential equation, such that V denotes d/dt. Since the polynomial in (6.113) is of degree $2r$, there are $2r$ homogeneous solutions denoted as $\phi_{h_k}(t)$, $k = 1, 2, \cdots, 2r$, for every eigenvalue λ. Once we obtain $\phi_{h_k}(t)$, $k = 1, 2, \cdots, 2r$, we form

$$\phi(t) = \sum_{k=1}^{2r} c_k \phi_{h_k}(t) \qquad (6.114)$$

where c_k's are constants, and substitute into the original integral equation to determine λ and c_k, as shown in Example 6.7.

Example 6.8

Let $X(t)$ be a zero-mean wide-sense stationary process with power spectrum density

$$S_X(\omega) = \frac{2\alpha P}{\omega^2 + \alpha^2} \qquad \text{for all } \omega$$

(a) Obtain the differential equation.
(b) Determine the eigenvalues λ.

Solution.
(a) The autocorrelation function is given by

$$R_{XX}(\tau) = Pe^{-\alpha|\tau|}$$

Assuming a symmetric observation interval, the integral equation

$$\begin{aligned}
\lambda\phi(t) &= \int_{-T}^{T} K_{XX}(t, u)\phi(u)du \\
&= \int_{-T}^{T} Pe^{-|t-u|}\phi(u)du
\end{aligned}$$

since $R_{XX}(\tau) = K_{XX}(\tau)$, where $\tau = t - u$. The above expression can be rewritten as

$$\lambda\phi(t) = \int_{-T}^{t} Pe^{-\alpha(t-u)}\phi(u)du + \int_{t}^{T} Pe^{-\alpha(u-t)}\phi(u)du$$

Differentiating with respect to t, we obtain

$$\lambda\frac{d\phi(t)}{dt} = -\alpha\int_{-T}^{t} Pe^{-\alpha}(t-u)\phi(u)du + \alpha\int_{t}^{T} Pe^{-\alpha(u-t)}\phi(u)du$$

Differentiating again results in

$$\begin{aligned}
\lambda\frac{d^2\phi(t)}{dt^2} &= \alpha^2\int_{-T}^{T} Pe^{-\alpha(t-u)}\phi(u)du - 2\alpha P\phi(t) + \alpha^2\int_{t}^{T} Pe^{-\alpha(u-t)}\phi(u)du \\
&= -2\alpha P\phi(t) + \alpha^2\int_{-T}^{T} Pe^{-|t-u|}\phi(u)du \\
&= -2\alpha P\phi(t) + \alpha^2\lambda\phi(t)
\end{aligned}$$

or

$$\frac{d^2\phi(t)}{dt^2} - \frac{\alpha^2}{\lambda}(\lambda - \frac{2P}{\alpha})\phi(t) = 0$$

(b) Since the eigenvalues must be nonnegative, we have four cases:
(i) $\lambda = 0$
(ii) $0 < \lambda < (2P)/\alpha$
(iii) $\lambda = (2P)/\alpha$
(iv) $\lambda > (2P)/\alpha$

Case (i): $\lambda = 0$ The differential equation is

$$-2\alpha\ p\phi(t) = 0$$

since $\alpha P \neq 0$, we only have the trivial solution $\phi(t) = 0$, and thus, $\lambda = 0$ is not an eigenvalue.

Case (ii): $0 < \lambda < (2P)/\alpha$. Let $\beta^2 = -(\alpha^2/\lambda)[\lambda - (2P/\alpha)] \implies 0 < \beta^2 < \infty$ and the differential equation becomes

$$\frac{d^2\phi(t)}{dt^2} + \beta^2\phi(t) = 0$$

This has a general solution of the form $\phi(t) = c_1 e^{j\beta t} + c_2 e^{-j\beta t}$ where c_1 and c_2 are constants to be determined. Substituting for $\phi(t)$ in the integral equation and integrating, we obtain

$$
\begin{aligned}
\lambda(c_1 e^{j\beta t} c_2 e^{-j\beta t}) \;=\; & c_1 P e^{j\beta t}\Big(\frac{1}{\alpha + j\beta} + \frac{1}{\alpha - j\beta}\Big) \\
& + c_2 P e^{-j\beta t}\Big(\frac{1}{\alpha - j\beta} + \frac{1}{\alpha + j\beta}\Big) \\
& - P e^{\alpha t}\Big(\frac{c_1 e^{-(\alpha + j\beta)T}}{\alpha + j\beta} + \frac{c_2 e^{-(\alpha - j\beta)T}}{\alpha - j\beta}\Big) \\
& - P e^{\alpha t}\Big(\frac{c_1 e^{-(\alpha - j\beta)T}}{\alpha - j\beta} + \frac{c_2 e^{-(\alpha + j\beta)T}}{\alpha + j\beta}\Big)
\end{aligned}
$$

By inspection of the above expression, we have

$$
\frac{c_1 e^{-(\alpha + j\beta)T}}{\alpha + j\beta} + \frac{c_2 e^{-(\alpha - j\beta)}}{\alpha - j\beta} = 0
$$

$$
\frac{c_1 e^{-(\alpha - j\beta)T}}{\alpha - j\beta} + \frac{c_2 e^{-(\alpha + j\beta)}}{\alpha + j\beta} = 0
$$

$$
\lambda c_1 = c_1 P\Big(\frac{1}{\alpha + j\beta} + \frac{1}{\alpha - j\beta}\Big)
$$

and

$$
\lambda c_2 = c_2 P\Big(\frac{1}{\alpha - j\beta} + \frac{1}{\alpha + j\beta}\Big)
$$

Solving for λ, c_1, and c_2 we obtain the eigenvalue:

$$
\lambda = \frac{2\alpha P}{\alpha^2 + \beta^2}
$$

and

$$
c_1^2 = c_2^2 \implies
\begin{cases}
c_1 = c_2 \\
\text{or} \\
c_1 = -c_2
\end{cases}
$$

When $c_1 = c_2$, the kth eigenfunction is

$$
\begin{aligned}
\phi_k(t) \;&=\; c_1(e^{j\beta_k t} + e^{-j\beta_k t}) \\
&=\; A_k \cos\beta_k t
\end{aligned}
$$

Since $\int_{-T}^{T} \phi_k^2(t) = 1$, we can solve for A_k as

$$A_k = \frac{1}{\sqrt{T(1 + \frac{\sin 2\beta_k T}{2\beta_k})}}$$

It should be noted that when $c_1 = c_2$, we have

$$(\alpha + j\beta)e^{j\beta T} = (-\alpha + j\beta)e^{-j\beta T}$$

or

$$\tan\frac{\beta}{T} = \frac{\alpha}{\beta}$$

The values of β satisfying the above equation can be determined graphically as shown in Figure 3.8 of Van Trees[28].

When $c_1 = -c_2$, we follow the same procedure as before to obtain

$$\phi_k(t) = B_k \sin\beta_k t$$

where

$$B_k = \frac{1}{\sqrt{T(1 - \frac{\sin 2\beta_k T}{2\beta_k T})}}$$

and

$$\tan\beta T = -\frac{\beta}{\alpha}$$

Case (iii): $\lambda = (2P)/\alpha$. In this case, the differential equation becomes

$$\frac{d^2\phi(t)}{dt^2} = 0 \implies \phi(t) = c_1 t + c_2$$

Substituting for $\phi(t)$ into the integral equation to determine the constants c_1 and c_2, we obtain

$$c_2 = c_1(T - \frac{1}{\alpha}) \quad \text{and} \quad c_2 = c_1(T - \frac{1}{\alpha})$$

For $T \neq \alpha \implies c_1 = c_2 = 0$, and hence we only have the trivial solution $\phi(t) = 0$; and $\lambda = 2P/\alpha$ is not an eigenvalue.

Case (iv): $\lambda > (2P)/\alpha$. Let $\gamma = (\alpha^2/\lambda)\{\lambda - [(2P)/\alpha]\} \implies 0 < \gamma^2 < \alpha^2$. For $|\gamma| < \alpha$, the differential equation is

$$\frac{d^2\phi(t)}{dt^2} - \gamma\phi(t) = 0$$

which has solution $\phi(t) = c_1 e^{\gamma t} + c_2 e^{-\gamma t}$. As in Case (ii), we obtain

$$\frac{c_2}{c_1} = -\frac{\alpha + \gamma}{\alpha - \gamma}e^{2\gamma T}$$

and

$$\frac{c_2}{c_2} = -\frac{\alpha - \gamma}{\alpha + \gamma}e^{-2\gamma T}$$

No solution satisfies the above equation, and hence, $\lambda > (2P)/\alpha$ is not an eigenvalue.

6.5.2. Wiener Process

In this subsection, we first show how the Wiener process (a nonstationary process) is obtained from the random walk and then derive its eigenvalues and the corresponding eigenfunctions in order to write the Karhunen-Loeve expansion.

Consider an experiment where starting at time $t = 0$ from a specific point, a particle moves from its current position with probability p, one step to the right, and with probability $q = 1 - p$, one step to the left. The particle moves every T seconds. A typical sketch of this process, known as *random walk*, is shown in Figure 6.9. If $p = q = 1/2$, the walk is called *symmetric*.

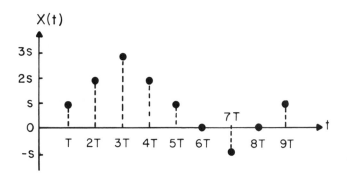

Figure 6.9 Random walk.

In n steps, assume the process takes k steps to the right, and $(n - k)$ steps to the left. That is,

$$
\begin{aligned}
X(nT) &= ks - (n - k)s \\
&= (2k - n)s
\end{aligned} \tag{6.115}
$$

Assuming the walk is symmetric, it follows that

$$
P[X(nT) = (2k - n)s] = \binom{n}{k} \frac{1}{2^k} \frac{1}{2^{n-k}} = \binom{n}{k} \frac{1}{2^k} \tag{6.116}
$$

and the probability density function of the process is

$$
f_X[x(nT)] = \sum_{k=0}^{n} \binom{n}{k} \frac{1}{2^n} \delta[x(nT) - (2k - n)s] \tag{6.117}
$$

Note that the increment $X(n_2T) - X(n_1T)$ depends only on the steps in the interval (n_1T, n_2T), for $n_2 > n_1$. Consequently, if $n_4 > n_3 \geq n_2 > n_1$, the increments $X(n_4T) - X(n_3T)$ and $X(n_2T) - X(n_1T)$ are independent.

The Wiener process, also known as the *Brownian motion*, is given by

$$W(t) = \lim_{\substack{T \to 0 \\ n \to \infty}} X(nT) \tag{6.118}$$

such that

$$\lim_{\substack{T \to 0 \\ n \to \infty}} nT = t \tag{6.119}$$

Using the central limit theorem, which states that if $X = X_1 + \cdots + X_n$ is the sum of n independent random variables, then as $n \to \infty$, the density function of X may be given by

$$f_X(x) \simeq \frac{1}{\sigma\sqrt{2\pi}} \, e^{-\frac{(x-m)^2}{2\sigma^2}} \tag{6.120}$$

where m is the mean and σ^2 is the variance of X. If X takes values ka, $k = 1, 2, \cdots$, with probability P_k, then

$$P(X = ka) \simeq \frac{1}{\sigma\sqrt{2\pi}} \, e^{-\frac{(ka-m)^2}{2\sigma^2}} \tag{6.121}$$

If X_k is a Bernoulli random variable taking values $+s$ and $-s$ with probability $1/2$, then $E[X_k] = 0$ and $\text{var}(X_k) = s^2$. Since the random walk is

$$X(nT) = X_1 + \cdots + X_n \tag{6.122}$$

where $X_k = 1, 2, \cdots, n$ are independent and symmetric Bernoulli random variables, then the mean and variance of $X(nT)$ are

$$E[X(nT)] = \sum_{k=1}^{n} E[X_k] = 0 \tag{6.123}$$

and

$$\text{var}[X(nT)] = \sum_{k=1}^{n} \text{var}[X_k] = ns^2 \tag{6.124}$$

Consequently, for n large we have

$$P[X(nT) = (2k - n)s] \simeq \frac{1}{s\sqrt{2\pi n}} \, e^{-\frac{(2k-n)^2}{2n}} \tag{6.125}$$

Using Equation (6.119), we have $\text{var}[X(nT)] = ns^2 = ts^2/T$ and thus, if

$$s^2 = \alpha T \tag{6.126a}$$

$$\omega = (2k - n)s \tag{6.126b}$$

the density function of the Wiener process $W(t)$ can be expressed as

$$f_{W(t)}[\omega(t)] = \frac{1}{\sqrt{2\pi\alpha t}} \, e^{-\frac{\omega^2(t)}{2\alpha t}} \tag{6.127}$$

Also, if $t_4 > t_3 \geq t_2 > t_1$, the increments $W(t_4) - W(t_3)$ and $W(t_2) - W(t_1)$ are independent.

We are now ready to determine the eigenvalues and eigenfunctions of the wiener process. Let $X(t)$ be the Wiener process as defined above with the properties as given in Equations $(6.123), (6.124)$, and (6.127). To determine the covariance function $K_{XX}(t, u) = E[X(t)X(u)]$ consider the increments $X(t) - X(u)$, and $X(u) - X(0)$, where $t \geq u > 0$. Since $X(u) - X(0) = X(u)$, $X(t) - X(u)$ and $X(u)$ are statistically independent. Consequently,

$$
\begin{aligned}
E\{[X(t) - X(u)]X(u)\} &= E[X(t)X(u)] - E[X^2(u)] \\
&= K_{XX}(t, u) - \alpha u = 0 \tag{6.128}
\end{aligned}
$$

since $E[X(u)] = 0$. Hence,

$$K_{XX}(t, u) = \alpha u \qquad , t \geq u \tag{6.129}$$

Similarly, for $u \geq t > 0$, we obtain

$$K_{XX}(t, u) = \alpha t \qquad , u \geq t \tag{6.130}$$

The covariance function of the Wiener process is

$$K_{XX}(t, u) = \alpha \min(u, t) = \begin{cases} \alpha u & , u \leq t \\ \alpha t & , t \leq u \end{cases} \tag{6.131}$$

To solve for the eigenfunctions, we use the integral equation:

$$
\begin{aligned}
\lambda\phi(t) &= \int_0^T K_{XX}(t, u)\phi(u)du \\
&= \alpha \int_0^t u\phi(u)du + \alpha t \int_0^T \phi(u)du \tag{6.132}
\end{aligned}
$$

Differentiating twice with respect to t, and using Leibnitz's rule, we obtain the differential equation:

$$\lambda \frac{d^2\phi(t)}{dt^2} + \alpha\phi(t) = 0 \tag{6.133}$$

we have two cases: (i) $\lambda = 0$ and (ii) $\lambda > 0$

Case (i): $\lambda = 0$. In this case $\phi(t) = 0$ which is the trivial solution.

Case (ii): $\lambda > 0$. Let $\beta^2 = \alpha/\lambda$, then the differential equation is

$$\frac{d^2\phi(t)}{dt^2} + \beta^2\phi(t) = 0 \tag{6.134}$$

where

$$\phi(t) = c_1\, e^{j\beta t} + c_2\, e^{-j\beta t} \tag{6.135}$$

substituting into the integral equation and solving for λ_k and $\phi_k(t)$, we obtain

$$\lambda_k = \frac{\alpha}{\beta_k} = \frac{\alpha T^2}{\pi^2(k - \frac{1}{2})^2} \tag{6.136}$$

where $\beta_k = (\pi/T)[k - (1/2)]$, and the normalized $\phi_k(t)$ is

$$\phi_k(t) = \sqrt{\frac{2}{T}}\, \sin[\frac{\pi}{T}(k - \frac{1}{2})t]\; ; \; 0 \le t \le T \tag{6.137}$$

Therefore, the Karhunen-Loeve expansion of the Wiener process is

$$\begin{aligned}
X(t) &= \sum_{k=1}^{\infty} X_k\phi_k(t) \\
&= \sum_{k=1}^{\infty} X_k\sqrt{\frac{2}{T}}\, \sin[\frac{\pi}{T}(k - \frac{1}{2})t]
\end{aligned} \tag{6.138}$$

where the mean-square value of the coefficient X_k is

$$E[X_k^2] = \lambda_k = \frac{\alpha T^2}{(k - \frac{1}{2})\pi^2} \tag{6.139}$$

6.5.3 White Noise Process

The white noise process can be derived from the Wiener process. Let $\alpha = \sigma^2$ and the K-term approximation of the Wiener process $X(t)$ be $X_K(t)$. That is,

$$X_K(t) = \sum_{k=1}^{K} X_k \sqrt{\frac{2}{T}} \sin[(k - \frac{1}{2})\frac{\pi}{T}t] \qquad (6.140)$$

Taking the derivative $X_K(t)$ with respect to t, we obtain

$$\frac{dX_K(t)}{dt} = \sum_{k=1}^{K} X_k(k - \frac{1}{2})\frac{\pi}{T}\sqrt{\frac{2}{T}} \cos[(k - \frac{1}{2})\frac{\pi}{T}t]$$

$$= \sum_{k=1}^{K} W_k \sqrt{\frac{2}{T}} \cos[(k - \frac{1}{2})\frac{\pi}{T}t \qquad (6.141a)$$

where

$$W_k = X_k(k - \frac{1}{2})\frac{\pi}{T} \qquad (6.141b)$$

and

$$E[W_k^2] = \sigma^2 \text{ for all } k \qquad (6.142)$$

Note also that the functions $\phi_k(t) = \sqrt{2/T} \cos\{[k - (1/2)](\pi/T)t\}$ are orthonormal in the observation interval $[0, T]$ and they are possible eigenfunctions to the derivative process.

To show that the set of functions $\{\phi_k(t)\}$ are eigenfunctions for the approximate integral equation corresponding to the white noise process, we need to define a white Gaussian noise process.

Definition: A white Gaussian process is a Gaussian process with co-variance function given by

$$\sigma^2 \delta(t - u) \qquad (6.143)$$

where δ is the delta function. The coefficients along each of the coordinate functions, are statistically independent Gaussian random variables with variance σ^2.

Now, considering the derivative of the covariance function of the Wiener process:

$$
\begin{aligned}
K_{\dot{X}\dot{X}}(t,u) &= E[\frac{dX(t)}{dt}\frac{dX(u)}{du}] = \frac{\partial^2}{\partial t \partial u} E[X(t)X(u)] \\
&= \frac{\partial^2}{\partial t \partial u} K_{XX}(t,u) = \frac{\partial^2}{\partial t \partial u}[\sigma^2 \min(u,t)] \\
&= \sigma^2 \delta(t-u)
\end{aligned}
\tag{6.144}
$$

The corresponding integral equation is

$$
\begin{aligned}
\lambda \phi(t) &= \int_0^T \sigma^2 \delta(t-u)\phi(u)du \\
&= \sigma^2 \phi(t)
\end{aligned}
\tag{6.145}
$$

and hence, the integral equation is satisfied for *any* set of orthonormal functions $\{\phi_k(t)\}$. In addition, we observe that

$$
\lambda_k = \sigma^2 \text{ for all } k
\tag{6.146}
$$

Note that the energy over the interval $[0,T]$ is not finite as $K \to \infty$, since

$$
\sum_{k=1}^{\infty} \lambda_k = \sum_{k=1}^{\infty} \sigma^2 \to \infty
\tag{6.147}
$$

Therefore, this derivative process is not realizable. Nevertheless, one possible representation, which is not unique, is

$$
W(t) = \frac{dX(t)}{dt} = \sum_{k=1}^{\infty} W_k \sqrt{\frac{2}{T}} \cos[(k-\frac{1}{2})\frac{\pi}{T}t]
\tag{6.148}
$$

6.6 SUMMARY

In this chapter, we have shown how a deterministic signal can be represented in series expansion of orthonormal functions. In doing this, we needed to cover the fundamental mathematical concepts of orthogonal functions, and generalized Fourier series. Then, we used the Gram-Schmidt orthogonalization procedure to show how a set of dependent or independent functions can be decomposed into another set of orthonormal, and independent functions.

We also showed how a random process can be represented by an orthonormal series expansion known as Karhunen-Loeve expansion. Specific processes such as: the rational power spectral densities, the Wiener process, and the white Gaussian noise process were considered. We showed how the Wiener process is obtained from the random walk and how the white Gaussian noise process is derived from the Wiener process. This required solving for eigenvalues and eigenvectors of linear transformations. We discussed Green's function and showed how integral equations can be reduced to linear differential equations in order to solve for eigenvalues, and their corresponding eigenfunctions.

The mathematical concepts covered such as solving for eigenvalues and eigenvectors/eigenfunctions, matrix diagonalization and series representation of signal; will be useful to us in the next two chapters, which deal with the general Gaussian problem, and detection in noise.

PROBLEMS

6.1 (a) Is the set of functions

$$\{1/\sqrt{T}, \sqrt{2/T} \cos(k\pi/T)t\} \qquad , k = 1, 2, \cdots$$

orthonormal in the interval $[0, T]$?
(b) Using the fact that the functions in (a) are orthonormal on the interval $[0, T]$, show that the set

$$\{1/\sqrt{2T}, (1/sqrtT) \cos(k\pi/T)t\} \qquad , k = 1, 2, \cdots$$

is orthonormal on the interval $[-T, T]$.

6.2 Let $s_1(t) = 1$ and $s_2(t) = t$ be defined on the interval $[-1, 1]$.
(a) Are $s_1(t)$ and $s_2(t)$ orthogonal in the given interval ?
(b) Determine the constants α and β such that $s_3(t) = 1 + \alpha t + \beta t^2$ is orthogonal to both $s_1(t)$, and $s_2((t)$ in the given interval.

6.3 (a) Find a set of orthonormal basis functions for the set of signals shown in Figure $P6.3$.
(b) Find the vector corresponding to each signal for the orthonormal basis set found in (a) and sketch the signal constellation.

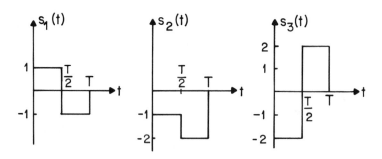

Figure P6.3 Set of signals.

6.4 Find the eigenvalues, the eigenvectors, and the Jordan form by the similarity transformation for the matrix $\mathbf{A}$.

(a) $\mathbf{A} = \begin{bmatrix} 2 & -2 & 3 \\ 1 & 1 & 1 \\ 1 & 3 & -1 \end{bmatrix}$ (b) $\mathbf{A} = \begin{bmatrix} 4 & -2 & 0 \\ 1 & 2 & 0 \\ 0 & 0 & 6 \end{bmatrix}$ (c) $\mathbf{A} = \begin{bmatrix} 4 & 2 & 1 \\ 0 & 6 & 1 \\ 0 & -4 & 2 \end{bmatrix}$

6.5 Find the Green's function for the differential system

$$\frac{d\phi}{dt} + \phi(t) = u(t) \quad ,0 \leq t \leq 1$$

$$\phi(0) + \phi(1) = \alpha$$

6.6 Consider the integral equation

$$\lambda\phi(t) = \int_{-\infty}^{\infty} K(t,u)\phi(u)du \qquad \text{for all } t$$

where

$$K(t,u) = \frac{1}{\sqrt{1-s^2}} e^{\frac{t^2+u^2}{2}} e^{-\frac{t^2+u^2-2sut}{1-s^2}}$$

and s, $0 < s < 1$, fixed.
Show that

$$\phi(t) = e^{-\frac{t^2}{2}}$$

is an eigenfunction corresponding to the eigenvalue $\lambda = \sqrt{\pi}$.

6.7 Determine the integral equation corresponding to the following second-order linear differential equation

$$\frac{d^2\phi(t)}{dt^2} + \lambda\phi(t) = 0$$

where λ is constant, $[d\phi(t)/dt]|_{t=0} = 0$ and $\phi(t)|_{t=1} = 0$

6.8 Consider the integral equation:

$$\lambda\phi(t) = \int_0^{\frac{\pi}{2}} K(u,t)\phi(u)du \ , \ 0 \le t \le \frac{\pi}{2}$$

where

$$K(u,t) = \begin{cases} u & ,u < t \\ t & ,u > t \end{cases}$$

Find all eigenvalues and eigenfunctions in the interval $[0, \pi/2]$.

6.9 Consider the integral equation

$$\lambda\phi(t) = \int_0^T K(u,t)\phi(u)du \ , \ 0 \le t \le T$$

where

$$K(u,t) = \begin{cases} T - t & ,u < t \\ T - u & ,u > t \end{cases}$$

Determine the eigenvalues and eigenfunctions in the interval $[0, T]$.

6.10 Consider the integral equation with the following coresponding linear differential equation:

$$\frac{d^2\phi(t)}{dt^2} + \lambda\phi(t) = 0 \qquad \begin{array}{c} 0 \le t \le 1 \\ , \ \alpha\phi(1) + \phi'(1) = 0 \\ \phi(0) = 0 \end{array}$$

where α is a positive constant. Determine all eigenvalues and eigenfunctions.

6.11 Determine all eigenvalues and eigenfunctions for the integral equation with the corresponding linear differential equation:

$$\frac{d^2\phi(t)}{dt^2} + \lambda\phi(t) = 0 \qquad \begin{array}{c} 0 \le t \le T \\ , \ \phi'(0) = \phi'(T) = 0 \end{array}$$

REFERENCES

1. Atkinson, F.V., *Discrete and Continuous Boundary Problems,* Academic Press, New York, 1964.

2. Bergman, S., and M. Schiffer, *Kernel Functions and Elliptic Differential Equations in Mathematical Physics,* Academic Press, New York, 1953.

3. Brogan, W.L., *Modern Control Theory,* Quantum, New York, 1974.

4. Churchill, R.V., and J.W. Brown, *Fourier Series and Boundary Value Problems,* McGraw-Hill, New York, 1978.

5. Cooper, G.R., and C.D. McGillem, *Modern Communications and Spread Spectrum,* McGraw-Hill, New York, 1986.

6. Davenport, W.B., Jr., and W.L. Root, *An Introduction to the Theory of Random Signals and Noise,* McGraw-Hill, New York, 1958.

7. Dorny, C.N., *A Vector Space Approach to Models and Optimization,* Krieger, New York, 1980.

8. Fox, L., *Numerical Solution of Ordinary and Partial Differential Equations,* Addison-Wesley, Reading, MA, 1962.

9. Golub, G.H., and C.F. Van Loan, *Matrix Computations,* The John Hopkins, Baltimore, MD, 1983.

10. Grimmett, G.R., and D.R. Stirzaker, *Probability and Random Processes,* Clarendon, Oxford, 1982.

11. Haykin, S., *Digital Communications,* John Wiley and Sons, New York, 1988.

12. Helstrom, C.W., *Statistical Theory of Signal Detection,* Pergamon, New York, 1960.

304

13. Hille, E., *Lectures on Ordinary Differential Equations*, Addison-Wesley, Reading, MA, 1969.

14. Hurewicz, W., *Lectures on Ordinary Differential Equations*, M.I.T. Press, Cambridge, MA, 1958.

15. Kreyszig, E., *Advanced Engineering Mathematics*, John Wiley and Sons, New York, 1972.

16. Lee, Y.W., *Statistical Theory of Communication*, John Wiley and Sons, New York, 1960.

17. Lipschutz, S., *Schaum's Outline Series: Linear Algebra*, McGraw-Hill, New York, 1968.

18. Mikhlin, S.G., and K.L. Smolitskiy, *Approximate Methods for Solution of Differential and Integral Equations*, American Elsevier, New York, 1967.

19. Noble, B., and J.W. Daniel, *Applied Linear Algebra*, Prentice-Hall, Englewood Cliffs, NJ, 1977.

20. Papoulis, A., *Probability, Random Variables, and Stochastic Processes*, McGraw-Hill, New York, 1984.

21. Proakis, J.G., *Digital Communications*, McGraw-Hill, New York, 1989.

22. Schwartz, M., *Information Transmission, Modulation, and Noise*, McGraw-Hill, New York, 1980.

23. Sneddon, I.N., *Elements of Partial Differential Equations*, McGraw-Hill, New York, 1957.

24. Spiegel, M.R., *Schaum's Outline Series: Fourier Analysis* , McGraw-Hill, New York, 1974.

25. Spiegel, M.R., *Schaum's Outline Series: Laplace Transforms*, McGraw-Hill, New York, 1965.

26. Stakgold, I., *Green's Functions and Boundary Value Problems*, John Wiley and Sons, New York, 1979.

27. Strang, G., *Introduction to Applied Mathematics*, Wellesley-Cambridge, Wellesley, MA, 1986.

28. Van Trees, H.L., *Detection, Estimation, and Modulation Theory, Part I*, John Wiley and Sons, New York, 1968.

29. Wayland, H., *Differential Equations Applied in Science and Engineering*, Van Nostrand, New York, 1957.

30. Wozencraft, J.M., and I.M. Jacobs, *Principles of Communication Engineering*, John Wiley and Sons, New York, 1965.

Chapter 7

The General Gaussian Problem

7.1 INTRODUCTION

In Chapters 1 and 2, we briefly mentioned Gaussian random variables and the Gaussian process. Due to the wide use of the Gaussian process, we now discuss it in some detail and state some of its important properties. In Section 7.3, we formulate the general Gaussian problem. In Section 7.4, we cover the general Gaussian problem with equal covariance matrix under either hypothesis H_1 or H_0. For nondiagonal covariance matrices we use an orthogonal transformation into a new coordinate system, so that the matrix is diagonalized. In Section 7.5, we also solve the general Gaussian binary hypothesis problems, but with mean vectors equal under both hypotheses. In Section 7.6, we consider symmetric hypotheses and obtain the likelihood ratio test (LRT).

7.2 GAUSSIAN PROCESS

In Chapter 2, we defined that a random process $Y(t)$ is Gaussian if the random variables $Y(t_1), Y(t_2), \cdots, Y(t_K)$ are jointly Gaussian for any choice of K, and distinct time instants $t_1, t_2, \cdots, t_K$. A linear combination of Gaussian random variables given by

$$Y = \sum_{k=1}^{K} \alpha_k Y_k$$

$$= \boldsymbol{\alpha}^T \mathbf{Y} \tag{7.1}$$

where

$$\mathbf{Y} = \begin{bmatrix} Y_1 \\ Y_2 \\ \vdots \\ Y_K \end{bmatrix} \quad , \quad \boldsymbol{\alpha} = \begin{bmatrix} \alpha_1 \\ \alpha_2 \\ \vdots \\ \alpha_K \end{bmatrix} \tag{7.2}$$

and all values of α_k, $k = 1, 2, \cdots, K$, is a Gaussian random variable. The K-fold joint probability density function with mean vector:

$$\mathbf{m} = E \begin{bmatrix} Y_1 \\ Y_2 \\ \vdots \\ Y_K \end{bmatrix} = \begin{bmatrix} m_1 \\ m_2 \\ \vdots \\ m_K \end{bmatrix} \tag{7.3}$$

such that $E[Y_k] = m_k$, $k = 1, 2, \cdots, K$, and covariance matrix:

$$\mathbf{C} = E[(\mathbf{Y} - \mathbf{m})^T (\mathbf{Y} - \mathbf{m})] = \begin{bmatrix} c_{11} & c_{12} & \cdots & c_{1K} \\ c_{21} & c_{22} & \cdots & c_{2K} \\ \vdots & \vdots & \vdots & \vdots \\ c_{K1} & c_{K2} & \cdots & c_{KK} \end{bmatrix} \tag{7.4}$$

such that

$$c_{jk} = E[(Y_j - m_j)(Y_k - m_k)] \quad j, k = 1, 2, \cdots, K \tag{7.5}$$

is given by

$$f_{\mathbf{Y}}(\mathbf{y}) = \frac{1}{(2\pi)^{\frac{K}{2}} \sqrt{|\mathbf{C}|}} e^{-\frac{1}{2}\{(\mathbf{y}-\mathbf{m})^T \mathbf{C}^{-1}(\mathbf{y}-\mathbf{m})\}} \tag{7.6}$$

where $|\mathbf{C}|$ is the determinant of the covariance matrix, and the superscript T denotes transpose. $\mathbf{C}^{-1}$ denotes inverse of the covariance matrix. If the set of random variables $Y(t_1), Y(t_2), \cdots, Y(t_K)$ obtained by sampling the Gaussian process $Y(t)$ at time $t_1, t_2, \cdots, t_K$ are uncorrelated; that is,

$$E\{[Y(t_j) - m_Y(t_j)][Y(t_k) - m_Y(t_k)]\} = 0 \quad , j \neq k \quad (7.7)$$

then these random variables are statistically independent. Note the subscript Y of m denotes the mean of Y $(m_Y(t) = E[Y(t)])$, and the notations $Y(t_j)$ and Y_j are equivalent. Since the random variables are statistically independent, the joint density function given in (7.6) may be written as the product of the individual density functions. Moreover, if the Gaussian process $Y(t)$ is stationary in the wide-sense, then it is stationary in the strict sense.

It can also be shown that the Gaussian process $Y(t)$ and its derivative $dY(t)/dt$ are jointly Gaussian. If $Y(t)$ is the output to a linear time-invariant system with impulse response $h(t)$, such that

$$Y(t) = \int_{-\infty}^{\infty} X(t - \tau)h(\tau)d\tau \quad (7.8)$$

where $X(t)$ is the input Gaussian process, then $Y(t)$ is also Gaussian.

The characteristic function of the vector $\mathbf{Y}$ composed of the set of the K jointly Gaussian random variables $Y_1, Y_2, \cdots, Y_K$ is given by

$$\Phi_{\mathbf{Y}}(j\boldsymbol{\omega}) = E[e^{j\boldsymbol{\omega}^T \mathbf{Y}}] \quad (7.9a)$$

$$= E[e^{j \sum_{k=1}^{K} y_k \omega_k}] \quad (7.9b)$$

$$= e^{j\boldsymbol{\omega}^T \mathbf{m} - \frac{1}{2} \boldsymbol{\omega}^T \mathbf{C} \boldsymbol{\omega}} \quad (7.9c)$$

$$= e^{j \sum_{k=1}^{K} \omega_k m_k - \frac{1}{2} \sum_{j=1}^{K} \sum_{k=1}^{K} \omega_j c_{jk} \omega_k} \quad (7.9d)$$

where

$$\boldsymbol{\omega} = \begin{bmatrix} \omega_1 \\ \omega_2 \\ \vdots \\ \omega_K \end{bmatrix} \quad (7.9e)$$

7.3 BINARY DETECTION

In this section, we formulate the general Gaussian problem for binary hypothesis testing. Consider the hypotheses:

$$H_1 : \quad \mathbf{Y} = \mathbf{X} + \mathbf{N}$$
$$H_0 : \quad \mathbf{Y} = \quad \mathbf{N} \tag{7.10}$$

where the vector observation $\mathbf{Y}$, the signal vector $\mathbf{X}$, and the noise vector $\mathbf{N}$ are given by

$$\mathbf{Y} = \begin{bmatrix} Y_1 \\ Y_2 \\ \vdots \\ Y_K \end{bmatrix} \quad , \quad \mathbf{X} = \begin{bmatrix} X_1 \\ X_2 \\ \vdots \\ X_K \end{bmatrix} \quad , \quad \mathbf{N} = \begin{bmatrix} N_1 \\ N_2 \\ \vdots \\ N_K \end{bmatrix} \tag{7.11}$$

the noise components are Gaussian random variables. By definition, a hypothesis testing problem is called a general Gaussian problem if the conditional density function $f_{\mathbf{Y}|H_j}(\mathbf{y}|H_j)$ for all j is a Gaussian density function. Similarly, an estimation problem is called a general Gaussian problem, if the conditional density function $f_{\mathbf{Y}|\Theta}(\mathbf{y}|\,\boldsymbol{\theta})$ has a Gaussian density for all $\boldsymbol{\theta}$; $\boldsymbol{\theta}$ is the parameter to be estimated.

Consider the binary hypothesis testing problem given in (7.10). Let the mean vectors $\mathbf{m}_1$ and $\mathbf{m}_0$ under hypotheses H_1 and H_0, respectively, be

$$\mathbf{m}_1 = E[\mathbf{Y}|H_1] \tag{712a}$$

and

$$\mathbf{m}_0 = E[\mathbf{Y}|H_0] \tag{7.12b}$$

The covariance matrices under each hypothesis are given by

$$\mathbf{C}_1 = E[(\mathbf{Y} - \mathbf{m}_1)(\mathbf{Y} - \mathbf{m}_1)^T |H_1] \tag{7.13a}$$

and

$$\mathbf{C}_0 = E[(\mathbf{Y} - \mathbf{m}_0)(\mathbf{Y} - \mathbf{m}_0)^T |H_0] \tag{7.13b}$$

In Chapter 3, we have seen that applying Bayes criterion to the binary hypothesis problem resulted in the likelihood ratio test; that is,

$$\Lambda(\mathbf{Y}) = \frac{f_{\mathbf{Y}|H_1}(\mathbf{y}|H_1)}{f_{\mathbf{Y}|H_0}(\mathbf{y}|H_0)} \overset{H_1}{\underset{H_0}{\gtrless}} \eta \qquad (7.14)$$

where

$$f_{\mathbf{Y}|H_j}(\mathbf{y}|H_j) = \frac{1}{(2\pi)^{\frac{K}{2}}|\mathbf{C}_j|^{\frac{1}{2}}} e^{-\frac{1}{2}\{(\mathbf{y}-\mathbf{m}_j)^T \mathbf{C}^{-1}(\mathbf{y}-\mathbf{m}_j)\}} \quad , \quad j = 0, 1 \qquad (7.15)$$

Substituting Equation (7.15) into (7.14) yields

$$\Lambda(\mathbf{y}) = \frac{|\mathbf{C}_0|^{\frac{1}{2}} e^{-\frac{1}{2}\{(\mathbf{y}-\mathbf{m}_1)^T \mathbf{C}_1^{-1}(\mathbf{y}-\mathbf{m}_1)\}}}{|\mathbf{C}_1|^{\frac{1}{2}} e^{-\frac{1}{2}\{(\mathbf{y}-\mathbf{m}_0)^T \mathbf{C}_0^{-1}(\mathbf{y}-\mathbf{m}_0)\}}} \overset{H_1}{\underset{H_0}{\gtrless}} \eta \qquad (7.16)$$

Taking the logarithm on both sides of the above equation, an equivalent test is

$$\frac{1}{2}(\mathbf{y} - \mathbf{m}_0)^T \mathbf{C}_0^{-1}(\mathbf{y} - \mathbf{m}_0) - \frac{1}{2}(\mathbf{y} - \mathbf{m}_1)^T \mathbf{C}_1^{-1}(\mathbf{y} - \mathbf{m}_1) \overset{H_1}{\underset{H_0}{\gtrless}} \gamma \qquad (7.17a)$$

where

$$\gamma = \ln\eta + \frac{1}{2}(\ln|\mathbf{C}_1| - \ln|\mathbf{C}_0|) \qquad (7.17b)$$

Thus, the likelihood ratio test reduces to the difference of two quadratic forms. The evaluation of such difference depends on several constraints on the mean vectors and the covariance matrices under each hypothesis.

7.4 SAME COVARIANCE

In this case, we assume the covariance matrices C_1 and C_0 under both hypotheses H_1 and H_0 are the same; that is,

$$C_1 = C_0 = C \qquad (7.18)$$

Substituting Equation (7.18) into (7.17a), the LRT can be written as

$$\frac{1}{2}(y - m_0)^T C^{-1}(y - m_0) - \frac{1}{2}(y - m_1)^T C^{-1}(y - m_1) \underset{H_0}{\overset{H_1}{\underset{<}{\gtrless}}} \gamma \qquad (7.19)$$

Expanding the above expression, we obtain

$$-\frac{1}{2}m_0^T C^{-1}y - \frac{1}{2}y^T C^{-1}m_0 + \frac{1}{2}m_0^T C^{-1}m_0 + \frac{1}{2}m_1^T C^{-1}y$$

$$+\frac{1}{2}y^T C^{-1}m_1 - \frac{1}{2}m_1^T C^{-1}m_1 \underset{H_0}{\overset{H_1}{\underset{<}{\gtrless}}} \gamma \qquad (7.20)$$

Using the fact that the inverse covariance matrix C^{-1} is symmetric, that is

$$C^{-1} = (C^{-1})^T \qquad (7.21)$$

and the fact that, the transpose of a scalar is equal to itself, that is

$$\begin{aligned} y^T C^{-1}m_j &= (y^T C^{-1}m_j)^T \\ &= m_j^T (C^{-1})^T y \\ &= m_j^T C^{-1}y \qquad , j = 0, 1 \end{aligned} \qquad (7.22)$$

Equation (7.20) reduces to the following test

$$m_1^T C^{-1}y - m_0^T C^{-1}y + \frac{1}{2}m_0^T C^{-1}m_0 - \frac{1}{2}m_1^T C^{-1}m_1 \underset{H_0}{\overset{H_1}{\underset{<}{\gtrless}}} \gamma \qquad (7.23)$$

Rearranging terms, an equivalent test is

$$(\mathbf{m}_1^T - \mathbf{m}_0^T)\mathbf{C}^{-1}\mathbf{y} \underset{H_0}{\overset{H_1}{\underset{<}{\overset{>}{\gtrless}}}} \gamma_1 \qquad (7.24a)$$

where

$$\gamma_1 = \gamma + \frac{1}{2}(\mathbf{m}_1^T\mathbf{C}^{-1}\mathbf{m}_1 - \mathbf{m}_0^T\mathbf{C}^{-1}\mathbf{m}_0) \qquad (7.24b)$$

Note that all terms in $\mathbf{y}$ are on one side and the others are on the other side. Hence, the sufficient statistic $T(\mathbf{Y})$ is

$$T(\mathbf{Y}) = (\mathbf{m}_1^T - \mathbf{m}_0^T)\mathbf{C}^{-1}\mathbf{Y} \qquad (7.25)$$

Let the difference mean vector be

$$\Delta\mathbf{m} = \mathbf{m}_1 - \mathbf{m}_0 \qquad (7.26)$$

Substituting Equation (7.26) into (7.24a), the LRT becomes

$$T(\mathbf{y}) = \Delta\mathbf{m}^T\mathbf{C}^{-1}\mathbf{y} \underset{H_0}{\overset{H_1}{\underset{<}{\overset{>}{\gtrless}}}} \gamma_1$$

$$= \mathbf{y}^T\mathbf{C}^{-1}\Delta\mathbf{m} \underset{H_0}{\overset{H_1}{\underset{<}{\overset{>}{\gtrless}}}} \gamma_1 \qquad (7.27)$$

We observe that $T(\mathbf{Y})$ is just a linear combination of jointly Gaussian random variables, and hence, by definition, it is a Gaussian random variable. Therefore, we only need to find the mean and variance of the sufficient statistic under each hypothesis and perform the test in (7.27) against the threshold γ_1 to determine the performance of this test. The mean and variance of $T(\mathbf{Y})$ are given by

$$\begin{aligned} E[T(\mathbf{Y})|H_j] &= E[\Delta\mathbf{m}^T\mathbf{C}^{-1}\mathbf{Y}|H_j] \\ &= \Delta\mathbf{m}^T\mathbf{C}^{-1}E[\mathbf{Y}|H_j] \\ &= \Delta\mathbf{m}^T\mathbf{C}^{-1}\mathbf{m}_j \qquad , j = 0,1 \qquad (7.28) \end{aligned}$$

314

and

$$
\begin{aligned}
\mathrm{var}[T(\mathbf{Y})|H_j] &= E\{(T(\mathbf{Y}) - E[T(\mathbf{Y})|H_j])^2|H_j\} \\
&= E\{[\Delta\mathbf{m}^T\mathbf{C}^{-1}\mathbf{Y} - \Delta\mathbf{m}^T\mathbf{C}^{-1}\mathbf{m}_j]^2|H_j\} \\
&= E\{[\Delta\mathbf{m}^T\mathbf{C}^{-1}\mathbf{Y} - \Delta\mathbf{m}^T\mathbf{C}^{-1}\mathbf{m}_j] \\
&\quad \cdot[\mathbf{Y}^T\mathbf{C}^{-1}\Delta\mathbf{m} - \mathbf{m}_j^T\mathbf{C}^{-1}\Delta\mathbf{m}]|H_j\} \\
&= E\{[\Delta\mathbf{m}^T\mathbf{C}^{-1}(\mathbf{Y} - \mathbf{m}_j)][(\mathbf{Y}^T - \mathbf{m}_j^T)\mathbf{C}^{-1}\Delta\mathbf{m}]|H_j\} \\
&= \Delta\mathbf{m}^T\mathbf{C}^{-1}E[(\mathbf{Y} - \mathbf{m}_j)(\mathbf{Y}^T - \mathbf{m}_j^T)|H_j]\mathbf{C}^{-1}\Delta\mathbf{m}
\end{aligned}
$$

$$(7.29)$$

but, from (7.13) and (7.18), we have,

$$E[(\mathbf{Y} - \mathbf{m}_j)(\mathbf{Y}^T - \mathbf{m}_j)|H_j] = \mathbf{C}_j = \mathbf{C} \quad ,j = 0, 1 \qquad (7.30)$$

Hence,

$$
\begin{aligned}
\mathrm{var}[T(\mathbf{Y})|H_j] &= \Delta\mathbf{m}^T\mathbf{C}^{-1}\mathbf{C}\mathbf{C}^{-1}\Delta\mathbf{m} \\
&= \Delta\mathbf{m}^T\mathbf{C}^{-1}\Delta\mathbf{m}
\end{aligned}
\qquad (7.31)
$$

since $\mathbf{C}\mathbf{C}^{-1} = \mathbf{I}$ is the identity matrix. Note that the variance is independent of any hypothesis. The performance of this test is affected by the choice $\mathbf{C}$ which we will study next.

7.4.1 Diagonal Covariance Matrix

Let the covariance matrix $\mathbf{C}$ be diagonal and given by

$$
\mathbf{C} =
\begin{bmatrix}
\sigma_1^2 & 0 & 0 & \cdots & 0 \\
0 & \sigma_2^2 & 0 & \cdots & 0 \\
\vdots & \vdots & \vdots & \vdots & \vdots \\
0 & 0 & 0 & \cdots & \sigma_K^2
\end{bmatrix}
\qquad (7.32)
$$

This means that the components Y_k, $k = 1, 2, \cdots, K$, are statistically independent. Two possible cases may arise. The variances of the components are either (i) equal; in this case $\sigma_1^2 = \sigma_2^2 = \cdots = \sigma_K^2 = \sigma^2$, or

(ii) unequal; in this case $\sigma_1^2 \neq \sigma_2^2 \neq \cdots \neq \sigma_K^2$.

Equal Variance

In this case, the covariance matrix is given by

$$\mathbf{C} = \begin{bmatrix} \sigma^2 & 0 & \cdots & 0 \\ 0 & \sigma^2 & \cdots & 0 \\ \vdots & \vdots & \vdots & \vdots \\ 0 & 0 & \cdots & \sigma^2 \end{bmatrix}$$

$$= \sigma^2 \mathbf{I} \tag{7.33}$$

that is,

$$E[(Y_j - m_j)(Y_k - m_k)] = \begin{cases} \sigma^2 & ,j = k \\ 0 & ,j \neq k \end{cases} \tag{7.34}$$

The inverse covariance matrix is

$$\mathbf{C}^{-1} = \frac{1}{\sigma^2}\mathbf{I} \tag{7.35}$$

Substituting Equation (7.35) into (7.25), we obtain the sufficient statistic to be

$$T(\mathbf{Y}) = \frac{1}{\sigma^2}\Delta\mathbf{m}^T\mathbf{Y} \tag{7.36}$$

which is simply the dot product between the mean difference vector $\Delta\mathbf{m}$ and the observation vector $\mathbf{Y}$.

Unequal Variance

In this case, the covariance matrix is as given in Equation (7.32) where $\sigma_1 \neq \sigma_2 \neq \cdots \neq \sigma_K$. The inverse covariance matrix is given by

$$\mathbf{C}^{-1} = \begin{bmatrix} \frac{1}{\sigma_1^2} & 0 & \cdots & 0 \\ 0 & \frac{1}{\sigma_2^2} & \cdots & 0 \\ \vdots & \vdots & \vdots & \vdots \\ 0 & 0 & \cdots & \frac{1}{\sigma_K^2} \end{bmatrix} \tag{7.37}$$

Consequently, after substitution into (7.25), the sufficient statistic becomes

$$T(\mathbf{Y}) = \Delta \mathbf{m}^T \mathbf{C}^{-1} \mathbf{Y}$$

$$= \sum_{k=1}^{K} \frac{\Delta m_k Y_k}{\sigma_k^2} \qquad (7.38)$$

7.4.2 Nondiagonal Covariance Matrix

In general, the covariance matrix $\mathbf{C}$ will not be a diagonal matrix and thus, the components of the received random vector $\mathbf{Y}$ are not statistically independent. In order to make the components independent, we need to find a new coordinate system in which the transformed components are independent. That is, the covariance matrix in the new coordinate system must be diagonal.

The concept of diagonalizing a matrix, which can be done by similarity transformation, was presented in the previous chapter and can now be used. Let the new coordinate system have coordinate axes denoted by the set of orthonormal vectors $\{\boldsymbol{\Phi}_k\}$, $k = 1, 2, \cdots, K$. Let $\mathbf{Y}$ be the original observation vector, and $\mathbf{Y}'$ its transformed vector in the new coordinate system. The vector $\mathbf{Y}'$ has also K components where the kth component, denoted Y_k', is just the projection of the observation vector $\mathbf{Y}$ onto the coordinate $\boldsymbol{\Phi}_k$ of the new system. This geometric interpretation mathematically represents the dot product between the vector $\mathbf{Y}$, and the vector $\boldsymbol{\Phi}_k$. That is,

$$Y_k' = \boldsymbol{\Phi}_k^T \mathbf{Y}$$

$$= \mathbf{Y}^T \boldsymbol{\Phi}_k \qquad (7.39)$$

Assuming we have a 3-dimensional vector, the transformation of the vector $\mathbf{Y}$ into $\mathbf{Y}'$ in the new coordinate system may be as shown in Figure 7.1.

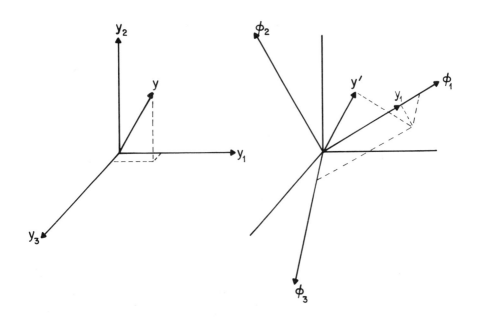

Figure 7.1 New coordinate system representing transformation of vector $\mathbf{Y}$ into $\mathbf{Y}'$.

The mean of $\mathbf{Y}'$ in the new coordinate system is

$$
\begin{aligned}
\mathbf{m}' &= E[\mathbf{Y}'] \\
&= E[\mathbf{\Phi}^T\mathbf{Y}] = \mathbf{\Phi}^T E[\mathbf{Y}] \\
&= \mathbf{\Phi}^T\mathbf{m} = \mathbf{m}^T\mathbf{\Phi}
\end{aligned}
\tag{7.40}
$$

The covariance matrix of $\mathbf{Y}'$ is diagonal since the components in this new coordinate system are now statistically independent; that is,

$$
E[(Y'_j - m_j)(Y'_k - m_k)] = \lambda_k \delta_{jk}
\tag{7.41}
$$

where $m_k = E[Y_k]$, and δ_{jk} is as given in (6.30). Using the fact that $Y'_k = \mathbf{\Phi}_k^T\mathbf{Y} = \mathbf{Y}^T\mathbf{\Phi}_k$ and $m_k = \mathbf{\Phi}_k^T\mathbf{m} = \mathbf{m}^T\mathbf{\Phi}_k$, Equation (7.41) can be rewritten as

$$E[\mathbf{\Phi}_j^T(\mathbf{Y} - \mathbf{m})(\mathbf{Y}^T - \mathbf{m}^T)\mathbf{\Phi}_k] = \lambda_k \delta_{jk} \tag{7.42}$$

or

$$\mathbf{\Phi}_j^T \mathbf{C} \mathbf{\Phi}_k = \lambda_k \delta_{jk} \tag{7.43}$$

Hence, Equation (7.43) is only true when

$$\mathbf{C}\mathbf{\Phi}_k = \lambda_k \mathbf{\Phi}_k \tag{7.44}$$

since

$$\mathbf{\Phi}_j^T \mathbf{\Phi}_k = \delta_{jk} \tag{7.45}$$

Consequently, the solution reduces to solving the eigenproblem

$$\mathbf{C}\mathbf{\Phi} = \lambda \mathbf{\Phi} \tag{7.46}$$

or

$$(\mathbf{C} - \mathbf{I}\lambda)\mathbf{\Phi} = 0 \tag{7.47}$$

The solution to the homogeneous Equations of (7.47) was studied in detail in Chapter 6. That is, we first obtain the nonzero eigenvalues from the equation $|\mathbf{C} - \mathbf{I}\lambda| = 0$. Then, using (7.46), we solve the set of K equations in K unknowns to obtain the eigenvectors $\mathbf{\Phi}_k$, $k = 1, 2, \cdots, K$, corresponding to the eigenvalues λ_k, $k = 1, 2, \cdots, K$. The eigenvectors are linearly independent. We form the modal matrix $\mathbf{M}$ given by

$$\mathbf{M} = [\mathbf{\Phi}_1 | \mathbf{\Phi}_2 | \cdots | \mathbf{\Phi}_K] \tag{7.48}$$

and then use the similarity transformation to diagonalize the covariance matrix $\mathbf{C}$. We obtain

$$\boldsymbol{\lambda} = \mathbf{M}^{-1}\mathbf{C}\mathbf{M} = \begin{bmatrix} \lambda_1 & 0 & \cdots & 0 \\ 0 & \lambda_2 & \cdots & 0 \\ \vdots & \vdots & \vdots & \vdots \\ 0 & 0 & \cdots & \lambda_K \end{bmatrix} \tag{7.49}$$

It should also be noted that since the covariance matrix $\mathbf{C}$ is real and symmetric, the inverse of the modal matrix $\mathbf{M}$ equals its transpose

$(\mathbf{M}^{-1} = \mathbf{M}^{T})$. Thus, the orthogonal transformation can be used to diagonalize $\mathbf{C}$. The vector $\mathbf{Y}'$ in the new coordinate system is given by

$$\mathbf{Y}' = \mathbf{M}^{T}\mathbf{Y} \tag{7.50}$$

or

$$\mathbf{Y} = \mathbf{M}\mathbf{Y}' \tag{7.51}$$

The above transformation corresponds to a rotation and hence, the norm of $\mathbf{Y}'$ in the new system is equal to the norm of $\mathbf{Y}$ in the original system.

Now, we can apply the LRT to the binary hypothesis problem in the new coordinate system. The sufficient statistic is still of the form given in (7.25). Let $\mathbf{m}_1$ and $\mathbf{m}_0$ be the mean vectors in the original coordinate system under H_1 and H_0, respectively, such that

$$\mathbf{m}_1 = \begin{bmatrix} m_{11} \\ m_{12} \\ \vdots \\ m_{1K} \end{bmatrix} \quad , \quad \mathbf{m}_0 = \begin{bmatrix} m_{01} \\ m_{02} \\ \vdots \\ m_{0K} \end{bmatrix} \tag{7.52}$$

The transformed mean vectors are as given by (7.40). Hence, the transformed mean difference vector $\Delta\mathbf{m}' = \mathbf{m}_1' - \mathbf{m}_0'$ is

$$\Delta\mathbf{m}' = \begin{bmatrix} \boldsymbol{\Phi}_1^{T}\Delta\mathbf{m} \\ \boldsymbol{\Phi}_2^{T}\Delta\mathbf{m} \\ \vdots \\ \boldsymbol{\Phi}_K^{T}\Delta\mathbf{m} \end{bmatrix} = \begin{bmatrix} \boldsymbol{\Phi}_1^{T} \\ \boldsymbol{\Phi}_2^{T} \\ \vdots \\ \boldsymbol{\Phi}_K^{T} \end{bmatrix} \Delta\mathbf{m}$$

$$= \mathbf{W}\Delta\mathbf{m} \tag{7.53}$$

where $\mathbf{W}$ is a $K \times K$ matrix with the vectors $\boldsymbol{\Phi}_k^{T}$, $k = 1, 2, \cdots, K$. That is,

$$\mathbf{W} = \mathbf{M}^{T} = \mathbf{M}^{-1} \tag{7.54}$$

Hence,

$$\begin{aligned} \Delta\mathbf{m} &= \mathbf{W}^{-1}\Delta\mathbf{m}' \\ &= \mathbf{M}\Delta\mathbf{m}' \end{aligned} \tag{7.55}$$

Substituting Equations (7.51) and (7.55) into (7.25), the sufficient statistic in the new coordinate system becomes

$$
\begin{aligned}
T(\mathbf{Y}) &= \Delta \mathbf{m}^T \mathbf{C}^{-1} \mathbf{Y} \\
&= (\mathbf{M} \Delta \mathbf{m}')^T \mathbf{C}^{-1} (\mathbf{M} \mathbf{Y}') \\
&= (\Delta \mathbf{m}')^T \mathbf{M}^T \mathbf{C}^{-1} \mathbf{M} \mathbf{Y}' \\
&= (\Delta \mathbf{m}')^T \mathbf{M}^{-1} \mathbf{C}^{-1} \mathbf{M} \mathbf{Y}'
\end{aligned}
\tag{7.56}
$$

Using Equation (7.49) into (7.56) the sufficient statistic reduces to

$$
T(\mathbf{Y}') = (\Delta \mathbf{m}')^T \boldsymbol{\lambda}^{-1} \mathbf{Y}'
$$

$$
= \sum_{k=1}^{K} \frac{\Delta m'_k Y'_k}{\lambda_k}
\tag{7.57}
$$

where

$$
\boldsymbol{\lambda}^{-1} =
\begin{bmatrix}
\frac{1}{\lambda_1} & 0 & \cdots & 0 \\
0 & \frac{1}{\lambda_2} & \cdots & 0 \\
\vdots & \vdots & \vdots & \vdots \\
0 & 0 & \cdots & \frac{1}{\lambda_K}
\end{bmatrix}
\tag{7.58}
$$

Example 7.1

Consider the binary hypothesis problem with specifications

$$
\mathbf{m}_0 = \mathbf{0} \quad , \quad
\mathbf{C} =
\begin{bmatrix}
1 & 0.5 & 0.25 \\
0.5 & 1 & 0.5 \\
0.25 & 0.5 & 1
\end{bmatrix}
\quad , \quad
\mathbf{m}_1 =
\begin{bmatrix}
m_{11} \\
m_{12} \\
m_{13}
\end{bmatrix}
$$

Obtain the sufficient statistic in the new coordinate system for which the components of the observation vector are independent.

Solution.

For the components of Y to be independent, the covariance matrix C in the new coordinate system must be diagonal. This can be achieved using the orthogonal transformation. First, we solve for the eigenvalues of C using

$$|C - I\lambda| = 0 \implies$$

$$\begin{vmatrix} 1 - \lambda & 0.5 & 0.25 \\ 0.5 & 1 - \lambda & 0.5 \\ 0.25 & 0.5 & 1 - \lambda \end{vmatrix} = -\lambda^3 + 3\lambda^2 - 2.4375\lambda + 0.5625$$

Therefore, $\lambda_1 = 0.4069$, $\lambda_2 = 0.75$ and $\lambda_3 = 1.8431$. To obtain the first eigenvector Φ_1, we solve

$$C\Phi_1 = \lambda_1 \Phi_1$$

$$\implies \begin{bmatrix} 1 & 0.5 & 0.25 \\ 0.5 & 1 & 0.5 \\ 0.25 & 0.5 & 1 \end{bmatrix} \begin{bmatrix} \phi_{11} \\ \phi_{12} \\ \phi_{13} \end{bmatrix} = 0.4069 \begin{bmatrix} \phi_{11} \\ \phi_{12} \\ \phi_{13} \end{bmatrix}$$

Solving for ϕ_{11}, ϕ_{12}, and ϕ_{13}, we obtain

$$\Phi_1 = \begin{bmatrix} 0.4544 \\ -0.7662 \\ 0.4544 \end{bmatrix}$$

Similarly, we solve for Φ_2 and Φ_3, using $C\Phi_2 = \lambda_2 \Phi_2$ and $C\Phi_3 = \lambda_3 \Phi_3$ to obtain

$$\Phi_2 = \begin{bmatrix} -0.7071 \\ 0.0000 \\ 0.7071 \end{bmatrix}, \quad \Phi_3 = \begin{bmatrix} 0.5418 \\ 0.6426 \\ 0.5418 \end{bmatrix}$$

Hence, the modal matrix $\mathbf{M}$ is

$$\mathbf{M} = [\boldsymbol{\Phi}_1|\boldsymbol{\Phi}_2|\boldsymbol{\Phi}_3]$$

$$= \begin{bmatrix} 0.4544 & -0.7071 & 0.5418 \\ -0.7662 & 0.0000 & 0.6426 \\ 0.4544 & 0.7071 & 0.5418 \end{bmatrix}$$

Since $\mathbf{Y} = \mathbf{M}\mathbf{Y}' \Longrightarrow \mathbf{Y}' = \mathbf{M}^{-1}\mathbf{Y} = \mathbf{M}^T\mathbf{Y}$, we have

$$\mathbf{Y}' = \begin{bmatrix} Y_1 \\ Y_2 \\ Y_3 \end{bmatrix} = \begin{bmatrix} 0.4544 & -0.7662 & 0.4544 \\ -0.7071 & 0.0000 & 0.7071 \\ 0.5418 & 0.6426 & 0.5418 \end{bmatrix} \begin{bmatrix} Y_{11} \\ Y_{12} \\ Y_{13} \end{bmatrix}$$

Also, since $\mathbf{m}_0 = \mathbf{0} \Longrightarrow \mathbf{m}_0' = \mathbf{M}^T\mathbf{m}_0 = \mathbf{0}$ and $\boldsymbol{\Delta}\mathbf{m}' = \mathbf{m}_1'$ where

$$\mathbf{m}_1' = \begin{bmatrix} m_1' \\ m_2' \\ m_3' \end{bmatrix} = \begin{bmatrix} 0.4544 & -0.7662 & 0.4544 \\ -0.7071 & 0.0000 & 0.7071 \\ 0.5418 & 0.6426 & 0.5418 \end{bmatrix} \begin{bmatrix} m_{11} \\ m_{12} \\ m_{13} \end{bmatrix}$$

Therefore, the sufficient statistic is

$$T(\mathbf{Y}') = \sum_{k=1}^{3} \frac{\Delta m_k' Y_k'}{\lambda_k} = \sum_{k=1}^{3} \frac{m_k' Y_k'}{\lambda_k}$$

Example 7.2

Consider the problem of Example 7.1, but $K = 2$, $\mathbf{m}_0 = \mathbf{0}$, and

$$\mathbf{C} = \begin{bmatrix} 1 & \rho \\ \rho & 1 \end{bmatrix}$$

Solution.

Following the same procedure as in Example 7.1, we solve for the eigenvalues using $|\mathbf{C} - \mathbf{I}\lambda| = 0$. That is,

$$\begin{vmatrix} 1 - \lambda & \rho \\ \rho & 1 - \lambda \end{vmatrix} = \lambda^2 - 2\lambda + 1 - \rho^2 = 0$$

Thus, $\lambda_1 = 1 + \rho$ and $\lambda_2 = 1 - \rho$. To obtain the eigenvector $\mathbf{\Phi}_1$, we have

$$\mathbf{C}\mathbf{\Phi}_1 = \lambda_1 \mathbf{\Phi}_1$$

or

$$\begin{bmatrix} 1 & \rho \\ \rho & 1 \end{bmatrix} \begin{bmatrix} \phi_{11} \\ \phi_{12} \end{bmatrix} = (1 + \rho) \begin{bmatrix} \phi_{11} \\ \phi_{12} \end{bmatrix}$$

Solving for ϕ_{11} and ϕ_{12}, such that $\mathbf{\Phi}_1^T \mathbf{\Phi}_1 = \phi_{11}^2 + \phi_{22}^2 = 1$, we obtain the normalized eigenvector

$$\mathbf{\Phi}_1 = \begin{bmatrix} \frac{1}{\sqrt{2}} \\ \frac{1}{\sqrt{2}} \end{bmatrix}$$

Similarly, we obtain

$$\mathbf{\Phi}_2 = \begin{bmatrix} \frac{1}{\sqrt{2}} \\ -\frac{1}{\sqrt{2}} \end{bmatrix}$$

The modal matrix is

$$\mathbf{M} = \begin{bmatrix} \mathbf{\Phi}_1 | \mathbf{\Phi}_2 \end{bmatrix}$$

$$= \begin{bmatrix} \frac{1}{\sqrt{2}} & \frac{1}{\sqrt{2}} \\ \frac{1}{\sqrt{2}} & -\frac{1}{\sqrt{2}} \end{bmatrix} = \frac{1}{\sqrt{2}} \begin{bmatrix} 1 & 1 \\ 1 & -1 \end{bmatrix}$$

Note that

$$\mathbf{M}^{-1}\mathbf{CM} = \frac{1}{2} \begin{bmatrix} 1 & 1 \\ 1 & -1 \end{bmatrix} \begin{bmatrix} 1 & \rho \\ \rho & 1 \end{bmatrix} \begin{bmatrix} 1 & 1 \\ 1 & -1 \end{bmatrix}$$

$$= \begin{bmatrix} 1+\rho & 0 \\ 0 & 1-\rho \end{bmatrix} = \begin{bmatrix} \lambda_1 & 0 \\ 0 & \lambda_2 \end{bmatrix} = \boldsymbol{\lambda}$$

The observation vector $\mathbf{Y}'$ in the new coordinate system is

$$\mathbf{Y}' = \mathbf{MY}$$

$$= \begin{bmatrix} \frac{1}{\sqrt{2}} & \frac{1}{\sqrt{2}} \\ \frac{1}{\sqrt{2}} & -\frac{1}{\sqrt{2}} \end{bmatrix} \begin{bmatrix} Y_1 \\ Y_2 \end{bmatrix}$$

$$\Longrightarrow Y_1' = \frac{Y_1 + Y_2}{\sqrt{2}} \quad \text{and} \quad Y_2' = \frac{Y_1 - Y_2}{\sqrt{2}}$$

Similarly, the mean vector $\mathbf{m}_1'$ is

$$\mathbf{m}_1' = \mathbf{M}^T\mathbf{m}_1$$

$$= \begin{bmatrix} \frac{1}{\sqrt{2}} & \frac{1}{\sqrt{2}} \\ \frac{1}{\sqrt{2}} & -\frac{1}{\sqrt{2}} \end{bmatrix} \begin{bmatrix} m_{11} \\ m_{12} \end{bmatrix}$$

$$\Longrightarrow m_{11}' = \frac{m_{11} + m_{12}}{\sqrt{2}} \quad \text{and} \quad m_{12}' = \frac{m_{11} - m_{12}}{\sqrt{2}}$$

The difference mean vector is $\Delta \mathbf{m}' = \mathbf{m}_1' - \mathbf{m}_0' = \mathbf{m}_1'$. Therefore, the sufficient statistic is given by

$$
\begin{aligned}
T(\mathbf{Y}') &= \sum_{k=1}^{2} \frac{\Delta m_k' Y_k'}{\lambda_k} = \frac{m_{11}' Y_1'}{1+\rho} + \frac{m_{12}' Y_2'}{1+\rho} \\
&= \frac{(m_{11} + m_{12})(Y_1 + Y_2)}{2(1+\rho)} + \frac{(m_{11} - m_{12})(Y_1 - Y_2)}{2(1-\rho)}
\end{aligned}
$$

7.5 SAME MEAN

In the previous section, the constraint was that the covariance matrix under both hypotheses was the same. Now, we consider the case with the constraint that the mean vectors under both hypotheses are equal. That is,

$$
\mathbf{m}_1 = \mathbf{m}_0 = \mathbf{m} \tag{7.59}
$$

Substituting (7.59) into the LRT given in (7.17), we obtain

$$
\frac{1}{2}(\mathbf{y} - \mathbf{m})^T (\mathbf{C}_0^{-1} - \mathbf{C}_1^{-1})(\mathbf{y} - \mathbf{m}) \underset{H_0}{\overset{H_1}{\underset{<}{>}}} \gamma \tag{7.60}
$$

Note that the mean vector $\mathbf{m}$ of the test in (7.60) does not affect the decision as to which hypothesis is true. Consequently, for simplicity and without loss of generality, let $\mathbf{m}_0 = \mathbf{0}$. The LRT reduces to

$$
T(\mathbf{y}) = \mathbf{y}^T (\mathbf{C}_0^{-1} - \mathbf{C}_1^{-1})\mathbf{y} \underset{H_0}{\overset{H_1}{\underset{<}{>}}} 2\gamma = \gamma_2 \tag{7.61}
$$

Furthermore, assume that this binary hypothesis problem can be characterized by

$$
\begin{aligned}
H_1 &: \quad Y_k = S_k + N_k \quad , k = 1, 2, \cdots, K \\
H_0 &: \quad Y_k = \qquad\quad N_k \quad , k = 1, 2, \cdots, K
\end{aligned} \tag{7.62}
$$

that is, we only have noise under hypothesis H_0, while under hypothesis H_1 we have signal plus noise. The signal and noise components are assumed statistically independent. In addition, the noise components are uncorrelated with equal variances σ_N^2, $k = 1, 2, \cdots, K$. Thus, the noise components under hypothesis H_0 are a multivariate Gaussian with covariance matrix

$$\mathbf{C}_N = \sigma_N^2 \mathbf{I} = \mathbf{C}_0 \qquad (7.63)$$

If the signal components are assumed independent of each other then, the covariance matrix $\mathbf{C}_S$ is diagonal. The signal components are also a multivariate Gaussian with covariance matrix $\mathbf{C}_S$. Since the signal and noise components are independent, the covariance matrix $\mathbf{C}_1$ under hypothesis H_1 is

$$
\begin{aligned}
\mathbf{C}_1 &= \mathbf{C}_S + \mathbf{C}_N \\
&= \mathbf{C}_S + \sigma_N^2 \mathbf{I}
\end{aligned}
\qquad (7.64)
$$

Substituting Equations (7.63) and (7.64) into (7.61), the LRT becomes

$$T(\mathbf{y}) = \mathbf{y}^T [\frac{1}{\sigma_N^2}\mathbf{I} - (\mathbf{C}_S + \sigma_N^2\mathbf{I})^{-1}]\mathbf{y} \underset{H_0}{\overset{H_1}{\gtrless}} \gamma_2 \qquad (7.65)$$

We note that, the LRT can be further reduced depending on the structure of the signal covariance matrix, which we consider next.

7.5.1 Uncorrelated Signal Components; Equal Variances

In this case, we assume that the signal components are uncorrelated, and identically distributed. Thus, the covariance matrix is diagonal with equal diagonal elements σ_S^2; that is,

$$\mathbf{C}_S = \sigma_S^2 \mathbf{I} \qquad (7.66)$$

Consequently, the LRT reduces to

$$T(\mathbf{y}) = \frac{\sigma_S^2}{\sigma_N^2(\sigma_S^2 + \sigma_N^2)}\mathbf{y}^T\mathbf{y} \underset{H_0}{\overset{H_1}{\underset{<}{>}}} \gamma_2$$

$$= \frac{\sigma_S^2}{\sigma_N^2(\sigma_S^2 + \sigma_N^2)}\sum_{k=1}^{K} y_k^2 \underset{H_0}{\overset{H_1}{\underset{<}{>}}} \gamma_2 \qquad (7.67)$$

where $\gamma_2 = 2\gamma$, and γ is given in (7.17b). Simplifying Equation (7.67) further, we obtain the equivalent test:

$$T(\mathbf{y}) = \sum_{k=1}^{K} y_k^2 \underset{H_0}{\overset{H_1}{\underset{<}{>}}} \gamma_3 \qquad (7.68a)$$

where

$$\gamma_3 = \frac{\sigma_N^2(\sigma_S^2 + \sigma_N^2)}{\sigma_S^2}\gamma_2 \qquad (7.68b)$$

Hence, the sufficient statistic is

$$T(\mathbf{Y}) = \sum_{k=1}^{K} Y_k^2 \qquad (7.69)$$

Since the Y_k's are independent, and identically distributed Gaussian random variables, $T(\mathbf{Y}) = Y_1^2 + Y_2^2 + \cdots + Y_K^2$ is a chi-square random variable with K degrees of freedom, as shown in Chapter 1. Consequently, we can carry the test further, and obtain an expression for P_D the probability of detection and P_F the probability of false alarm. Note that once we obtain P_D and P_F, we can plot the ROC.

Using the concepts of transformation of random variables developed in Chapter 1, it is easily shown that the density function of the random variable $Y = X^2$, where X is Gaussian with mean zero and variance σ^2, is

$$f_Y(y) = \begin{cases} \frac{1/2\sigma^2}{\Gamma(\frac{1}{2})}(\frac{1}{2\sigma^2}y)^{\frac{1}{2}}e^{-\frac{y}{2\sigma^2}} & ,y > 0 \\ \\ 0 & , \text{ otherwise} \end{cases} \tag{7.70}$$

where, from Equation (1.79) $r = 1/2$ and $\alpha = 1/2\sigma^2$. Hence, the mean and variance of Y are

$$E[Y] = \frac{r}{\alpha} = \sigma^2 \tag{7.71}$$

and

$$\text{var}(Y) = \frac{r}{\alpha^2} = 2\sigma^4 \tag{7.72}$$

Also, from Equation (1.80) the characteristic function of $Y = X^2$ is

$$\Phi_X(j\omega) = E[e^{j\omega X}] = (1 - \frac{j\omega}{\alpha})^{-r} \tag{7.73}$$

Generalizing the result in (7.73) to $Y = Y_1 + Y_2 + \cdots + Y_K$, the sum of K independent random variables, we obtain

$$\begin{aligned} \Phi_Y(j\omega) &= E[e^{j\omega Y}] \\ &= E[e^{j\omega(Y_1+Y_2+\cdots+Y_K)}] \\ &= E[e^{j\omega Y_1}]E[e^{j\omega Y_2}]\cdots E[e^{j\omega Y_K}] \\ &= \Phi_{Y_1}(j\omega)\Phi_{Y_2}(j\omega)\cdots\Phi_{Y_K}(j\omega) \\ &= (1 - \frac{j\omega}{\alpha})^{-(r_1+\cdots+r_K)} \end{aligned} \tag{7.74}$$

and hence, the density function of Y is

$$f_Y(y) = \begin{cases} \frac{\alpha}{\Gamma(r_1+\cdots+r_K)}(\alpha y)^{r_1+\cdots+r_K-1}e^{-\alpha y} & ,y > 0 \\ \\ 0 & , \text{ otherwise} \end{cases} \tag{7.75}$$

Using $r = 1/2$ and $\alpha = 1/(2\sigma^2)$, we obtain the density function of the sufficient statistic to be

$$f_Y(y) = \begin{cases} \frac{1}{2^{\frac{K}{2}}\sigma^K\Gamma(\frac{K}{2})}y^{\frac{K}{2}-1}e^{-\frac{y}{2\sigma^2}} & ,y > 0 \\ \\ 0 & , \text{ otherwise} \end{cases} \tag{7.76}$$

Note that the variance σ^2 of Y_k, $k = 1, 2, \cdots, K$, denotes σ_N^2 under hypothesis H_0 and $(\sigma_S^2 + \sigma_N^2)$ under hypothesis H_1. That is, the density function of the sufficient statistic $T(\mathbf{Y})$ under each hypothesis is

$$f_{T|H_0}(t|H_0) = \begin{cases} \dfrac{1}{2^{\frac{K}{2}}\sigma_0^K \Gamma(\frac{K}{2})} t^{\frac{K}{2}-1} e^{-\frac{t}{2\sigma_0^2}} & , t > 0 \\ \\ 0 & , \text{otherwise} \end{cases} \qquad (7.77)$$

and

$$f_{T|H_1}(t|H_1) = \begin{cases} \dfrac{1}{2^{\frac{K}{2}}\sigma_1^K \Gamma(\frac{K}{2})} t^{\frac{K}{2}-1} e^{-\frac{t}{2\sigma_1^2}} & , t > 0 \\ \\ 0 & , \text{otherwise} \end{cases} \qquad (7.78)$$

where $\sigma_0^2 = \sigma_N^2$ and $\sigma_1^2 = \sigma_S^2 + \sigma_N^2$. Knowing the conditional density functions $f_{T|H_1}(t|H_1)$ and $f_{T|H_0}(t|H_0)$, we can obtain expressions for P_D and P_F. From (7.68a), the probabilities of detection and false alarm are

$$\begin{aligned} P_F &= \int_{\gamma_3}^{\infty} f_{T|H_0}(t|H_0)dt \\ \\ &= \frac{1}{2^{\frac{K}{2}}\sigma_0^K \Gamma(\frac{K}{2})} \int_{\gamma_3}^{\infty} t^{\frac{K}{2}-1} e^{-\frac{t}{2\sigma_0^2}} dt \end{aligned} \qquad (7.79)$$

and

$$\begin{aligned} P_D &= \int_{\gamma_3}^{\infty} f_{T|H_1}(t|H_1)dt \\ \\ &= \frac{1}{2^{\frac{K}{2}}\sigma_1^K \Gamma(\frac{K}{2})} \int_{\gamma_3}^{\infty} t^{\frac{K}{2}-1} e^{-\frac{t}{2\sigma_1^2}} dt \end{aligned} \qquad (7.80)$$

7.5.2 Uncorrelated Signal Components; Unequal Variances

In this case, we assume that the signal components are uncorrelated and thus, the covariance matrix $\mathbf{C}_S$ is diagonal. We also assume that the variances of the different components are not equal; that is,

$$\mathbf{C}_S = \begin{bmatrix} \sigma_{S_1}^2 & 0 & \cdots & 0 \\ 0 & \sigma_{S_2}^2 & \cdots & 0 \\ \vdots & \vdots & \ddots & \vdots \\ 0 & 0 & \cdots & \sigma_{S_K}^2 \end{bmatrix} \tag{7.81}$$

From the LRT in Equation (7.65), let the term in brackets be denoted $\mathbf{H}$; that is,

$$\mathbf{H} = \frac{1}{\sigma_N^2}\mathbf{I} - (\mathbf{C}_S + \sigma_N^2\mathbf{I})^{-1} \tag{7.82}$$

Substituting Equation (7.81) into Equation (7.82) and rearranging terms, the $\mathbf{H}$ matrix reduces to

$$\mathbf{H} = \begin{bmatrix} \frac{\sigma_{S_1}^2}{\sigma_N^2(\sigma_{S_1}^2 + \sigma_N^2)} & 0 & \cdots & 0 \\ 0 & \frac{\sigma_{S_2}^2}{\sigma_N^2(\sigma_{S_2}^2 + \sigma_N^2)} & \cdots & 0 \\ \vdots & \vdots & \ddots & \vdots \\ 0 & 0 & \cdots & \frac{\sigma_{S_K}^2}{\sigma_N^2(\sigma_{S_K}^2 + \sigma_N^2)} \end{bmatrix} \tag{7.83}$$

and, consequently, the LRT becomes

$$T(\mathbf{y}) = \mathbf{y}^T\mathbf{H}\mathbf{y}$$

$$= \frac{1}{\sigma_N^2}\sum_{k=1}^K \frac{\sigma_{S_k}^2}{(\sigma_{S_k}^2 + \sigma_N^2)}y_k^2 \mathop{\gtrless}_{H_0}^{H_1} \gamma_2 \tag{7.84}$$

We observe that the above expression is not as simple as the one in the previous subsection and, consequently, it may not be easy to obtain expressions for P_D and P_F.

Remark: If the signal components are not independent, and thus the signal covariance matrix is not diagonal, we can diagonalize the matrix using an orthogonal transformation following the procedure given in Section 7.4.2.

Example 7.3

Consider the binary hypothesis problem

$$H_1 : \quad Y_k \quad = \quad S_k \quad + \quad N_k \quad , k = 1, 2$$

$$H_0 : \quad Y_k \quad = \quad \qquad \qquad N_k \quad , k = 1, 2$$

where the noise components are zero-mean, uncorrelated Gaussian random variables with variance σ_N^2, $k = 1, 2$. The signal components are also independent, and zero-mean with variance σ_S^2, $k = 1, 2$. The signal and noise components are independent.
(a) Obtain the optimum decision rule.
(b) Determine expressions for the probability of detection and the probability of false alarm.

Solution.

(a) This is the case where the noise components are independent and identically distributed, and the signal components are also independent and identically distributed. Both covariance matrices $\mathbf{C}_S$, and $\mathbf{C}_N$ of the signal and noise are diagonal. The optimum decision rule is given by (7.68a) to be

$$T(\mathbf{y}) = y_1^2 + y_2^2 \underset{H_0}{\overset{H_1}{\underset{<}{\gtrless}}} \gamma_3$$

where

$$\gamma_3 = \frac{\sigma_N^2(\sigma_S^2 + \sigma_N^2)}{\sigma_S^2}\gamma_2$$

$$\gamma_2 = 2\gamma$$

and

$$\gamma = \ln\eta + \frac{1}{2}(\ln|C_1| - \ln|C_0|)$$

The covariance matrices C_1 and C_0 under hypotheses H_1 and H_0 are

$$C_1 = C_S + C_N = \begin{bmatrix} \sigma_S^2 + \sigma_N^2 & 0 \\ 0 & \sigma_S^2 + \sigma_N^2 \end{bmatrix}$$

and

$$C_0 = C_N = \begin{bmatrix} \sigma_N^2 & 0 \\ 0 & \sigma_N^2 \end{bmatrix}$$

Rearranging terms, the decision rule becomes

$$T(\mathbf{y}) = y_1^2 + y_2^2 \underset{H_0}{\overset{H_1}{\underset{<}{>}}} 2\frac{\sigma_N^2(\sigma_S^2 + \sigma_N^2)}{\sigma_S^2}(\ln\eta + \ln\frac{\sigma_S^2 + \sigma_N^2}{\sigma_N^2}) = \gamma_3$$

Consequently, the sufficient statistic is $T(\mathbf{Y}) = Y_1^2 + Y_2^2$.

(b) Using the results derived in Equations (7.77) and (7.78), the conditional probability density function of the sufficient statistic under each hypothesis is

$$f_{T|H_0}(t|H_0) = \begin{cases} \frac{1}{2\sigma_0^2}e^{-\frac{t}{2\sigma_0^2}} &, t > 0 \\ \\ 0 & \text{otherwise} \end{cases}$$

and

$$f_{T|H_1}(t|H_1) = \begin{cases} \frac{1}{2\sigma_1^2}e^{-\frac{t}{2\sigma_1^2}} &, t > 0 \\ \\ 0 & \text{otherwise} \end{cases}$$

where $\sigma_0^2 = \sigma_N^2$ and $\sigma_1^2 = \sigma_S^2 + \sigma_N^2$. Consequently, the probability of detection and the probability of false alarm are

$$P_D = \frac{1}{2\sigma_1^2}\int_{\gamma_3}^{\infty} e^{-\frac{t}{2\sigma_1^2}}\, dt = e^{-\frac{\gamma_3}{2\sigma_1^2}}$$

and

$$P_F = \frac{1}{2\sigma_0^2} \int_{\gamma_3}^{\infty} e^{-\frac{t}{2\sigma_0^2}} dt = e^{-\frac{\gamma_3}{2\sigma_0^2}}$$

7.6 SAME MEAN; SYMMETRIC HYPOTHESES

Consider the binary symmetric hypothesis problem given by

$$H_1 : \begin{array}{llll} Y_k & = & S_k & + & N_k & , k = 1, 2, \cdots, K \\ Y_k & = & & & N_k & , k = (K+1), (K+2), \cdots, 2K \end{array}$$

$$H_0 : \begin{array}{llll} Y_k & = & & N_k & , k = 1, 2, \cdots, K \\ Y_k & = & S_k + & N_k & , k = (K+1), (K+2), \cdots, 2K \end{array}$$

$$(7.85)$$

We assume, as before, that the mean vectors $\mathbf{m}_1 = \mathbf{m}_0 = \mathbf{0}$ and that the noise components are uncorrelated with variance σ_N^2. Thus, the noise covariance matrix is $\mathbf{C}_N = \sigma_N^2 \mathbf{I}$. Let $\mathbf{C}_S$ denote the signal covariance matrix. Then, the $2K \times 2K$ covariance matrices $\mathbf{C}_0$ and $\mathbf{C}_1$ under hypotheses H_0 and H_1, respectively, can be partitioned into $K \times K$ submatrices. That is,

$$\mathbf{C}_1 = \left[\begin{array}{c:c} \mathbf{C}_S + \mathbf{C}_N & \mathbf{0} \\ \hdashline \mathbf{0} & \mathbf{C}_N \end{array} \right] = \left[\begin{array}{c:c} \mathbf{C}_S + \sigma_N^2 \mathbf{I} & \mathbf{0} \\ \hdashline \mathbf{0} & \sigma_N^2 \mathbf{I} \end{array} \right]$$

$$(7.86)$$

and

$$\mathbf{C}_0 = \left[\begin{array}{c:c} \mathbf{C}_N & \mathbf{0} \\ \hdashline \mathbf{0} & \mathbf{C}_S + \mathbf{C}_N \end{array} \right] = \left[\begin{array}{c:c} \sigma_N^2 \mathbf{I} & \mathbf{0} \\ \hdashline \mathbf{0} & \mathbf{C}_S + \sigma_N^2 \mathbf{I} \end{array} \right]$$

$$(7.87)$$

Let the difference of the inverse covariance matrices of $\mathbf{C}_0$ and $\mathbf{C}_1$ be denoted by

$$\mathbf{\Delta C}^{-1} = \mathbf{C}_0^{-1} - \mathbf{C}_1^{-1} \qquad (7.88)$$

Thus,

$$\Delta \mathbf{C}^{-1} = \left[\begin{array}{ccc} \frac{1}{\sigma_N^2}\mathbf{I} & \vdots & \mathbf{0} \\ \cdots & \vdots & \cdots\cdots\cdots\cdots \\ \mathbf{0} & \vdots & (\mathbf{C}_S + \sigma_N^2\mathbf{I})^{-1} \end{array} \right] - \left[\begin{array}{ccc} (\mathbf{C}_S + \sigma_N^2\mathbf{I})^{-1} & \vdots & \mathbf{0} \\ \cdots\cdots\cdots\cdots & \vdots & \cdots \\ \mathbf{0} & \vdots & \frac{1}{\sigma_N^2}\mathbf{I} \end{array} \right]$$

$$= \left[\begin{array}{ccc} \frac{1}{\sigma_N^2}\mathbf{I} - (\mathbf{C}_S + \sigma_N^2\mathbf{I})^{-1} & \vdots & \mathbf{0} \\ \cdots\cdots\cdots\cdots\cdots\cdots & \vdots & \cdots\cdots\cdots\cdots\cdots\cdots \\ \mathbf{0} & \vdots & (\mathbf{C}_S + \sigma_N^2\mathbf{I})^{-1} - \frac{1}{\sigma_N^2}\mathbf{I} \end{array} \right]$$

$$(7.89)$$

Partitioning the $2K \times 1$ vector $\mathbf{Y}$ into two $K \times 1$ vectors, such that

$$\mathbf{Y} = \left[\begin{array}{c} \mathbf{Y}_1 \\ \cdots \\ \mathbf{Y}_2 \end{array} \right] \qquad (7.90)$$

and substituting Equations (7.88) and (7.89) into Equation (7.61), the LRT becomes

$$T(\mathbf{y}) = \mathbf{y}^T \Delta \mathbf{C}^{-1} \mathbf{y}$$

$$= \left[\begin{array}{c} \mathbf{y}_1^T \\ \cdots \\ \mathbf{y}_2^T \end{array} \right] \left[\begin{array}{ccc} \frac{1}{\sigma_N^2}\mathbf{I} - (\mathbf{C}_S + \sigma_N^2\mathbf{I})^{-1} & \vdots & \mathbf{0} \\ \cdots\cdots\cdots\cdots\cdots & \vdots & \cdots\cdots\cdots\cdots \\ \mathbf{0} & \vdots & (\mathbf{C}_S + \sigma_N^2\mathbf{I})^{-1} - \frac{1}{\sigma_N^2}\mathbf{I} \end{array} \right] \left[\begin{array}{c} \mathbf{y}_1 \\ \cdots \\ \mathbf{y}_2 \end{array} \right]$$

$$= \mathbf{y}_1^T [\tfrac{1}{\sigma_N^2}\mathbf{I} - (\mathbf{C}_S + \sigma_N^2\mathbf{I})^{-1}]\mathbf{y}_1 + \mathbf{y}_2^T [(\mathbf{C}_S + \sigma_N^2\mathbf{I})^{-1} - \tfrac{1}{\sigma_N^2}\mathbf{I}]\mathbf{y}_2 \underset{H_0}{\overset{H_1}{\underset{<}{\overset{>}{}}}} \gamma_2$$

$$(7.91)$$

Again, depending on the structure of the signal covariance matrix $\mathbf{C}_S$, the above expression may be reduced as in the previous section.

7.6.1 Uncorrelated Signal Components; Equal Variances

In order to carry the test in (7.91) further, let the signal components be uncorrelated, and identically distributed. That is,

$$\mathbf{C}_S = \sigma_S^2 \mathbf{I} \tag{7.92}$$

Substituting the above value of the signal covariance matrix into Equation (7.91), the LRT test is obtained to be

$$
\begin{aligned}
T(\mathbf{y}) &= \mathbf{y}_1^T[\frac{1}{\sigma_N^2}\mathbf{I} - (\sigma_S^2\mathbf{I} + \sigma_N^2\mathbf{I})^{-1}]\mathbf{y}_1 + \mathbf{y}_2^T[(\sigma_S^2\mathbf{I} + \sigma_N^2\mathbf{I})^{-1} - \frac{1}{\sigma_N^2}\mathbf{I}]\mathbf{y}_2 \\
&= \frac{\sigma_S^2}{\sigma_N^2(\sigma_S^2 + \sigma_N^2)}\mathbf{I}(\mathbf{y}_1^T\mathbf{y}_1 - \mathbf{y}_2^T\mathbf{y}_2) \underset{H_0}{\overset{H_1}{\underset{<}{>}}} \gamma_2
\end{aligned}
\tag{7.93}
$$

or

$$T(\mathbf{y}) = \sum_{k=1}^{K} y_k^2 - \sum_{k=K+1}^{2K} y_k^2 \underset{H_0}{\overset{H_1}{\underset{<}{>}}} \gamma_3 \tag{7.94}$$

where γ_3 is defined in (7.68b).

We can have more insight into this problem by assuming that we have a minimum probability of error criterion, and that both hypotheses are equally likely. Thus, the threshold η equals one, and γ_2 and γ_3 become zero. We observe that the determinants of both covariance matrices are equal ($|\mathbf{C}_1| = |\mathbf{C}_0|$), since the hypotheses are symmetrical. Consequently, the LRT reduces to

$$T_1(\mathbf{y}) = \sum_{k=1}^{K} y_k^2 \underset{H_0}{\overset{H_1}{\underset{<}{>}}} \sum_{k=K+1}^{2K} y_k^2 = T_0(\mathbf{y}) \tag{7.95}$$

The probability of error is defined as

$$P(\varepsilon) = P(\varepsilon|H_0)P(H_0) + P(\varepsilon|H_1)P(H_1)$$

$$= \frac{1}{2}[P(\varepsilon|H_0) + P(\varepsilon|H_1)] \qquad (7.96)$$

Since the test is symmetrical with respect to both hypotheses, we have

$$P(\varepsilon|H_0) = P(\varepsilon|H_1) \qquad (7.97)$$

Thus, the probability of error is just

$$P(\varepsilon) = P(\varepsilon|H_0) = P(T_0 < T_1|H_0)$$

$$= P(\varepsilon|H_1) = P(T_1 < T_0|H_1) \qquad (7.98)$$

From Figure 7.2, we see that the probability of error is given by

$$
\begin{aligned}
P(\varepsilon) \;&=\; P(\varepsilon|H_0) = \int_0^\infty \int_0^{t_1} f_{T_1 T_0}(t_1, t_0|H_0)\,dt_0\,dt_1 \\
&=\; P(\varepsilon|H_1) = \int_0^\infty \int_{t_1}^\infty f_{T_1 T_0}(t_1, t_0|H_1)\,dt_0\,dt_1 \qquad (7.99)
\end{aligned}
$$

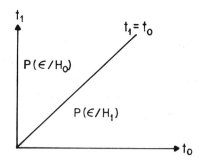

Figure 7.2 Regions of integration for $P(\varepsilon)$.

From (7.76), $T_1(\mathbf{Y})$ and $T_2(\mathbf{Y})$ are statistically independent and chi-square distributed; that is,

$$f_{T_1}(t_1) = \frac{1}{2^{\frac{K}{2}} \sigma_1^K \Gamma(\frac{K}{2})} t_1^{\frac{n}{2}-1} e^{-\frac{t_1}{2\sigma_1^2}} \qquad (7.100)$$

$$f_{T_0}(t_0) = \frac{1}{2^{\frac{K}{2}} \sigma_0^K \Gamma(\frac{K}{2})} t_0^{\frac{n}{2}-1} e^{-\frac{t_0}{2\sigma_0^2}} \qquad (7.101)$$

where $\sigma_0^2 = \sigma_N^2$ and $\sigma_1^2 = \sigma_S^2 + \sigma_N^2$. Substituting Equations (7.100) and (7.101) into (7.99) and solving the integral, it can be shown that the probability of error reduces to

$$P(\varepsilon) = \left(\frac{\sigma_N^2}{\sigma_1^2 + \sigma_N^2}\right)^{\frac{K}{2}} \sum_{k=0}^{\frac{K}{2}-1} \left(\begin{array}{c} \frac{K}{2}+k-1 \\ k \end{array}\right)\left(1 - \frac{\sigma_N^2}{\sigma_1^2 + \sigma_N^2}\right)^k \qquad (7.102)$$

7.6.2 Uncorrelated Signal Components; Unequal Variances

In this case, we assume that the signal components are uncorrelated, but their corresponding variances are not equal. That is, the signal covariance matrix is still diagonal, but with unequal elements. Thus, we have

$$\mathbf{C}_S = \begin{bmatrix} \sigma_{S_1}^2 & 0 & \cdots & 0 \\ 0 & \sigma_{S_2}^2 & \cdots & 0 \\ \vdots & \vdots & \ddots & \vdots \\ 0 & 0 & \cdots & \sigma_{S_K}^2 \end{bmatrix} \qquad (7.103)$$

Substituting (7.103) into (7.88) and rearranging terms, we obtain

338

$$\Delta \mathbf{C}^{-1} =$$

$$
\begin{bmatrix}
\frac{\sigma_{S_1}^2}{\sigma_N^2(\sigma_{S_1}^2+\sigma_N^2)} & \cdots & 0 & 0 & 0 & 0 \\
0 & \cdots & 0 & 0 & 0 & 0 \\
\vdots & \ddots & \vdots & \vdots & \vdots & \vdots \\
0 & \cdots & \frac{\sigma_{S_K}^2}{\sigma_N^2(\sigma_{S_K}^2+\sigma_N^2)} & 0 & 0 & 0 \\
0 & 0 & 0 & -\frac{\sigma_{S_1}^2}{\sigma_{S_1}^2+\sigma_N^2} & \cdots & 0 \\
0 & 0 & 0 & 0 & \cdots & 0 \\
\vdots & \vdots & \vdots & \vdots & \ddots & \vdots \\
0 & 0 & 0 & 0 & \cdots & -\frac{\sigma_{S_K}^2}{\sigma_{S_K}^2+\sigma_N^2}
\end{bmatrix}
\quad (7.104)
$$

It follows that the test in (7.91) becomes

$$
T(\mathbf{y}) = \frac{\sigma_1^2}{\sigma_N^2} \left\{ \sum_{k=1}^{K} \frac{\sigma_{S_k}^2}{\sigma_{S_k}^2 + \sigma_N^2} y_k^2 - \sum_{k=K+1}^{2K} \frac{\sigma_{S(k-N)}^2}{\sigma_{S(k-N)}^2 + \sigma_N^2} y_k^2 \right\} \underset{H_0}{\overset{H_1}{\underset{<}{>}}} \gamma_2 \quad (7.105)
$$

This expression is too complicated to proceed any further with the test.

7.7 SUMMARY

In this chapter we have discussed the general Gaussian problem. We defined the Gaussian process and stated some of its most important properties. Then, we considered the binary hypothesis problem. Due to the characteristics of the Gaussian process and Gaussian random variables, the general Gaussian problem was considered in terms of the covariance matrices and the mean vectors under each hypothesis. First, we considered the case of equal covariance matrix for both hypotheses. The noise samples were always assumed uncorrelated, and thus statistically independent, with equal variances. The signal components considered, however, were either independent or not independent. When the signal components were independent and of equal variances, the problem was relatively simple since the covariance matrix is diagonal with equal value elements. When the signal component variances were not equal, the expressions were more difficult and in this case we were able to solve for the sufficient statistic only.

In the case when the covariance matrices are general, we transformed the problem from one coordinate system into another coordinate system such that the covariance matrix is diagonal. We solved for the eigenvalues and eigenvectors, and then used an orthogonal transformation to diagonalize the covariance matrix. In Sections 7.5 and 7.6, we considered the case of equal mean vectors and obtained the LRT.

PROBLEMS

7.1 For the binary hypothesis problem with $\mathbf{m}_0 = \mathbf{0}$. Let the covariance matrix $\mathbf{C}$ be

(a) $\mathbf{C} = \begin{bmatrix} 1 & \frac{1}{2} \\ \frac{1}{2} & 1 \end{bmatrix}$ (b) $\mathbf{C} = \begin{bmatrix} 1 & 0.1 \\ 0.1 & 1 \end{bmatrix}$ (c) $\mathbf{C} = \begin{bmatrix} 1 & 0.9 \\ 0.9 & 1 \end{bmatrix}$

Determine the LRT for the three cases above.

7.2 Repeat Problem 7.1, assuming that the covariance matrix $\mathbf{C}$ is

$$\mathbf{C} = \begin{bmatrix} 1 & 0.9 \\ 0.9 & 2 \end{bmatrix}$$

7.3 Consider the binary hypothesis problem:

$$H_1: \ Y_k \ = \ S_k \ + \ N_k \ , k = 1, 2$$

$$H_0: \ Y_k \ = \ \qquad N_k \ , k = 1, 2$$

where the noise components are zero-mean, uncorrelated Gaussian random variables with variance $\sigma_N^2 = 1$, $k = 1, 2$. The signal components are also independent and zero-mean with variances $\sigma_S^2 = 2$, $k = 1, 2$. The signal and noise components are independent.
(a) Obtain the optimum decision rule.
(b) Determine the minimum probability error. Assume $P(H_0) = P(H_1) = 1/2$.

7.4 Repeat Problem 7.3 with $k = 1, 2, 3, 4$.

7.5 Plot the ROC of Problem 7.3 with the ratio σ_S^2/σ_N^2 as a parameter.

7.6 Consider the problem given in 7.3 with signal covariance matrix:

$$\mathbf{C}_S = \begin{bmatrix} \sigma_s^2 & 0 \\ 0 & 2\sigma_s^2 \end{bmatrix}$$

Design an optimum test.

7.7 Consider the symmetrical binary hypothesis problem

$$H_1 : \begin{array}{llll} Y_k & = & S_k + N_k & ,k = 1, 2 \\ Y_k & = & N_k & ,k = 3, 4 \end{array}$$

$$H_0 : \begin{array}{llll} Y_k & = & N_k & ,k = 1, 2 \\ Y_k & = & S_k + N_k & ,k = 3, 4 \end{array}$$

Let the mean vectors under each hypothesis be zero for both hypotheses H_1 and H_0. The noise components are identically distributed Gaussian random variables with variance 1. The signal components are also independent, and identically distributed with variance 2. The signal and noise components are independent.
(a) Design an optimum test.
(b) Determine the probability of error assuming minimum probability of error criterion and $P_0 = P_1 = \frac{1}{2}$.

7.8 Repeat Problem 7.1, if the covariance matrix is given by

(a) $\mathbf{C} = \begin{bmatrix} 1 & 0.9 & 0.5 \\ 0.9 & 1 & 0.1 \\ 0.5 & 0.1 & 1 \end{bmatrix}$ (b) $\mathbf{C} = \begin{bmatrix} 1 & 0.8 & 0.6 & 0.2 \\ 0.8 & 1 & 0.8 & 0.6 \\ 0.6 & 0.8 & 1 & 0.8 \\ 0.2 & 0.6 & 0.8 & 1 \end{bmatrix}$

REFERENCES

1. Brogan, W.L., *Modern Control Theory*, Quantum, New York, 1974.

2. Cox, D.R., and H.D. Miller, *The Theory of Stochastic Processes*, John Wiley and Sons, New York, 1965.

3. Davenport, Jr., W.B., and W.L. Root, *An Introduction to the Theory of Random Signals and Noise*, McGraw-Hill, New York, 1958.

4. Dorny, C.N., *A Vector Space Approach to Models and Optimization*, Robert E. Krieger, New York, 1980.

5. Haykin, S., *Communication Systems*, John Wiley and Sons, New York, 1983.

6. Melsa, J.L., and D.L. Cohn, *Decision and Estimation Theory*, McGraw-Hill, New York, 1978.

7. Mohanty, N., *Signal Processing: Signals, Filtering, and Detection*, Van Nostrand Reinhold, New York, 1987.

8. Noble, B., and J.W. Daniel, *Applied Linear Algebra*, Prentice-Hall, Englewood Cliffs, NJ, 1977.

9. Papoulis, A., *Probability, Random Variables, and Stochastic Processes*, McGraw-Hill, New York, 1984.

10. Van Trees, H.L., *Detection, Estimation, and Modulation Theory: Part I*, John Wiley and Sons, New York, 1968.

11. Wozencraft, J.M., and I.M. Jacobs, *Principles of Communication Engineering*, John Wiley and Sons, New York, 1965.

Chapter 8

Detection and Parameter Estimation

8.1 INTRODUCTION

In Chapters 1 and 2, we presented the fundamentals of probability theory and stochastic processes. In Chapters 3 and 4, we developed the basic principles needed for solving decision and estimation problems. The observations considered were represented by random variables. In Chapter 5, we presented the orthogonality principle and its application in the optimum linear mean-square estimation. In Chapter 6, we presented some mathematical principles such as the Gram-Schmidt orthogonalization procedure, diagonalization of a matrix and similarity transformation, integral equations, and generalized Fourier series. The concept of generalized Fourier series was then used to represent random processes by an orthogonal series expansion, referred to as Karhunen–Loeve expansion. Chapter 6, gave us the basic mathematical background for Chapters 7 and 8. In Chapter 7, we covered the general detection Gaussian problem.

In this chapter, we extend the concepts of decision and estimation problems to time varying waveforms. If a signal is transmitted, the received waveform is composed of the transmitted signal and an additive noise process. If no signal is transmitted, then the received waveform is noise only. The goal is to design an optimum receiver (detector) according to some criterion. In Section 8.2, we discuss the general and simple binary detection of known signals corrupted by an additive white Gaussian noise process with mean zero, and power spectral density $N_0/2$. The received waveforms are observed over the interval of time $t\epsilon[0, T]$. In Section 8.3, we extend the concepts of binary detection to M-ary detection. In Section 8.4, we assume that the received signals in the presence of the additive white Gaussian noise process have some unknown parameters, which need to be estimated. Some linear estimation techniques are used to estimate these unknown parameters, which may be either random or nonrandom. Nonlinear estimation is presented in Section 8.5. In Section 8.6, we consider the general binary detection with unwanted parameters in additive white Gaussian noise. In this case, the received waveform is not completely known *a priori* as in the previous sections. The unknown parameters of the signal are referred to *unwanted parameters*. We conclude this chapter by a section on detection in colored noise. Specifically, we consider the general binary detection in nonwhite Gaussian noise. Two different approaches, using the Karhunen–Loeve expansion and whitening, are suggested to solve this problem.

8.2 BINARY DETECTION

In a binary communication problem, the transmitter may send a deterministic signal $s_0(t)$ under the null hypothesis H_0, or a deterministic signal $s_1(t)$ under the alternate hypothesis H_1. At the receiver, the signal is corrupted by $W(t)$ an additive white Gaussian noise process. Assume that the additive noise is zero-mean, and has a double-sided power spectral density of $N_0/2$. The goal is to design an optimum receiver which observes the received signal $Y(t)$ over the interval $t\epsilon[0, T]$, and decides whether hypothesis H_0, or hypothesis H_1, is true.

8.2.1 Simple Binary Detection

In a simple binary detection problem, the transmitted signal under hypothesis H_1 is $s(t)$, and no signal is transmitted under the null hypothesis H_0. At the receiver, we have

$$H_1 : Y(t) = s(t) + W(t) \quad , 0 \le t \le T$$
$$H_0 : Y(t) = \qquad W(t) \quad , 0 \le t \le T \tag{8.1}$$

Note that the signal is a continuous time function. In order to obtain a set of countable random variables so that we may apply the concepts developed in Chapter 3, we need to take K samples, where K may be infinite. However, in Chapter 6 we saw that a continuous time signal may be represented by the Karhunen–Loeve expansion using a set K of complete orthonormal functions. The coefficients in the series expansion are the desired set of random variables.

The energy of the known deterministic signal is

$$\mathcal{E} = \int_0^T s^2(t)dt \tag{8.2}$$

Thus, let the first normalized function $\phi_1(t)$ be

$$\phi_1(t) = \frac{s(t)}{\sqrt{\mathcal{E}}} \tag{8.3}$$

or

$$s(t) = \sqrt{\mathcal{E}}\phi_1(t) \tag{8.4}$$

Consequently, the first coefficient in the Karhunen–Loeve expansion of $Y(t)$ is

$$
\begin{aligned}
Y_1 &= \int_0^T Y(t)\phi_1(t)dt \\
&= \begin{cases} H_1 : \int_0^T [s(t) + W(t)]\phi_1(t)dt = \sqrt{\mathcal{E}} + W_1 \\ H_0 : \int_0^T W(t)\phi_1(t)dt = W_1 \end{cases}
\end{aligned} \tag{8.5}
$$

where W_1 is the first coefficient in the series expansion of $W(t)$. The rest of the coefficients Y_k, $k = 2, 3, \cdots$, are obtained by using arbitrary orthogonal functions ϕ_k, $k = 2, 3, \cdots$. The ϕ_k's are orthogonal to $\phi_1(t)$ ($\int_0^T \phi_k(t)\phi_1(t)dt = 0$). Thus,

$$Y_k = \begin{cases} H_1 : \int_0^T [s(t) + W(t)]\phi_k(t)dt = W_k \\ \\ H_0 : \int_0^T W(t)\phi_k(t)dt = W_k \end{cases} \tag{8.6}$$

since $W(t)$ is a Gaussian process, the random variables W_k, $k = 2, 3, \cdots$, are Gaussian. We observe from Equation (8.6), that the coefficients Y_k, $k = 2, 3, \cdots$, are the coefficients of white Gaussian process (W_k's), and do not depend on which hypothesis is true. Only the coefficient Y_1 depends on the hypotheses H_1 and H_0.

We need to find a sufficient statistic for this infinite number of random variables in order to make a decision as to which hypothesis is true. Since the coefficients W_j, W_k, $j \neq k$, of the Karhunen–Loeve expansion are uncorrelated, that is,

$$E[W_j W_k | H_0] = E[W_j W_k | H_1] = E[W_j W_k] = 0 \quad , j \neq k \tag{8.7}$$

and are jointly Gaussian, they are statistically independent. Thus, all Y_k, $k = 2, 3, \cdots$, are statistically independent of Y_1, and have no effect on the decision. Hence, the sufficient statistic is only Y_1; that is,

$$T(\mathbf{Y}) = Y_1 \tag{8.8}$$

Equation (8.8) tells us that the infinite observation space has been reduced to a one-dimensional decision space. Thus, the equivalent problem to Equation (8.1) is

$$\begin{aligned} H_1 &: Y_1 = \sqrt{\mathcal{E}} + W_1 \\ \\ H_0 &: Y_1 = \phantom{\sqrt{\mathcal{E}} +} W_1 \end{aligned} \tag{8.9}$$

where W_1 is Gaussian with means:

$$\begin{aligned} E[W_1 | H_1] &= E[W_1 | H_0] = E[\int_0^T \phi_1(t)W(t)dt] \\ &= \int_0^T \phi_1(t)E[W(t)]dt = 0 \end{aligned} \tag{8.10}$$

and variances:

$$
\begin{aligned}
E[W_1^2|H_1] &= E[W_1^2|H_0] \\
&= E[\int_0^T \int_0^T \phi_1(t)\phi_1(u)W(t)W(u)dtdu] \\
&= \int_0^T \int_0^T \phi_1(t)\phi_1(u)E[W(t)W(u)]dtdu \quad (8.11)
\end{aligned}
$$

We are given that the power spectral density of $W(t)$ is $N_0/2$ for all frequency f, and thus, its autocorrelation function $R_{WW}(t,u)$ is

$$
\begin{aligned}
E[W(t)W(u)] &= R_{WW}(t,u) \\
&= \frac{N_0}{2}\delta(t-u) \\
&= C_{WW}(t,u) \quad (8.12)
\end{aligned}
$$

where $C_{WW}(t,u)$ is the covariance function. Substituting Equation (8.12) into Equation (8.11), we obtain the variance of W_1 to be

$$
\begin{aligned}
E[W_1^2] &= \frac{N_0}{2}\int_0^T \int_0^T \phi_1(t)\phi_1(u)\delta(t-u)dtdu \\
&= \frac{N_0}{2}\int_0^T \phi_1^2(t)dt = \frac{N_0}{2} \quad (8.13)
\end{aligned}
$$

We observe that Problem (8.9) is the same as the one solved in Example 3.1 with $m = \sqrt{\mathcal{E}}$ and $\sigma^2 = N_0/2$. Consequently, the optimum decision rule is

$$
T(y) = y_1 \mathop{\gtrless}_{H_0}^{H_1} \frac{N_0}{2\sqrt{\mathcal{E}}}\ln\eta + \frac{\sqrt{\mathcal{E}}}{2} = \gamma \quad (8.14)
$$

The probabilities of detection and false alarm are

$$
P_D = Q(2\frac{\gamma - \sqrt{\mathcal{E}}}{N_0}) \quad (8.15)
$$

and

$$
P_F = Q(\frac{2\gamma}{N_0}) \quad (8.16)
$$

348

where $Q(\cdot)$ is the Q-function. Thus, the only factors affecting the performance of such receiver are the signal energy $\mathcal{E}$ and the noise power spectral density $N_0/2$. From Chapter 6, we note that the optimum receiver is either a *correlation receiver* or a *matched filter receiver*. The receivers are illustrated in Figures 8.1 and 8.2.

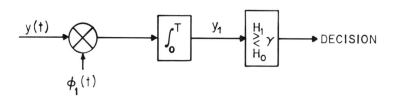

Figure 8.1 Correlation receiver.

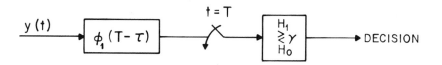

Figure 8.2 Matched filter receiver.

Note that the impulse response $h(\tau)$ of the matched filter is

$$h(\tau) = \begin{cases} \phi_1(T - \tau) & ,0 \leq t \leq T \\ 0 & ,\text{otherwise} \end{cases} \qquad (8.17)$$

We now derive the optimum receiver without resorting to the concept of sufficient statistics. Given a complete set $\{\phi_k(t)\}$, of K orthonormal functions, the Karhunen–Loeve expansion of the received process $Y(t)$ is

$$Y(t) = \sum_{k=1}^{K} Y_k \phi_k(t) \qquad , 0 \le t \le T \qquad (8.18)$$

where

$$Y_k = \int_0^T Y(t) \phi_k(t) dt \qquad , k = 1, 2, \cdots, K \qquad (8.19)$$

The observation vector is $\mathbf{Y} = [Y_1, Y_2, \cdots . Y_k]^T$. Under hypothesis H_0, Y_k is expressed as

$$
\begin{aligned}
Y_k &= \int_0^T W(t) \phi_k(t) dt \\
&= W_k
\end{aligned}
\qquad (8.20)
$$

while under hypothesis H_1, Y_k is

$$
\begin{aligned}
Y_k &= \int_0^T [s(t) + W(t)] \phi_k(t) dt \\
&= \int_0^T s(t) \phi_k(t) dt + \int_0^T W(t) \phi_k(t) dt \\
&= s_k + W_k
\end{aligned}
\qquad (8.21)
$$

Y_k indicates Gaussian random variables, and thus, we only need to find the means and variances under each hypothesis to have a complete description of the conditional density functions. The means and variances of Y_k are

$$E[Y_k | H_0] = E[W_k] = 0 \qquad (8.22)$$

$$E[Y_k | H_1] = E[s_k + W_k] = s_k \qquad (8.23)$$

$$
\begin{aligned}
\operatorname{var}(Y_k | H_0) &= E[Y_k^2 | H_0] = E[W_k^2 | H_0] \\
&= R_{WW}(0) = \frac{N_0}{2}
\end{aligned}
\qquad (8.24)
$$

and

$$\begin{aligned} \mathrm{var}(Y_k|H_1) &= E[(Y_k - s_k)^2|H_1] \\ &= E[W_k^2|H_1] \\ &= R_{WW}(0) = \frac{N_0}{2} \end{aligned} \qquad (8.25)$$

Since uncorrelated Gaussian random variables are statistically independent, the conditional density functions are

$$f_{\mathbf{Y}|H_1}(\mathbf{y}|H_1) = \prod_{k=1}^{K} \frac{1}{\sqrt{\pi N_0}} e^{-\frac{(y_k - s_k)^2}{N_0}} \qquad (8.26)$$

and

$$f_{\mathbf{Y}|H_0}(\mathbf{y}|H_0) = \prod_{k=1}^{K} \frac{1}{\sqrt{\pi N_0}} e^{-\frac{y_k^2}{N_0}} \qquad (8.27)$$

Also, from Equations $(6.15), (6.23)$, and (6.25), we have

$$s(t) = \lim_{K \to \infty} s_K(t) \qquad (8.28)$$

where

$$s_K(t) = \sum_{k=1}^{K} s_k \phi_k(t) \qquad (8.29)$$

Consequently, the likelihood ratio is

$$\Lambda[y(t)] = \lim_{K \to \infty} \Lambda[y_K(t)] = \frac{f_{\mathbf{Y}|H_1}(\mathbf{y}|H_1)}{f_{\mathbf{Y}|H_0}(\mathbf{y}|H_0)}$$

$$= \frac{\prod_{k=1}^{K} \frac{1}{\sqrt{\pi N_0}} e^{-\frac{(y_k - s_k)^2}{N_0}}}{\prod_{k=1}^{K} \frac{1}{\sqrt{\pi N_0}} e^{-\frac{y_k^2}{N_0}}} \qquad (8.30)$$

where $\Lambda[y_K(t)]$ is the K-term likelihood ratio. Taking the logarithm and simplifying, Equation (8.30) may be rewritten as

$$\lim_{K \to \infty} \ln\Lambda[Y_K(t)] = \lim_{K \to \infty} \left\{ \frac{2}{N_0} \sum_{k=1}^{K} Y_k s_k - \frac{1}{N_0} \sum_{k=1}^{K} s_k^2 \right\} \qquad (8.31)$$

where

$$\lim_{K \to \infty} \sum_{k=1}^{K} Y_k s_k = \int_0^T Y_K(t) s_K(t) dt \qquad (8.32)$$

and

$$\lim_{K \to \infty} \sum_{k=1}^{K} s_k^2 = \int_0^T s_K^2(t) dt \qquad (8.33)$$

The likelihood ratio, letting $K \to \infty$, is

$$\ln\Lambda[Y(t)] = \frac{2}{N_0} \int_0^T Y(t) s(t) dt - \frac{1}{N_0} \int_0^T s^2(t) dt \qquad (8.34)$$

and the decision rule is given by

$$\ln\Lambda[y(t)] \underset{H_0}{\overset{H_1}{\underset{<}{>}}} \ln\eta \qquad (8.35)$$

Substituting Equation (8.2) into Equation (8.4), and then into Equation (8.35), we obtain

$$\ln\Lambda[y(t)] = \frac{2}{N_0} \int_0^T y(t) s(t) dt - \frac{\mathcal{E}}{N_0} \underset{H_0}{\overset{H_1}{\underset{<}{>}}} \ln\eta \qquad (8.36)$$

Since $s(t) = \sqrt{\mathcal{E}} \phi_1(t)$, the test reduces to

$$\frac{2\sqrt{\mathcal{E}}}{N_0} \int_0^T y(t) \phi_1(t) dt \underset{H_0}{\overset{H_1}{\underset{<}{>}}} \ln\eta + \frac{\mathcal{E}}{N_0} \qquad (8.37)$$

or

$$\int_0^T y(t) \phi_1(t) dt \underset{H_0}{\overset{H_1}{\underset{<}{>}}} \gamma \qquad (8.38)$$

which is the optimum receiver derived earlier in (8.14), using the sufficient statistic.

Example 8.1

Consider the digital communication system shown in Figure 8.3.

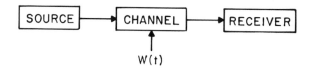

Figure 8.3 Digital communication system.

The information source is binary and produces zeros and ones with equal probability. The communication system uses ASK so that the received signals under hypotheses H_1 and H_0 are

$$H_1 : Y(t) = As(t) \; + \; W(t) \quad ,0 \le t \le T$$

$$H_0 : Y(t) = \qquad\qquad W(t) \quad ,0 \le t \le T$$

The attenuation A produced by the communication channel is a Gaussian random variable with mean zero and variance σ_a^2. The signal $s(t)$ is deterministic, and $W(t)$ is an additive white Gaussian noise with mean zero and power spectral density $N_0/2$. Determine the optimum receiver assuming minimum probability of error criterion.

Solution.

Assuming $s(t)$ has signal energy $\mathcal{E}$, from Equation (8.3), the first normalized function $\phi_1(t)$ is

$$\phi_1(t) = \frac{s(t)}{\sqrt{\mathcal{E}}}$$

Following the same procedure described from Equations (8.3) to (8.9), the problem reduces to

$$H_1 : Y_1 = A + W_1$$

$$H_0 : Y_1 = \qquad W_1$$

The conditional density functions are

$$f_{Y_1|H_0}(y_1|H_0) = f_{W_1}(w_1) = \frac{1}{\sqrt{\pi N_0}} \, e^{-\frac{y_1^2}{N_0}}$$

and

$$f_{Y_1|H_1}(y_1|H_1) = f_A(a) * f_{W_1}(w_1)$$

where $*$ denotes convolution. The convolution of two Gaussian density functions is Gaussian with mean:

$$E[Y_1|H_1] = E[A + W_1|H_1] = 0$$

and variance:

$$\begin{aligned}
\text{var}(Y_1|H_1) &= \text{var}(A + W_1|H_1) \\
&= E[(A + W_1)^2|H_1] \\
&= \text{var}(A) + \text{var}(W_1) \\
&= \sigma_a^2 + \frac{N_0}{2}
\end{aligned}$$

since the random variables A and W_1 are independent, and each with mean zero. Thus, the conditional density function under hypothesis H_1 is

$$f_{Y_1|H_1}(y_1|H_1) = \frac{1}{\sqrt{2\pi}\sqrt{\sigma_a^2 + \frac{N_0}{2}}} \, e^{-\frac{y_1^2}{2(\sigma_a^2 + \frac{N_0}{2})}}$$

Applying the likelihood ratio test, we have

$$\Lambda(y_1) = \frac{f_{Y_1|H_1}(y_1|H_1)}{f_{Y_1|H_0}(y_1|H_0)} = \frac{\frac{1}{\sqrt{\pi(2\sigma_a^2+N_0)}}\, e^{-\frac{y_1^2}{2\sigma_a^2+N_0}}}{\frac{1}{\sqrt{\pi N_0}}\, e^{-\frac{y_1^2}{N_0}}} \underset{H_0}{\overset{H_1}{\underset{<}{>}}} \eta$$

Taking the logarithm and rearranging terms, an equivalent test is

$$y_1^2 \left[\frac{2\sigma_a^2}{N_0(2\sigma_a^2+N_0)} \right] \underset{H_0}{\overset{H_1}{\underset{<}{>}}} \ln\eta - \frac{1}{2}\ln\frac{N_0}{2\sigma_a^2+N_0}$$

or

$$y_1^2 \underset{H_0}{\overset{H_1}{\underset{<}{>}}} \frac{N_0(2\sigma_a^2+N_0)}{2\sigma_a^2}\left[\ln\eta - \frac{1}{2}\ln\frac{N_0}{2\sigma_a^2+N_0} \right]$$

For minimum probability of error, $C_{00} = C_{11} = 0$, and $C_{01} = C_{10} = 1$, we have,

$$\eta = \frac{P_0(C_{10} - C_{00})}{P_1(C_{01} - C_{11})} = \frac{P_0}{P_1} = 1$$

since the hypotheses are equally likely. Thus, $\ln\eta = 0$ and the optimum decision rule becomes

$$y_1^2 \underset{H_0}{\overset{H_1}{\underset{<}{>}}} \frac{N_0(2\sigma_a^2+N_0)}{4\sigma_a^2}\ln\frac{2\sigma_a^2+N_0}{N_0} = \gamma$$

The sufficient statistic is

$$T(\mathbf{Y}) = Y_1^2$$

and the optimum receiver is shown in Figure 8.4.

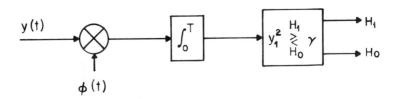

Figure 8.4 Optimum receiver for Example 8.1.

8.2.2 General Binary Detection

In this case, the transmitter sends the signal $s_1(t)$ under hypothesis H_1 and the signal $s_0(t)$ under hypothesis H_0. At the receiver, we have

$$H_1 : Y(t) = s_1(t) + W(t) \quad , 0 \leq t \leq T$$

$$H_0 : Y(t) = s_0(t) + W(t) \quad , 0 \leq t \leq T$$

(8.39)

Let the signal $s_0(t)$ and $s_1(t)$ have energies

$$\mathcal{E}_0 = \int_0^T s_0^2(t)dt$$

(8.40)

and

$$\mathcal{E}_1 = \int_0^T s_1^2(t)dt$$

(8.41)

and correlation coefficient ρ, $-1 \leq \rho \leq 1$, such that

$$\rho = \frac{1}{\sqrt{\mathcal{E}_0 \mathcal{E}_1}} \int_0^T s_0(t)s_1(t)dt$$

(8.42)

Following the same procedure as in the previous subsection, we use the Gram-Schmidt orthogonalization procedure to obtain a complete set of orthonormal functions. The first basis function is

$$\phi_1(t) = \frac{s_1(t)}{\sqrt{\int_0^T s_1^2(t)dt}} = \frac{s_1(t)}{\sqrt{\mathcal{E}_1}} \tag{8.43}$$

The second basis function $\phi_2(t)$ orthogonal to $\phi_1(t)$ is

$$\phi_2(t) = \frac{f_2(t)}{\sqrt{\int_0^T f_2^2(t)dt}} \tag{8.44}$$

where

$$f_2(t) = s_0(t) - s_{01}\phi_1(t) \tag{8.45}$$

and

$$s_{01} = \int_0^T s_0(t)\phi_1(t)dt \tag{8.46}$$

Substituting Equation (8.43) into Equation (8.46), we obtain

$$
\begin{aligned}
s_{01} &= \frac{1}{\sqrt{\mathcal{E}_1}} \int_0^T s_0(t)s_1(t)dt \\
&= \rho\sqrt{\mathcal{E}_0}
\end{aligned}
\tag{8.47}
$$

Thus,

$$f_2(t) = s_0(t) - \rho\sqrt{\mathcal{E}_0}\phi_1(t) \tag{8.48}$$

and

$$\phi_2(t) = \frac{1}{\sqrt{\mathcal{E}_0(1-\rho^2)}}[s_0(t) - \rho\sqrt{\mathcal{E}_0}\phi_1(t)] \tag{8.49}$$

The remaining $\phi_k(t)$, $k = 3, 4, \cdots$, needed to complete the orthonormal set, can be selected from any set orthogonal to $\phi_1(t)$ and $\phi_2(t)$. In terms of the basis functions, $s_1(t)$, and $s_0(t)$ are

$$s_1(t) = \sqrt{\mathcal{E}_1}\phi_1(t) \tag{8.50}$$

$$s_0(t) = [\rho\sqrt{\mathcal{E}_0}\phi_1(t) + \sqrt{\mathcal{E}_0(1-\rho^2)}\phi_2(t)] \tag{8.51}$$

The general binary hypothesis problem is now given by

$$H_1 : Y(t) = \sqrt{\mathcal{E}_1}\phi_1(t) + W(t) \qquad\qquad ,0 \leq t \leq T$$

$$H_0 : Y(t) = [\rho\sqrt{\mathcal{E}_0}\phi_1(t) + \sqrt{\mathcal{E}_0(1-\rho^2)}\phi_2(t)] + W(t) \quad ,0 \leq t \leq T$$

$$(8.52)$$

To obtain the random variables Y_k, $k = 1, 2, \cdots$, we need to determine the Karhunen–Loeve coefficients of $Y(t)$. Thus,

$$Y_1 = \begin{cases} H_1 : \int_0^T \{Y(t)\}\phi_1(t)dt = \int_0^T \sqrt{\mathcal{E}_1}\phi_1^2(t)dt + \int_0^T W(t)\phi_1(t)dt \\[2mm] H_0 : \int_0^T \{[\rho\sqrt{\mathcal{E}_0}\phi_1(t) + \sqrt{\mathcal{E}_0(1-\rho^2)}\phi_2(t)] + W(t)\}\phi_1(t)dt \end{cases}$$

$$(8.53)$$

or

$$Y_1 = \begin{cases} H_1 : \sqrt{\mathcal{E}_1} + W_1 \\[2mm] H_0 : \rho\sqrt{\mathcal{E}_0} + W_1 \end{cases}$$

$$(8.54)$$

since, $\int_0^T \phi_1(t)\phi_2(t)dt = 0$ and $\int_0^T \phi_1^2(t)dt = 1$. Also,

$$Y_2 = \begin{cases} H_1 : \int_0^T \{Y(t)\}\phi_2(t)dt = \int_0^T \sqrt{\mathcal{E}_1}\phi_1(t)\phi_2(t)dt + \int_0^T W(t)\phi_2(t)dt \\[2mm] H_0 : \int_0^T \{[\rho\sqrt{\mathcal{E}_0}\phi_1(t) + \sqrt{\mathcal{E}_0(1-\rho^2)}\phi_2(t)] + W(t)\}\phi_2(t)dt \end{cases}$$

$$(8.55)$$

or

$$Y_2 = \begin{cases} H_1 : W_2 \\[2mm] H_0 : \sqrt{\mathcal{E}_0(1-\rho^2)} + W_2 \end{cases}$$

$$(8.56)$$

The random variables Y_k's for $k > 2$ are dependent on the choice of the hypotheses, and thus, they have no effect on the decision. They are

$$
Y_k = \begin{cases} H_1 : \int_0^T \{\sqrt{\mathcal{E}_1}\phi_1(t) + W(t)\}\phi_k(t)dt = W_k \\[3mm] H_0 : \int_0^T \{[\rho\sqrt{\mathcal{E}_0}\phi_1(t) + \sqrt{\mathcal{E}_0(1-\rho^2)}\phi_2(t)] + W(t)\}\phi_k(t)dt = W_k \end{cases}
$$

$$(8.57)$$

since the W_k's, $k = 1, 2, \cdots$, are the coefficients of the Karhunen–Loeve expansion of the white Gaussian process, with mean zero, and power spectral density $N_0/2$, they are statistically independent Gaussian random variables with mean zero, and variance $N_0/2$.

The equivalent problem to (8.39) is now two-dimensional and is given by

$$
H_1 : \begin{cases} Y_1 = \sqrt{\mathcal{E}_1} + W_1 \\[3mm] Y_2 = \qquad\quad W_2 \end{cases}
$$

$$(8.58a)$$

$$
H_0 : \begin{cases} Y_1 = \qquad \rho\sqrt{\mathcal{E}_0} + W_1 \\[3mm] Y_2 = \sqrt{\mathcal{E}_0(1-\rho^2)} + W_2 \end{cases}
$$

$$(8.58b)$$

In vector form, the received vector $\mathbf{Y}$, and the signal vectors $\mathbf{s}_1$, and $\mathbf{s}_0$, are

$$
\mathbf{Y} = \begin{bmatrix} Y_1 \\ Y_2 \end{bmatrix} \quad , \quad \mathbf{s}_1 = \begin{bmatrix} s_{11} \\ s_{12} \end{bmatrix} \quad \text{and} \quad \mathbf{s}_0 = \begin{bmatrix} s_{01} \\ s_{02} \end{bmatrix} \quad (8.59)
$$

Y_1 and Y_2 are statistically independent Gaussian random variables with mean vector $\mathbf{m}_1$, under hypothesis H_1, and mean vector $\mathbf{m}_0$, under hypothesis H_0 given by

$$
\mathbf{m}_1 = \begin{bmatrix} m_{11} \\ m_{12} \end{bmatrix} = E[\mathbf{Y}|H_1] = \begin{bmatrix} \sqrt{\mathcal{E}_1} \\ 0 \end{bmatrix} = \begin{bmatrix} s_{11} \\ s_{12} \end{bmatrix} = \mathbf{s}_1 \quad (8.60)
$$

and

$$\mathbf{m}_0 = \begin{bmatrix} m_{01} \\ m_{02} \end{bmatrix} = E[\mathbf{Y}|H_0] = \begin{bmatrix} \rho\sqrt{\mathcal{E}_0} \\ \sqrt{\mathcal{E}_0(1-\rho^2)} \end{bmatrix} = \begin{bmatrix} s_{01} \\ s_{02} \end{bmatrix} = \mathbf{s}_2 \quad (8.61)$$

Since the components of $\mathbf{Y}$ are uncorrelated, the covariance matrix of $\mathbf{Y}$ under each hypothesis is diagonal and is given by

$$\mathbf{C}_1 = \begin{bmatrix} \frac{N_0}{2} & 0 \\ 0 & \frac{N_0}{2} \end{bmatrix} = \mathbf{C}_0 = \mathbf{C} \quad (8.62)$$

Thus, using the results in Equation (7.24) for diagonal equal covariances, the decision rule is

$$T(\mathbf{y}) = (\mathbf{m}_1^T - \mathbf{m}_0^T)\mathbf{C}^{-1}\mathbf{y} \underset{H_0}{\overset{H_1}{\underset{<}{>}}} \gamma \quad (8.63)$$

where

$$\gamma = \ln\eta + \frac{1}{2}(\mathbf{m}_1^T\mathbf{C}^{-1}\mathbf{m}_1 - \mathbf{m}_0^T\mathbf{C}^{-1}\mathbf{m}_0) \quad (8.64)$$

and

$$\mathbf{C}^{-1} = \begin{bmatrix} \frac{2}{N_0} & 0 \\ 0 & \frac{2}{N_0} \end{bmatrix} \quad (8.65)$$

since $\mathbf{C}^{-1}$ is also diagonal, the decision rule reduces to

$$T(\mathbf{y}) = \mathbf{y}^T(\mathbf{m}_1 - \mathbf{m}_0) \underset{H_0}{\overset{H_1}{\underset{<}{>}}} \frac{N_0}{2}\ln\eta + \frac{1}{2}(|\mathbf{m}_1|^2 - |\mathbf{m}_0|^2) = \gamma_1 \quad (8.66)$$

The sufficient statistic is

$$T(\mathbf{Y}) = \mathbf{Y}^T(\mathbf{m}_1 - \mathbf{m}_0) \quad (8.67)$$

Substituting Equations (8.59) to (8.61) in Equation (8.67), the sufficient statistic can be written as

$$
\begin{aligned}
T(\mathbf{Y}) &= Y_1(m_{11} - m_{01}) + Y_2(m_{12} - m_{02}) \\
&= Y_1(\sqrt{\mathcal{E}_1} - \rho\sqrt{\mathcal{E}_0}) - Y_2\sqrt{\mathcal{E}_0(1 - \rho^2)} \\
&= (\sqrt{\mathcal{E}_1} - \rho\sqrt{\mathcal{E}_0})\int_0^T Y(t)\phi_1(t)dt - \sqrt{\mathcal{E}_0(1 - \rho^2)}\int_0^T Y(t)\phi_2(t)dt
\end{aligned}
\tag{8.68}
$$

The optimum correlation receiver is shown in Figure 8.5.

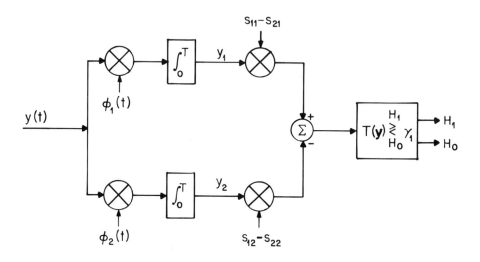

Figure 8.5 Optimum receiver for general binary detection.

This optimum receiver can be implemented in terms of a single correlator. Substituting for the values of $\phi_1(t)$ and $\phi_2(t)$ into Equation (8.68), we have

$$T(Y) = (\sqrt{\mathcal{E}_1} - \rho\sqrt{\mathcal{E}_0}) \int_0^T Y(t) \frac{s_1(t)}{\sqrt{\mathcal{E}_1}} dt$$

$$-\sqrt{\mathcal{E}_0(1-\rho^2)} \int_0^T Y(t) \frac{s_0(t) - \rho\sqrt{\mathcal{E}_0}\phi_1(t)}{\sqrt{\mathcal{E}_0(1-\rho^2)}} dt$$

$$= \int_0^T Y(t)[s_1(t) - s_0(t)]dt$$

$$= \int_0^T Y(t)s_\Delta(t)dt \qquad (8.69)$$

where
$$s_\Delta(t) = s_1(t) - s_0(t) \qquad (8.70)$$

The decision, in this case, is

$$\int_0^T y(t)\{s_1(t) - s_0(t)\}dt \underset{H_0}{\overset{H_1}{\underset{<}{\gtrless}}} \gamma_1 \qquad (8.71)$$

where
$$\gamma_1 = \frac{N_0}{2}\ln\eta + \frac{1}{2}\int_0^T \{s_1^2(t) - s_0^2(t)\}dt \qquad (8.72)$$

The corresponding optimum receiver is shown in Figure 8.6.

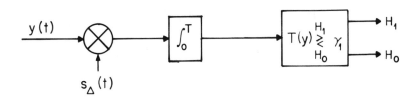

Figure 8.6 Optimum receiver for general binary detection problem with one correlator.

We now study the performance of this detector. Since the sufficient statistic is Gaussian, we only need to solve for the means and variances under each hypothesis, to have a complete description of the conditional density functions. Solving for the means, we have

$$
\begin{aligned}
E[T(\mathbf{Y})|H_1] &= (\sqrt{\mathcal{E}_1} - \rho\sqrt{\mathcal{E}_0})E[Y_1|H_1] - \sqrt{\mathcal{E}_0(1-\rho^2)}E[Y_2|H_1] \\
&= (\sqrt{\mathcal{E}_1} - \rho\sqrt{\mathcal{E}_0})\sqrt{\mathcal{E}_1} = \mathcal{E}_1 - \rho\sqrt{\mathcal{E}_0\mathcal{E}_1}
\end{aligned}
\tag{8.73}
$$

and

$$
\begin{aligned}
E[T(\mathbf{Y})|H_0] &= (\sqrt{\mathcal{E}_1} - \rho\sqrt{\mathcal{E}_0})E[Y_1|H_0] - \sqrt{\mathcal{E}_0(1-\rho^2)}E[Y_2|H_0] \\
&= (\sqrt{\mathcal{E}_1} - \rho\sqrt{\mathcal{E}_0})\rho\sqrt{\mathcal{E}_0} - \sqrt{\mathcal{E}_0(1-\rho^2)}\sqrt{\mathcal{E}_0(1-\rho^2)} \\
&= \rho\sqrt{\mathcal{E}_0\mathcal{E}_1} - \mathcal{E}_0
\end{aligned}
\tag{8.74}
$$

The variances are

$$
\begin{aligned}
\text{var}[T(\mathbf{Y})|H_1] &= \text{var}[T(\mathbf{Y})|H_0] \\
&= (\sqrt{\mathcal{E}_1} - \rho\sqrt{\mathcal{E}_0})^2 \, \text{var}(Y_1|H_1) \\
&\quad +(\sqrt{\mathcal{E}_0(1-\rho^2)})^2 \, \text{var}(Y_2|H_1) \\
&= \{(\sqrt{\mathcal{E}_1} - \rho\sqrt{\mathcal{E}_0})^2 + (\sqrt{\mathcal{E}_0(1-\rho^2)})^2\}\frac{N_0}{2} \\
&= (\mathcal{E}_1 + \mathcal{E}_0 - 2\rho\sqrt{\mathcal{E}_0\mathcal{E}_1})\frac{N_0}{2} = \sigma^2
\end{aligned}
\tag{8.75}
$$

Therefore, the probabilities of detection and false alarm are

$$
\begin{aligned}
P_D &= \int_{\gamma_1}^{\infty} f_{T|H_1}(t|H_1)dt \\
&= \int_{\gamma_1}^{\infty} \frac{1}{\sqrt{2\pi}\sigma} e^{-\frac{1}{2}\frac{(t-\mathcal{E}_1+\rho\sqrt{\mathcal{E}_0\mathcal{E}_1})^2}{\sigma^2}} dt \\
&= Q(\frac{\gamma_1 - \mathcal{E}_1 + \rho\sqrt{\mathcal{E}_0\mathcal{E}_1}}{\sigma})
\end{aligned}
\tag{8.76}
$$

where

$$\gamma_1 = \frac{1}{2}(N_0 \ln\eta + \mathcal{E}_1 - \mathcal{E}_0) \tag{8.77}$$

and

$$
\begin{aligned}
P_F &= \int_{\gamma_1}^{\infty} f_{T|H_0}(t|H_0)dt \\
&= \int_{\gamma_1}^{\infty} \frac{1}{\sqrt{2\pi}\sigma} e^{-\frac{1}{2}\frac{(t-\rho\sqrt{\mathcal{E}_0\mathcal{E}_1}+\mathcal{E}_0)^2}{\sigma^2}} dt \\
&= Q\left(\frac{\gamma_1 + \mathcal{E}_0 - \rho\sqrt{\mathcal{E}_0\mathcal{E}_1}}{\sigma}\right) \tag{8.78}
\end{aligned}
$$

We get more insight into the performance of this system if we assume that the hypotheses are equally likely, and that we use minimum probability of error criterion. In this case,

$$\gamma_1 = \frac{1}{2}(\mathcal{E}_1 - \mathcal{E}_0) \tag{8.79}$$

Define the constant:

$$\alpha = \mathcal{E}_1 + \mathcal{E}_0 - 2\rho\sqrt{\mathcal{E}_1\mathcal{E}_0} \tag{8.80}$$

Substituting (8.79) and (8.80) into (8.77) and (8.78), and rearranging terms, we obtain

$$P_F = Q\left(\frac{1}{2}\sqrt{\frac{2\alpha}{N_0}}\right) \tag{8.81}$$

and

$$P_D = Q\left(-\frac{1}{2}\sqrt{\frac{2\alpha}{N_0}}\right) = 1 - Q\left(\frac{1}{2}\sqrt{\frac{2\alpha}{N_0}}\right) \tag{8.82}$$

Since the probability of miss $P_M = 1 - P_D$, then the probability of error is

$$P(\varepsilon) = P_F = P_M = Q(\frac{1}{2}\sqrt{\frac{2\alpha}{N_0}}) \qquad (8.83)$$

We observe that the probability of error decreases as α increases while N_0 is fixed. Thus, from (8.80), the optimum system is obtained when the correlation coefficient $\rho = -1$. In this case, $s_1(t) = -s_0(t)$, and we say that the signals are *antipodal*. If, in addition, the signal energies are equal, $\mathcal{E}_0 = \mathcal{E}_1 = \mathcal{E}$, the likelihood ratio test is

$$T(\mathbf{y}) = \mathbf{y}^T(\mathbf{m}_1 - \mathbf{m}_0) \mathop{\gtrless}_{H_0}^{H_1} 0 \qquad (8.84)$$

or

$$T(\mathbf{y}) = \sqrt{\mathcal{E}}(1 - \rho) \int_0^T y(t)\phi_1(t)dt - \sqrt{\mathcal{E}(1 - \rho^2)} \int_0^T y(t)\phi_2(t)dt \mathop{\gtrless}_{H_0}^{H_1} 0$$

$$(8.85)$$

Substituting for the values of $\phi_1(t)$ and $\phi_2(t)$ in terms of $s_1(t)$ and $s_0(t)$, into Equation (8.84) and simplifying, an equivalent test is,

$$\int_0^T y(t)s_1(t)dt \mathop{\gtrless}_{H_0}^{H_1} \int_0^T y(t)s_0(t)dt \qquad (8.86)$$

The corresponding receiver is shown in Figure 8.7.

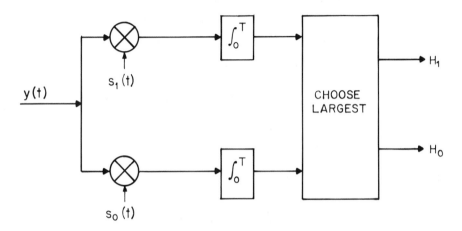

Figure 8.7 Optimum receiver representing Equation (8.85).

The decision rule of Equation (8.86) means that the receiver chooses the signal that has the largest correlation coefficient with the received one.

Example 8.2

Consider a communication system with binary transmission during each duration $T_b = 2\pi/\omega_b$ seconds. The transmitted signal under each hypothesis is

$$H_1 : s_1(t) = A \sin\omega_b t \quad ,0 \le t \le T_b$$

$$H_0 : s_0(t) = A \sin 2\omega_b t \quad ,0 \le t \le T_b$$

The hypotheses are equally likely. During transmission, the channel superimposes on the signals a white Gaussian noise process, with mean zero, and power spectral density $N_0/2$.

Determine the optimum receiver and calculate the probability of error. Assume minimum probability of error criterion.

Solution.

The received signal is characterized by

$$H_1 : Y(t) = s_1(t) + W(t) \qquad , 0 \le t \le T_b$$

$$H_0 : Y(t) = s_0(t) + W(t) \qquad , 0 \le t \le T_b$$

We observe that the signals $s_1(t)$ and $s_0(t)$ are orthogonal with energy

$$\mathcal{E}_1 = \frac{A^2 T_b}{2} = \frac{A^2 \pi}{\omega_b} = \mathcal{E}_0 = \mathcal{E}$$

Thus, the orthonormal basis functions are

$$\phi_1 = \frac{s_1(t)}{\sqrt{\mathcal{E}}} = \sqrt{\frac{2}{T_b}} \sin\omega_b t$$

and

$$\phi_2 = \frac{s_0(t)}{\sqrt{\mathcal{E}}} = \sqrt{\frac{2}{T_b}} \sin 2\omega_b t$$

Using Equations (8.53) and (8.55), we obtain the equivalent decision problem

$$Y_1 = \begin{cases} H_1 : \int_0^T \{s_1(t) + W(t)\}\phi_1(t)dt = \sqrt{\mathcal{E}} + W_1 \\ \\ H_0 : \int_0^T \{s_0(t) + W(t)\}\phi_1(t)dt = W_1 \end{cases}$$

and

$$Y_2 = \begin{cases} H_1 : \int_0^T \{s_1(t) + W(t)\}\phi_2(t)dt = W_2 \\ \\ H_0 : \int_0^T \{s_0(t) + W(t)\}\phi_2(t)dt = \sqrt{\mathcal{E}} + W_2 \end{cases}$$

Correspondingly, the coefficients of the signal vectors s_1 and s_0 are

$$s_1 = \begin{bmatrix} \sqrt{\mathcal{E}} \\ 0 \end{bmatrix} \quad \text{and} \quad s_2 = \begin{bmatrix} 0 \\ \sqrt{\mathcal{E}} \end{bmatrix}$$

Applying the decision rule of Equation (8.66), we have

$$T(\mathbf{y}) = \mathbf{y}^T(s_1 - s_0) \underset{H_0}{\overset{H_1}{\underset{<}{>}}} \frac{1}{2}(|s_1|^2 - |s_0|^2)$$

where $\ln \eta$ is zero, since we are using minimum probability of error criterion, and $P_0 = P_1$. Substituting for the values of $\mathbf{y}, s_1$ and s_0, the test reduces to

$$T(\mathbf{y}) = y_1 - y_2 \underset{H_0}{\overset{H_1}{\underset{<}{>}}} 0$$

or

$$y_1 \underset{H_0}{\overset{H_1}{\underset{<}{>}}} y_2$$

To determine the probability of error, we need to solve for the mean and variance of the sufficient statistic $T(\mathbf{Y}) = Y_1 - Y_2$. Since Y_1 and Y_2 are uncorrelated Gaussian random variables, $T(\mathbf{Y}) = Y_1 - Y_2$ is also Gaussian with means:

$$\begin{aligned} E[T(\mathbf{Y})|H_1] &= E[Y_1 - Y_2|H_1] \\ &= \sqrt{\mathcal{E}} \end{aligned}$$

$$\begin{aligned} E[T(\mathbf{Y})|H_0] &= E[Y_1 - Y_2|H_0] \\ &= -\sqrt{\mathcal{E}} \end{aligned}$$

and variances:

$$\begin{aligned} \text{var}[T(\mathbf{Y})|H_1] &= \text{var}[T(\mathbf{Y})|H_0] \\ &= \text{var}(Y_1|H_j) + \text{var}(Y_2|H_j) \quad , j = 0, 1 \end{aligned}$$

The variance of Y_1 under hypothesis H_0 is

$$\begin{aligned} \text{var}(Y_1|H_0) &= E[Y_1^2|H_0] \\ \\ &= E[\int_0^T \int_0^T W(t)\phi_1(t)W(u)\phi_1(u)dtdu] \\ \\ &= \int_0^T \int_0^T \phi_1(t)\phi_1(u)E[W(t)W(u)]dtdu \end{aligned}$$

where

$$\begin{aligned} E[W(t)W(u)] &= R_{WW}(t,u) = C_{WW}(t,u) \\ &= \frac{N_0}{2}\delta(t-u) \end{aligned}$$

Thus,

$$\begin{aligned} \text{var}[Y_1|H_0] &= \frac{N_0}{2}\int_0^T \int_0^T \phi_1(t)\phi_1(u)\delta(t-u)dtdu \\ \\ &= \frac{N_0}{2}\int_0^T \phi_1^2(t)dt = \frac{N_0}{2} \end{aligned}$$

and

$$\text{var}[T(\mathbf{Y})|H_1] = \text{var}[T(\mathbf{Y})|H_0] = N_0$$

The conditional density functions of the sufficient statistic are

$$f_{T|H_1}(t|H_1) = \frac{1}{\sqrt{2\pi N_0}} e^{-\frac{1}{2}\frac{(t-\sqrt{\mathcal{E}})^2}{N_0}}$$

$$f_{T|H_0}(t|H_0) = \frac{1}{\sqrt{2\pi N_0}} e^{-\frac{1}{2}\frac{(t+\sqrt{\mathcal{E}})^2}{N_0}}$$

The probability of error in this case is

$$P(\varepsilon) = P(\varepsilon|H_1)P(H_1) + P(\varepsilon|H_0)P(H_0)$$

$$= P(\varepsilon|H_1) = P(\varepsilon|H_0)$$

$$= \frac{1}{\sqrt{2\pi N_0}} \int_0^\infty e^{-\frac{1}{2}\frac{(t+\sqrt{\mathcal{E}})^2}{N_0}} dt$$

$$= Q(\sqrt{\frac{\mathcal{E}}{N_0}}) = Q(\sqrt{\frac{A^2 T_b}{2N_0}})$$

The optimum receiver is shown in Figure 8.8.

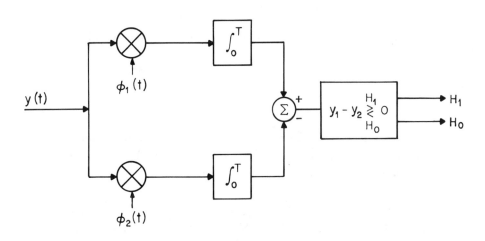

Figure 8.8 Optimum receiver for Example 8.2.

8.3 M-ARY DETECTION

We now generalize the concepts developed for binary hypothesis to M hypotheses. In this case, the decision space consists of at most $(M-1)$ dimensions. The problem may be characterized as

$$H_k : Y(t) = s_k(t) + W(t) \quad , \begin{array}{l} 0 \le t \le T \\ k = 1, 2, \cdots, M \end{array} \tag{8.87}$$

where $s_k(t)$ is a known deterministic signal with energy $\mathcal{E}_k$ such that

$$\mathcal{E}_k = \int_0^T s_k^2(t) dt \quad , k = 1, 2, \cdots, M \tag{8.88}$$

and $W(t)$ is an additive white Gaussian noise process with mean zero, and power spectral density $N_0/2$, or of covariance (autocorrelation) function:

$$C_{WW}(t, u) = R_{WW}(t, u) = \frac{N_0}{2} \delta(t - u) \tag{8.89}$$

The M signals may be dependent and correlated with correlation coefficients

$$\rho_{jk} = \frac{1}{\sqrt{\mathcal{E}_j \mathcal{E}_k}} \int_0^T s_j(t) s_k(t) dt \quad , j, k = 1, 2, \cdots, M \tag{8.90}$$

As before, we need to find a set of orthonormal basis functions in order to expand the received process $Y(t)$, and thus, $W(t)$ into a Karhunen–Loeve expansion, since $C_{YY}(t, u) = C_{WW}(t, u)$.

Using the Gram-Schmidt orthogonalization procedure, we can find a set of K basis functions, $K \le M$, if only K signals $\{s_k(t)\}$ are linearly independent out of the original M signals. Once the complete set of K orthonormal functions $\{\phi_j(t)\}$, $j = 1, 2, \cdots, K$, are obtained, we generate the corresponding coefficients by

$$Y_j = \int_0^T Y(t) \phi_j(t) \quad , j = 1, 2, \cdots, K \tag{8.91}$$

From Equation (8.28), the signals $s_k(t)$, $k = 1, 2, \cdots, M$, may be written as

$$s_k(t) = \sum_{j=1}^{K} s_{kj} \phi_j(t) \quad , \quad \begin{array}{l} 0 \le t \le T \\ k = 1, 2, \cdots, M \end{array} \tag{8.92}$$

where s_{kj} is as defined in Equation (6.29). Substituting Equation (8.92) into Equation (8.91) the equivalent M-ary decision problem becomes

$$
\begin{aligned}
H_k : Y_k &= \int_0^T \{s(t) + W(t)\} \phi_k(t) dt \\
&= \int_0^T \{\sum_{j=1}^{K} s_{kj} + W(t)\} \phi_k(t) dt \\
&= \begin{cases} s_{kj} + W_k & , k = 1, 2, \cdots, K \\ \\ W_k & , k = K+1, K+2, \cdots \end{cases}
\end{aligned} \tag{8.93}
$$

We observe that Y_k's are statistically independent Gaussian random variables with variance $N_0/2$, and that only the first K terms affect the decision, since for $k > K$, the coefficients are W_k irrespect of the hypothesis considered. That is, we have reduced the decision space to $K, K \le M$. The mean of the first K coefficients under each hypothesis is

$$E[Y_k|H_j] = m_{kj} = s_{kj} \quad , \quad \begin{array}{l} j = 1, 2, \cdots, M \\ k = 1, 2, \cdots, K \end{array} \tag{8.94}$$

whereas, for $k > K$, the mean is

$$E[Y_k|H_k] = E[W_k] = 0 \tag{8.95}$$

From Equation (3.50), we have seen that the optimum decision is based on the computation of the *a posteriori* probability $P(H_j|\mathbf{Y})$. A decision is made in favor of the hypothesis corresponding to the largest *a posteriori* probability. Since the set of K statistically independent random variables is described by the joint density function

$$
\begin{aligned}
f_{\mathbf{Y}|H_j}(\mathbf{y}|H_j) &= \prod_{k=1}^{K} \frac{1}{\sqrt{\pi N_0}} e^{-\frac{(\mathbf{y}-\mathbf{m}_k)^2}{N_0}} \\
&= \frac{1}{(\pi N_0)^{\frac{K}{2}}} e^{-\frac{1}{N_0} \sum_{k=1}^{K} (y_k - m_{kj})^2}
\end{aligned} \tag{8.96}
$$

and the *a posteriori* probability on which the decision is based is given by

$$P(H_j|\mathbf{Y}) = \frac{P(H_j)f_{\mathbf{Y}|H_j}(\mathbf{y}|H_j)}{f_{\mathbf{Y}}(\mathbf{y})} \tag{8.97}$$

the sufficient statistic can be expressed as

$$T_j(\mathbf{y}) = P_j f_{\mathbf{Y}|H_j}(\mathbf{y}|H_j) \qquad ,j = 1, 2, \cdots, M \tag{8.98}$$

Note that $f_{\mathbf{Y}}(\mathbf{y})$ which is the denominator of Equation (8.97), is common to all signals, and hence, it does not affect the decision and need not be included in the computation. Substituting Equation (8.96) into Equation (8.98) and taking the logarithm, an equivalent sufficient statisic is

$$T_j^1(\mathbf{y}) = \ln P_j - \frac{1}{N_0}\sum_{k=1}^{K}(y_k - m_{kj})^2 \qquad ,j = 1, 2, \cdots, M \tag{8.99}$$

where

$$T_j^1(\mathbf{y}) = T_j(\mathbf{y}) + \ln(\pi N_0)^{\frac{K}{2}} \tag{8.100}$$

the 1 of $T_j^1(\mathbf{y})$ is superscript. Also, from Equations (8.94) and (8.95), the signal vector is equal to the mean vector. That is,

$$\mathbf{s}_j = \begin{bmatrix} s_{1j} \\ s_{2j} \\ \vdots \\ s_{Kj} \end{bmatrix} = E[\mathbf{Y}|H_j] = \mathbf{m}_j = \begin{bmatrix} m_{1j} \\ m_{2j} \\ \vdots \\ m_{Kj} \end{bmatrix} \qquad ,j = 1, 2, \cdots, M \tag{8.101}$$

We observe that if the hypotheses are equally likely, $P_j = 1/M$ for all j, then Equation (8.98) means compute $f_{\mathbf{Y}|H_j}(\mathbf{y}|H_j)$ and select the maximum. That is, the MAP criterion is reduced to the ML criterion. Also, the sufficient statistic reduces to

$$
\begin{aligned}
T_j^2(\mathbf{y}) &= -\sum_{k=1}^{K}(y_k - m_{kj})^2 = -\sum_{k=1}^{K}(y_k - s_{kj})^2 \\
&= |\mathbf{y} - \mathbf{s}_j|^2 \qquad , j = 1, 2, \cdots, M
\end{aligned}
$$

(8.102)

where

$$
T_j^2(\mathbf{y}) = N_0[T_j^1(\mathbf{y}) + \ln M] \tag{8.103}
$$

and the 2 of $T_j^2(\mathbf{y})$ is superscript. In other words, the receiver decides in favor of the signal that maximizes the metric. Dropping the minus sign in Equation (8.102), means that the receiver computes $\sum_{k=1}^{K}(y_k - s_{kj})^2$ and decides in favor of the signal with the smallest distance.

The computation of the decision random variables given by the sufficient statistic in Equation (8.103) can be simplified if the signals transmitted have equal energy. The equivalent sufficient statistic is just (see problem 8.7)

$$
\begin{aligned}
T_j^3(\mathbf{Y}) &= \mathbf{s}_k^T \mathbf{Y} \\
&= \int_0^T s_k(t)Y(t) \qquad , j = 1, 2, \cdots, M
\end{aligned}
$$

(8.104)

where the 3 of $T_j^3(\mathbf{Y})$ is superscript. The optimum receiver computes the decision variables from Equation (8.104) and decides in favor of one. This receiver is referred to as the "largest of" receiver and is shown in Figure 8.9.

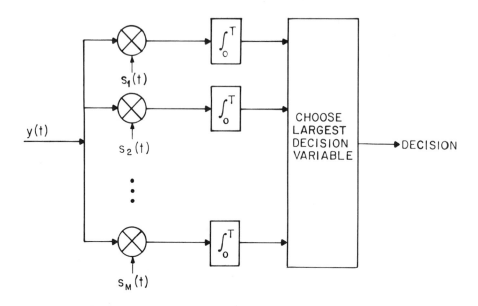

Figure 8.9 "Largest of" receiver.

Probability of Error of M-orthogonal signals

We have seen that when all hypotheses are equally likely and that all signals have equal energy $\mathcal{E}$, the optimum receiver is the "largest of" receiver, as shown in Figure 8.9, which computes the sufficient statistics given in Equation (8.104) and decides in favor of the hypothesis with largest T_j. The probability of error is given by

$$P(\varepsilon) = P_1 P(\varepsilon|H_1) + \cdots + P_M P(\varepsilon|H_M)$$
$$= P(\varepsilon|H_1) \tag{8.105}$$

due to symmetry. It is easier to calculate $P(\varepsilon)$ using the complement. Thus,

$$P(\varepsilon) = 1 - P_c$$
$$= 1 - P(\text{ all } T_k < T_1, k = 2, 3, \cdots, M|H_1) \tag{8.106}$$

where P_c is the probability of a correct decision. A correct decision for H_1 means that the receiver decides H_1 ($T_1 > T_k$ for all $k \neq 1$) when H_1 is transmitted.

Since the variables Y_k, $k = 1, 2, \cdots, M$, are Gaussian and uncorrelated, the sufficient statistics are also Gaussian and uncorrelated, and thus, statistically independent. The mean of T_k, $k = 1, 2, \cdots, M$, under hypothesis H_1 are

$$E[T_k|H_1] = E[s_k^T \mathbf{Y}|H_1] = E[\int_0^T s_k(t)Y(t)dt|H_1]$$

$$= \begin{cases} \mathcal{E} & , k = 1 \\ 0 & , k \neq 1 \end{cases} \tag{8.107}$$

and

$$\text{var}[T_k|H_1] = \frac{N_0}{2} \text{ for all } k \tag{8.108}$$

Hence, the conditional density functions of the sufficient statistics are

$$f_{T_1|H_1}(t_1|H_1) = \frac{1}{\sqrt{\pi N_0}} e^{-\frac{(t_1-\mathcal{E})^2}{N_0}} \tag{8.109}$$

and

$$f_{T_k|H_1}(t_k|H_1) = \frac{1}{\sqrt{\pi N_0}} e^{-\frac{(t_k)^2}{N_0}} \quad , k = 2, 3, \cdots, M \tag{8.110}$$

The probability of error is given by

$$P(\varepsilon) = 1 - P_c \tag{8.111}$$

where P_c is given by

$$P_c = P(T_2 < T_1, T_3 < T_1, \cdots, T_M < T_1|H_1)$$

$$= P(T_2 < T_1|H_1)P(T_3 < T_1|H_1) \cdots P(T_M < T_1|H_1) \tag{8.112}$$

Given a value of the random variable T_1, we have

$$P(T_k < t_1, k = 2, 3, \cdots, M | H_1) = [\int_{-\infty}^{t_1} f_{T_k|H_1}(t_k|H_1)dt_k]^{M-1} \quad (8.113)$$

Averaging all possible values of T_1, the probability of a correct decision is

$$P_c = \int_{-\infty}^{\infty} f_{T_1|H_1}(t_1|H_1)[\int_{-\infty}^{t_1} f_{T_k|H_1}(t_k|H_1)dt_k]^{M-1}dt_1 \quad (8.114)$$

Thus, $P(\varepsilon)$ is obtained to be

$$P_c = \frac{1}{\sqrt{\pi N_0}} \int_{-\infty}^{\infty} \left[1 - Q(t_1\sqrt{\tfrac{2}{N_0}}) \right]^{M-1} e^{-\frac{(t_1-\varepsilon)^2}{N_0}} dt_1 \quad (8.115)$$

Example 8.3

A signal source generates the following waveforms:

$$s_1(t) = \cos\omega_c t \qquad , 0 \le t \le T$$

$$s_2(t) = \cos(\omega_c t + \tfrac{2\pi}{3}) \quad , 0 \le t \le T$$

$$s_3(t) = \cos(\omega_c t - \tfrac{2\pi}{3}) \quad , 0 \le t \le T$$

where $\omega_c = (2\pi)/T$. During transmission, the channel superimposes on the signal a Gaussian noise with mean zero, and power spectral density $N_0/2$. Determine the optimum receiver, and show the decision regions on the signal space. Assume that the signals are equally likely, and minimum probability of error criterion.

Solution.
   We observe that the three signals $s_1(t), s_2(t)$, and $s_3(t)$ have equal energy $\mathcal{E} = T/2$. Let the first basis function be

$$\phi_1(t) = \frac{s_1(t)}{\sqrt{\mathcal{E}}} = \sqrt{\frac{2}{T}} \cos\omega_c t \qquad , 0 \le t \le T$$

Using trigonometric identities, $s_2(t)$, and $s_3(t)$ can be written as

$$s_2(t) = \cos(\omega_c t + \frac{2\pi}{3}) = \cos(\omega_c t)\cos(\frac{2\pi}{3}) - \sin(\omega_c t)\sin(\frac{2\pi}{3})$$

$$s_3(t) = \cos(\omega_c t - \frac{2\pi}{3}) = \cos(\omega_c t)\cos(\frac{2\pi}{3}) + \sin(\omega_c t)\sin(\frac{2\pi}{3})$$

where $\cos[(2\pi)/3] = -1/2$ and $\sin[(2\pi)/3] = \sqrt{3/2}$. By inspection, $k = 2$ orthonormal functions are needed to span the signal set. Hence,

$$\phi_1(t) = \sqrt{\frac{2}{T}}\ \cos\omega_c t \quad , 0 \le t \le T$$

$$\phi_2(t) = \sqrt{\frac{2}{T}}\ \sin\omega_c t \quad , 0 \le t \le T$$

The optimum receiver is the "largest of" receiver as shown in Figure 8.10.

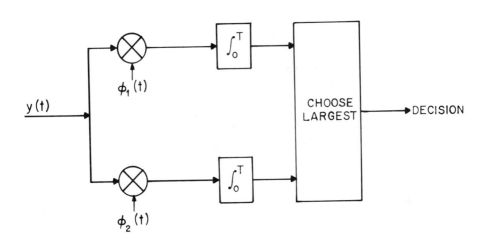

Figure 8.10 Optimum receiver for Example 8.3.

In terms of the basis functions, the signal set $\{s_k(t)\}$ may be expressed as

$$s_1(t) = \sqrt{\frac{T}{2}}\, \phi_1(t)$$

$$s_2(t) = -\frac{1}{2}\sqrt{\frac{T}{2}}\, \phi_1(t) - \frac{1}{2}\sqrt{\frac{3T}{2}}\, \phi_2(t)$$

$$s_3(t) = -\frac{1}{2}\sqrt{\frac{T}{2}}\, \phi_1(t) + \frac{1}{2}\sqrt{\frac{3T}{2}}\, \phi_2(t)$$

The signal constellation and the decision regions are shown in Figure 8.11.

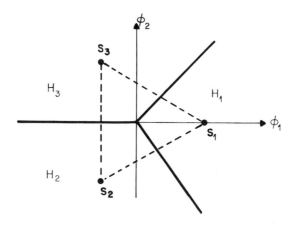

Figure 8.11 Decision regions for Example 8.3.

Example 8.4

Consider the problem given in Example 8.3 assuming the signal set:

$$s_k(t) = A \sin[\omega_c t + (k-1)\frac{\pi}{2}] \quad , \quad \begin{matrix} 0 \le t \le T \\ k = 1,2,3,4 \end{matrix}$$

Solution.

Using trigonometric identities, $s_k(t)$ can be written as

$$s_k(t) = A \sin(\omega_c t) \cos[(k-1)\frac{\pi}{2}] + A \cos(\omega_c t) \sin[(k-1)\frac{\pi}{2}] \quad , k = 1,2,3,4$$

or

$$s_1(t) = A \sin\omega_c t \qquad s_3(t) = -A \sin\omega_c t$$
$$s_2(t) = A \cos\omega_c t \qquad s_4(t) = -A \cos\omega_c t$$

The signals have equal energy $\mathcal{E} = (A^2 T)/2$. By inspection, $K = 2$ orthonormal functions are needed to span the signal set $\{s_k(t)\}$, $k = 1,2,3,4$. Thus, we have

$$\phi_1(t) = \sqrt{\frac{2}{T}} \cos\omega_c t \quad , 0 \le t \le T$$

$$\phi_2(t) = \sqrt{\frac{2}{T}} \sin\omega_c t \quad , 0 \le t \le T$$

Again, since the signals have equal energy and are equally likely, the optimum receiver is the "largest of" receiver and the decision regions, which are based on the "nearest neighbor" rule, are shown in Figure 8.12.

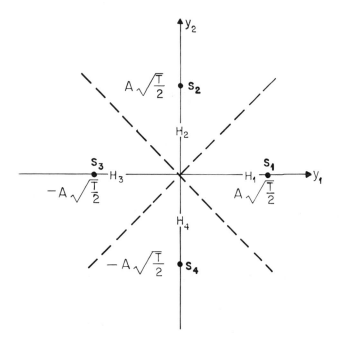

Figure 8.12 Decison space for Example 8.12.

Note that a rotation of the signal set does not affect the probability of error. For convenience, let the new signal set be as shown in Figure 8.13.

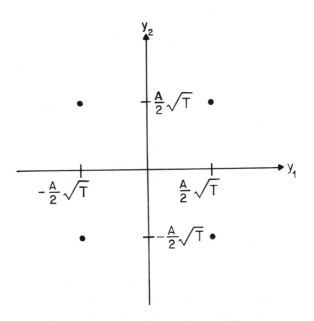

Figure 8.13 Signal set for Example 8.4 after rotation.

Assuming that the signal $s_1(t)$ is transmitted, the probability of error is

$$P(\varepsilon|H_1) = P(\mathbf{Y} \text{ falls outside first quadrant}|H_1)$$

Due to symmetry and the fact $P_j = 1/4$ for $j = 1, 2, 3, 4,$

$$
\begin{aligned}
P(\varepsilon|H_1) &= P(\varepsilon|H_2) = P(\varepsilon|H_3) = P(\varepsilon|H_4) \\
&= P(\varepsilon)
\end{aligned}
$$

Y_1 and Y_2 are statistically independent Gaussian random variables with means:

$$E[Y_1|H_1] = E[Y_2|H_1] = \frac{A}{2}\sqrt{T}$$

and variances:

$$\text{var}(Y_1|H_1) = \text{var}(Y_2|H_1) = \frac{N_0}{2}$$

Therefore,

$$
\begin{aligned}
P_c &= P(0 \le Y_1 \le \infty)P(0 \le Y_2 \le \infty) \\
&= \{\int_0^\infty \frac{1}{\sqrt{\pi N_0}} e^{-\frac{(y-\frac{A}{2}\sqrt{T})^2}{N_0}} dy\}^2 \\
&= \{Q(-\frac{A}{2}\sqrt{\frac{2T}{N_0}})\}^2 = \{1 - Q(\frac{A}{2}\sqrt{\frac{2T}{N_0}})\}^2
\end{aligned}
$$

and the probability of error is

$$P(\varepsilon) = 1 - \{1 - Q(\frac{A}{2}\sqrt{\frac{2T}{N_0}})\}^2$$

8.4 LINEAR ESTIMATION

In Chapter 4, we studied some techniques for parameter estimation in some optimum way, based on a finite number of samples of the signal. In this section, we consider parameter estimation of the signal, but in the presence of an additive white Gaussian noise process with mean zero, and power spectral density $N_0/2$. The received waveform is of the form

$$Y(t) = s(t, \theta) + W(t) \quad , 0 \le t \le T \qquad (8.116)$$

where θ is the unknown parameter to be estimated, and $s(t)$ is a deterministic signal with energy $\mathcal{E}$. The parameter θ may be either random or nonrandom. If it is random, we use Bayes estimation; otherwise, we use the maximum likelihood estimation. We assume that $s(t, \theta)$, a mapping of the parameter θ into a time function, is linear. That is, the superposition principle holds such that

$$s(t, \theta_1 + \theta_2) = s(t, \theta_1) + s(t, \theta_2) \qquad (8.117)$$

The estimator of the above mentioned problem is *linear* as will be shown later, and thus, we refer to the problem as *a linear estimation* problem.

Systems that use linear mappings are known as *linear signaling* or *linear modulation* systems. For such signaling the received waveform may be expressed as

$$Y(t) = \theta s(t) + W(t) \qquad ,0 \le t \le T \qquad (8.118)$$

We now consider the cases where the parameter is nonrandom and random.

8.4.1 ML Estimation

In this case, θ is a nonrandom parameter. $Y(t)$ may be expressed in a series of orthonormal functions such that

$$Y(t) = \lim_{K \to \infty} \sum_{k=1}^{K} Y_k \phi_k(t) \qquad (8.119)$$

where

$$Y_k = \int_0^T Y(t)\phi_k(t)dt \qquad (8.120)$$

and the functions ϕ_k's form a complete set of orthonormal functions. Thus, the first basis function is

$$\phi_1(t) = \frac{s(t)}{\sqrt{\mathcal{E}}} \qquad (8.121)$$

Substituting Equation (8.121) into Equation (8.120), and let $k > 1$, we obtain

$$Y_k = \int_0^T \{\theta\sqrt{\mathcal{E}}\phi_1(t) + W(t)\}\phi_k(t)dt = W_k \qquad (8.122)$$

which does not depend on the parameter to be estimated, whereas

$$Y_1 = \int_0^T \{\theta\sqrt{\mathcal{E}}\phi_1(t) + W(t)\}\phi_1(t)dt = \theta\sqrt{\mathcal{E}} + W_1 \qquad (8.123)$$

depends on θ. Consequently, Y_1 is a sufficient statistic. Y_1 is a Gaussian random variable with mean $\theta\sqrt{\mathcal{E}}$ and variance $N_0/2$.

The likelihood function is

$$L(\theta) = f_{Y_1|\Theta}(y_1|\theta) = \frac{1}{\sqrt{\pi N_0}}\, e^{-\frac{(y_1 - \theta\sqrt{\mathcal{E}})^2}{N_0}} \tag{8.124}$$

From Equation (4.3), we know that to obtain the ML estimate $\hat{\theta}$ is to solve the likelihood equation. That is,

$$\frac{\partial}{\partial\theta} \ln L(\theta) = \frac{\partial}{\partial\theta}[-\frac{1}{2} \ln \pi N_0 - \frac{(y_1 - \theta\sqrt{\mathcal{E}})^2}{N_0}]$$

$$= \frac{2\sqrt{\mathcal{E}}}{N_0}(y_1 - \theta\sqrt{\mathcal{E}}) = 0 \tag{8.125}$$

or

$$\hat{\theta}_{ml} = \frac{Y_1}{\sqrt{\mathcal{E}}} \tag{8.126}$$

Therefore, this optimum estimator is just a correlation of the received signal with the signal $s(t)$ normalized, as shown in Figure 8.14.

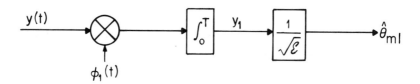

Figure 8.14 Optimum ML estimator.

To check if an estimate is "good", we need to compute its bias, error variance or Cramer–Rao bound, and determine its consistency. We observe that $\hat{\theta}_{ml}$ is unbiased, since $E[Y_1] = \theta\sqrt{\mathcal{E}}$, and thus, from (8.126):

$$E[\hat{\theta}_{ml}(Y_1)] = \frac{1}{\sqrt{\mathcal{E}}} E[Y_1] = \theta \tag{8.127}$$

Also, for an unbiased estimate, the variance of the error is equal to lower bound of the Cramer–Rao inequality provided it is efficient. Using equations (4.46) and (8.125), we have

$$\frac{\partial \ln f_{Y_1|\Theta}(y_1|\theta)}{\partial \theta} = \frac{2\sqrt{\mathcal{E}}}{N_0}(y_1 - \theta\sqrt{\mathcal{E}})$$

$$= \frac{2\mathcal{E}}{N_0}(\frac{y_1}{\sqrt{\mathcal{E}}} - \theta) = k(\theta)[\hat{\theta}(y_1) - \theta]|_{\theta=\hat{\theta}_{ml}} \qquad (8.128)$$

which means that $\text{var}(\hat{\theta}_{ml} - \theta)$ equals the lower bound of the Cramer–Rao inequality given in (4.28).

8.4.2 MAP Estimation

Following the same procedure as in Section 8.4.1, we obtain the sufficient statistic Y_1. However, since θ is a random variable, the MAP estimate is obtained by solving the MAP equation in (4.26). Assume that θ is Gaussian with mean zero and variance σ_θ^2; that is,

$$f_\Theta(\theta) = \frac{1}{\sqrt{2\pi}\sigma_\theta} e^{-\frac{\theta^2}{2\sigma_\theta^2}} \qquad (8.129)$$

The MAP equation is

$$\frac{\partial \ln f_{\Theta|Y_1}(\theta|y_1)}{\partial \theta} = \frac{\partial \ln f_{Y_1|\Theta}(y_1|\theta)}{\partial \theta} + \frac{\partial \ln f_\Theta(\theta)}{\partial \theta}$$

$$= \frac{2\sqrt{\mathcal{E}}}{N_0}(y_1 - \theta\sqrt{\mathcal{E}}) + \frac{\theta}{\sigma_\theta^2} = 0 \qquad (8.130)$$

Solving for θ, we obtain the MAP estimate to be

$$\theta_{map}(Y_1) = \frac{(2\sqrt{\mathcal{E}})/N_0}{[(2\mathcal{E})/N_0] + (1/\sigma_\theta^2)} Y_1$$

$$= \alpha\hat{\theta}_{ml} \qquad (8.131)$$

where

$$\alpha = \frac{(2\sqrt{\mathcal{E}})/N_0}{[(2\mathcal{E})/N_0] + (1/\sigma_\theta^2)} \qquad (8.132)$$

386

Also, it is easily shown that the mean-square error of the MAP estimate is equal to the lower bound of the Cramer–Rao inequality; that is,

$$\text{var}\{[\theta_{map}(Y_1) - \theta]^2\} = -\frac{1}{E[\frac{\partial^2 \ln f_{Y_1|\Theta}(y|\theta)}{\partial \theta^2}]}$$

$$= \frac{\sigma_\theta^2 N_0}{2\sigma_\theta^2 \mathcal{E} + N_0} \qquad (8.133)$$

The optimum MAP estimator is shown in Figure 8.15.

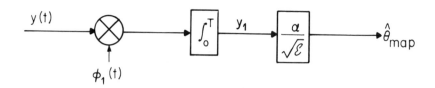

Figure 8.15 Optimum MAP estimator.

8.5 NONLINEAR ESTIMATION

The function $s(t, \theta)$ is now a nonlinear function in θ. Again, θ may be random or nonrandom.

8.5.1 ML Estimation

Let $\{\phi_k(t)\}$ be a set of K orthonormal basis functions. Since we require an infinite number of basis functions to represent $Y(t)$, we approximate the received signal $Y(t)$ as

$$Y(t) = \sum_{k=1}^{K} Y_k \phi_k(t) \qquad (8.134)$$

where

$$Y_k = \int_0^T Y(t)\phi_k(t)dt \tag{8.135}$$

Substituting Equation (8.116) into Equation (8.135), we have

$$
\begin{aligned}
Y_k &= \int_0^T \{s(t,\theta) + W(t)\}\phi_k(t)dt \\
&= \int_0^T s(t,\theta)\phi_k(t)dt + \int_0^T W(t)\phi_k(t)dt \\
&= s_k(\theta) + W_k \qquad\qquad k = 1, 2, \cdots, K
\end{aligned}
\tag{8.136}
$$

where

$$s_k(\theta) = \int_0^T s(t,\theta)\phi_k(t)dt \tag{8.137}$$

The Y_k's are statistically independent Gaussian random variables with mean $s_k(\theta)$, and variance $N_0/2$. Thus, the likelihood function, from Equation (4.2), is

$$L(\theta) = f_{\mathbf{Y}|\Theta}(\mathbf{y}|\theta) = \frac{1}{(\pi N_0)^{\frac{K}{2}}} \prod_{k=1}^{K} e^{-\frac{\{y_k - s_k(\theta)\}^2}{N_0}} \tag{8.138}$$

As $K \to \infty$, Equation (8.138) is not well defined. In fact,

$$\lim_{K \to \infty} f_{\mathbf{Y}|\Theta}(\mathbf{y}|\theta) = \begin{cases} 0 & \text{for } \pi N_0 > 1 \\ \infty & \text{for } \pi N_0 < 1 \end{cases} \tag{8.139}$$

Since the likelihood function is not affected if it is divided by any function that does not depend on θ, we avoid the convergence difficulty of Equation (8.137) by dividing $L(\theta)$ by

$$f_{\mathbf{Y}}(\mathbf{y}) = \prod_{k=1}^{K} \frac{1}{\sqrt{\pi N_0}} e^{-\frac{y_k^2}{N_0}} \tag{8.140}$$

Consequently, we define $\Lambda'[\mathbf{y}, \theta]$ as

$$
\begin{aligned}
\Lambda'[\mathbf{y}, \theta] &\triangleq \frac{f_{\mathbf{Y}|\Theta}(\mathbf{y}|\theta)}{f_{\mathbf{Y}}(\mathbf{y})} \\
&= e^{\frac{2}{N_0} \sum_{k=1}^{K} y_k s_k(\theta) - \frac{1}{N_0} \sum_{k=1}^{K} s_k^2(\theta)}
\end{aligned}
\tag{8.141}
$$

The ML estimate is the value of θ for which $\Lambda_k[\mathbf{Y}, \theta]$ is maximum. Using Parseval's theorem and the fact that $\lim_{K \to \infty} y_k(t) = y(t)$ and $\lim_{k \to \infty} s_k(t, \theta) = s(t, \theta)$, we obtain

$$\lim_{K \to \infty} \sum_{k=1}^{K} y_k s_k(\theta) = \int_0^T y(t) s(t, \theta) dt \qquad (8.142)$$

and

$$\lim_{K \to \infty} \sum_{k=1}^{K} s_k^2(\theta) = \int_0^T s^2(t, \theta) dt \qquad (8.143)$$

Using Equations (8.142) and (8.143), and taking the logarithm as $K \to \infty$, the likelihood function is

$$\ln\Lambda'[Y(t), \theta] = \frac{2}{N_0} \int_0^T Y(t) s(t, \theta) dt - \frac{1}{N_0} \int_0^T s^2(t, \theta) dt \qquad (8.144)$$

To obtain the ML estimate $\hat{\theta}_{ml}$, which maximizes the log likelihood function, we differentiate Equation (8.144) with respect to θ, and set the result to zero. We find that the ML estimate $\hat{\theta}$ is the solution to the equation:

$$\int_0^T [Y(t) - s(t, \hat{\theta})] \frac{\partial s(t, \hat{\theta})}{\partial \hat{\theta}} dt = 0 \qquad (8.145)$$

Since $\hat{\theta}_{ml}$ is an unbiased estimate, it can be shown that the error variance from the inequality:

$$\text{var}\{\hat{\theta}[Y(t)] - \theta\} \geq \frac{N_0}{2 \int_0^T [\frac{\partial s(t,\theta)}{\partial \theta}]^2 dt} \qquad (8.146)$$

equals the lower bound of the Cramer-Rao, if and only if, as $K \to \infty$,

$$\frac{\partial \ln\Lambda'[Y(t), \theta]}{\partial \theta} = k(\theta)\{\hat{\theta}[Y(t)] - \theta\} \qquad (8.147)$$

Example 8.5

Consider a known signal of the form:

$$s(t,\theta) = A_c \sin(2\pi f_c t + \theta) \quad , 0 \le t \le T$$

where the amplitude A_c and the frequency f_c, $\omega_c = (k\pi)/T$ and k integer, are known. We wish to estimate the unknown phase θ.

Solution.
The ML estimate $\hat{\theta}_{ml}$ is the solution to Equation (8.145). That is,

$$\int_0^T [Y(t) - A_c \sin(2\pi f_c t + \hat{\theta})] \cos(2\pi f_c t + \hat{\theta}) dt = 0$$

or

$$\int_0^T Y(t) \cos(2\pi f_c t + \hat{\theta}) dt = 0$$

Since $\int_0^T A_c \sin(2\pi f_c t + \hat{\theta}) \cos(2\pi f_c t + \hat{\theta}) dt = 0$. Using trigonometric identities, we can express the above integral as

$$\cos\hat{\theta} \int_0^T Y(t) \cos(2\pi f_c t) dt - \sin\hat{\theta} \int_0^T Y(t) \sin(2\pi f_c t) dt = 0$$

Solving for $\hat{\theta}$, we obtain the ML estimate to be

$$\hat{\theta} = \tan^{-1} \frac{\int_0^T Y(t) \cos(2\pi f_c t) dt}{\int_0^T Y(t) \sin(2\pi f_c t) dt}$$

8.5.2 MAP Estimation

Now θ is a random variable with density function $f_\Theta(\theta)$. Following the same approach as in Section 8.5.1 and using the fact that $\hat{\theta}_{map}$ is that value of θ for which the conditional density function $f_{\Theta|\mathbf{Y}}(\theta|\mathbf{y})$ is maximum,

$$\begin{aligned} \hat{\theta}_{map} &= \frac{\partial \ln\Lambda'[Y(t),\theta]}{\partial \theta} \\ &= \frac{2}{N_0} \int_0^T \{Y(t) - s(t,\theta)\} \frac{\partial s(t,\theta)}{\partial \theta} dt + \frac{d}{d\theta} \ln f_\Theta(\theta) \end{aligned} \qquad (8.148)$$

If θ is Gaussian with mean zero and variance σ_θ^2, then $d[\ln f_\Theta(\theta)]/dt = -\theta/\sigma_\theta^2$. The MAP estimate becomes

$$\hat{\theta}_{map} = \frac{2\sigma_\theta^2}{N_0} \int_0^T \{Y(t) - s(t,\theta)\} \frac{\partial s(t,\theta)}{\partial \theta} dt|_{\theta=\hat{\theta}_{map}} \qquad (8.149)$$

8.6 GENERAL BINARY DETECTION WITH UNWANTED PARAMETERS

In this section, we consider the general binary detection of signals in an additive white Gaussian noise process with mean zero, and power spectral density $N_0/2$. However, the received waveform is not completely known in advance as in the previous section, where we assume that the only uncertainties are due to additive white Gaussian noise. These signals, which are not completely known in advance, arise in many applications due to fading, random of phase in an echo pulse, etc. The unknown parameters of the signal are known as *unwanted parameters*.

Consider the general binary detection problem where the received signal under hypothesis H_1 and H_0 is given by

$$H_1 : Y(t) = s_1(t, \boldsymbol{\theta}_1) + W(t) \quad , 0 \leq t \leq T$$
$$H_0 : Y(t) = s_0(t, \boldsymbol{\theta}_0) + W(t) \quad , 0 \leq t \leq T$$
$$(8.150)$$

where $\boldsymbol{\theta}_1$ and $\boldsymbol{\theta}_0$ are the unknown random vectors. Note that if $\boldsymbol{\theta}_1$ and $\boldsymbol{\theta}_0$ are known, the signal $s_1(t, \boldsymbol{\theta}_1)$ and $s_0(t, \boldsymbol{\theta}_0)$ are deterministics, and thus, they are completely specified.

The unknown parameter $\boldsymbol{\theta}_j$, $j = 0, 1$, may be either random or nonrandom. In our case, we assume that $\boldsymbol{\theta}_j$, $j = 0, 1$, is a random vector with known *a priori* density function. That is, the joint density function of the components of $\boldsymbol{\theta}_j$, $j = 0, 1$, is known. The approach to solve this problem is to obtain a set of K orthonormal basis functions $\{\phi_k(t)\}$, approximate $Y(t)$ with the K-term series expansion and let $K \rightarrow \infty$. We form the K-term approximate to the likelihood ratio and let $K \to \infty$ to obtain

$$\Lambda[Y(t)] = \lim_{K \to \infty} \Lambda[Y_k(t)]$$

$$= \frac{f_{\mathbf{Y}|H_1}(\mathbf{y}|H_1)}{f_{\mathbf{Y}|H_0}(\mathbf{y}|H_0)} \qquad (8.151)$$

where

$$f_{\mathbf{Y}|H_1}(\mathbf{y}|H_1) = \int_{\mathcal{X}_{\boldsymbol{\theta}_1}} f_{\mathbf{Y},\boldsymbol{\Theta}_1|H_1}(\mathbf{y},\boldsymbol{\theta}_1|H_1)d\boldsymbol{\theta}_1$$

$$= \int_{\mathcal{X}_{\boldsymbol{\theta}_1}} f_{\mathbf{Y}|\boldsymbol{\Theta}_1,H_1}(\mathbf{y}|\boldsymbol{\theta}_1,H_1)f_{\boldsymbol{\Theta}_1|H_1}(\boldsymbol{\theta}_1|H_1)d\boldsymbol{\theta}_1 \qquad (8.152)$$

and

$$f_{\mathbf{Y}|H_0}(\mathbf{y}|H_0) = \int_{\mathcal{X}_{\boldsymbol{\theta}_0}} f_{\mathbf{Y},\boldsymbol{\Theta}_0|H_0}(\mathbf{y},\boldsymbol{\theta}_0|H_0)d\boldsymbol{\theta}_0$$

$$= \int_{\mathcal{X}_{\boldsymbol{\theta}_0}} f_{\mathbf{Y}|\boldsymbol{\Theta}_0,H_0}(\mathbf{y}|\boldsymbol{\theta}_0,H_0)f_{\boldsymbol{\Theta}_0|H_0}(\boldsymbol{\theta}_0|H_0)d\boldsymbol{\theta}_0 \qquad (8.153)$$

$\mathcal{X}_{\boldsymbol{\theta}_j}$, $j = 0, 1$, denotes the space of the parameter $\boldsymbol{\theta}_j$.

We now solve for $f_{\mathbf{Y}|\boldsymbol{\theta}_j,H_j}$, $j = 0, 1$, under the given conditions. Let

$$Y_K(t) = \sum_{k=1}^{K} Y_k \phi_k(t) \qquad (8.154)$$

where

$$Y_k = \int_0^T Y_k \phi_k(t)dt \qquad (8.155)$$

The observation vector is

$$\mathbf{Y}_K = \begin{bmatrix} Y_1 \\ Y_2 \\ \vdots \\ Y_K \end{bmatrix} \qquad (8.156)$$

Substituting Equation (8.150) into Equation (8.155), we obtain that Y_k under H_1 is

$$
\begin{aligned}
Y_k &= \int_0^T s_1(t, \boldsymbol{\theta}_1) \phi_k(t) dt + \int_0^T W(t) \phi_k(t) dt \\
&= s_{k1} + W_k
\end{aligned}
\tag{8.157}
$$

while

$$
\begin{aligned}
Y_k &= \int_0^T s_0(t, \boldsymbol{\theta}_0) \phi_k(t) dt + \int_0^T W(t) \phi_k(t) dt \\
&= s_{k0} + W_k
\end{aligned}
\tag{8.158}
$$

under H_0. Given $\boldsymbol{\theta}_j$, $j = 0, 1$, the Y_k's are statistically independent Gaussian random variables with means:

$$
E[Y_k | \boldsymbol{\theta}_1, H_1] = s_{k1}
\tag{8.159}
$$

$$
E[Y_k | \boldsymbol{\theta}_0, H_0] = s_{k0}
\tag{8.160}
$$

and variances:

$$
\text{var}(Y_k | \boldsymbol{\theta}_1, H_1) = \text{var}(Y_k | \boldsymbol{\theta}_0, H_0) = \frac{N_0}{2}
\tag{8.161}
$$

Thus, the conditional density functions are

$$
f_{\mathbf{Y} | \Theta_1, H_1}(\mathbf{y} | \boldsymbol{\theta}_1, H_1) = \prod_{k=1}^{K} \frac{1}{\sqrt{\pi N_0}} \, e^{-\frac{(y_k - s_{k1})^2}{N_0}}
\tag{8.162}
$$

$$
f_{\mathbf{Y} | \Theta_0, H_0}(\mathbf{y} | \boldsymbol{\theta}_0, H_0) = \prod_{k=1}^{K} \frac{1}{\sqrt{\pi N_0}} \, e^{-\frac{(y_k - s_{k0})^2}{N_0}}
\tag{8.163}
$$

We observe that $\Lambda[Y_k(t)]$ is the ratio of Equations (8.162) and (8.163). In the limit as $K \to \infty$, the terms in the exponent of Equations (8.162) and (8.163), which can be expressed as summations, become

$$
\lim_{K \to \infty} \sum_{k=1}^{K} (y_k - s_{k1})^2 = \int_0^T \{y(t) - s_1(t, \boldsymbol{\theta}_1)\}^2 dt
\tag{8.164}
$$

and

$$
\lim_{K \to \infty} \sum_{k=1}^{K} (y_k - s_{k0})^2 = \int_0^T \{y(t) - s_0(t, \boldsymbol{\theta}_0)\}^2 dt
\tag{8.165}
$$

Substituting Equations (8.164) and (8.165) into Equations (8.158) and (8.153) respectively, we obtain

$$f_{\mathbf{Y}|H_1}(\mathbf{y}|H_1) = \int_{\mathcal{X}_{\boldsymbol{\theta}_1}} f_{\Theta_1|H_1}(\boldsymbol{\theta}_1|H_1) e^{-\frac{1}{N_0}\int_0^T \{y(t)-s_1(t,\boldsymbol{\theta}_1)\}^2 dt} d\boldsymbol{\theta}_1 \quad (8.166)$$

and

$$f_{\mathbf{Y}|H_0}(\mathbf{y}|H_0) = \int_{\mathcal{X}_{\boldsymbol{\theta}_0}} f_{\Theta_0|H_0}(\boldsymbol{\theta}_0|H_0) e^{-\frac{1}{N_0}\int_0^T \{y(t)-s_0(t,\boldsymbol{\theta}_0)\}^2 dt} d\boldsymbol{\theta}_0 \quad (8.167)$$

Hence, the likelihood ratio is just the ratio of Equations (8.166) and (8.167).

Example 8.6

Consider a digital communications system with a source using ON-OFF signaling. The channel superimposes on the transmitted signal an additive white Gaussian noise process with mean zero, and power spectral density $N_0/2$. The received waveforms are given by

$$H_1 : Y(t) = A\cos(\omega_c t + \Theta) + W(t) \quad ,0 \le t \le T$$

$$H_0 : Y(t) = \qquad\qquad\qquad W(t) \quad ,0 \le t \le T$$

where the amplitude A and the phase Θ are independet random variables, with known density functions. Assume that $\omega_c = (2n\pi)/T$ where n is an integer, Θ is uniformly distributed over the interval $[0, 2\pi]$, and A is Rayleigh-distributed with density function:

$$f_A(a) = \begin{cases} \dfrac{a}{\sigma_a} e^{-\frac{a^2}{2\sigma_a^2}} & ,a \ge 0 \\ \\ 0 & \text{otherwise} \end{cases}$$

Solution.

Let $s_1(t) = \cos\omega_c t$, then the signal energy is $\mathcal{E} = T/2$, and the first basis function is $\phi_1(t) = \sqrt{2/T}\,\cos\omega_c t$. Consequently, the first coefficient in the Karhunen–Loeve expansion of $Y(t)$ is

$$
\begin{aligned}
Y_1 &= \int_0^T Y(t)\phi_1(t)dt \\
&= \begin{cases} H_1 : \int_0^T \{A\,\cos(\omega_c t + \theta) + W(t)\}\phi_1(t)dt \\[2mm] H_0 : \int_0^T W(t)\phi_1(t)dt \end{cases}
\end{aligned}
$$

Then, we select a suitable set of functions $\phi_k(t)$, $k = 2, 3, \cdots$, orthogonal to $\phi_1(t)$. We observe that for $k > 2$, we always obtain W_k independently of the hypothesis. Only Y_1 depends on which hypothesis is true. Thus, Y_1 is a sufficient statistic. Y_1 is a Gaussian random variable with means:

$$
\begin{aligned}
E[Y_1|a, \theta, H_1] &= E[a\sqrt{\frac{T}{2}}\,\cos\theta] \\
&= a\sqrt{\mathcal{E}}\,\cos\theta
\end{aligned}
$$

$$
E[Y_1|a, \theta, H_0] = E[W_1] = 0
$$

and variances:

$$
\text{var}(Y_1|a, \theta, H_1) = \text{var}(Y_1|a, \theta, H_0) = \frac{N_0}{2}
$$

We define the conditional likelihood ratio as

$$
\begin{aligned}
\Lambda[y(t)|a, \theta] &= \frac{f_{Y_1|a,\theta,H_1}(y_1|a, \theta, H_1)}{f_{Y_1|a,\theta,H_0}(y_1|a, \theta, H_0)} \\
&= e^{\frac{2a}{N_0}\sqrt{\mathcal{E}}\,\cos\theta y_1} e^{-\frac{a^2}{N_0}\mathcal{E}\,\cos^2\theta}
\end{aligned}
$$

where

$$
\Lambda[y(t)] = \int_{X_a}\int_{X_\theta} \Lambda[y(t)|a, \theta]f_{A,\Theta}(a, \theta)dad\theta
$$

and $\mathcal{X}_a, \mathcal{X}_\theta$ denote range a and θ, respectively. Substituting for the value of $\Lambda[y(t)|a, \theta]$ and $f_{A,\Theta}(a, \theta)$ into the above integral, it can be shown that the decision rule reduces to

$$\Lambda[y(t)] = \frac{N_0}{2\sigma_a^2 + N_0} e^{\frac{2\sigma_a^2}{N_0(2\sigma_a^2 + N_0)} y_1^2} \underset{H_0}{\overset{H_1}{\underset{<}{>}}} \eta$$

or

$$y_1^2 \underset{H_0}{\overset{H_1}{\underset{<}{>}}} \gamma$$

where

$$\gamma = \frac{N_0(2\sigma_a^2 + N_0)}{2\sigma_a^2} \ln\eta \frac{(2\sigma_a^2 + N_0)}{N_0}$$

8.7 BINARY DETECTION IN COLORED NOISE

In the previous sections, we assumed that the additive Gaussian noise is zero mean and white. However, in many applications this assumption is not valid. We now consider detection of signals in nonwhite Gaussian noise. Consequently, the power spectral density is not constant in the filter bandwidth. The noise samples are no longer uncorrelated, and thus, they are statistically dependent. One way to deal with this problem is to extend the concepts using Karhunen–Loeve expansion for white Gaussian noise to colored Gaussian noise. Another way may be to use some preliminary processing for the noise, referred to as whitening, to make the colored noise white, and then, use the Karhunen–Loeve expansion.

The problem under consideration is to design a receiver to test for the general binary detection given by

$$H_1 : Y(t) = s_1(t) + N(t) \quad ,0 \le t \le T$$

$$H_0 : Y(t) = s_0(t) + N(t) \quad ,0 \le t \le T$$

(8.168)

where $Y(t)$ is the received waveform, $s_1(t)$ and $s_0(t)$ are known deterministic signals, and $N(t)$ is the additive colored Gaussian with mean zero, and covariance function $C_{NN}(t, u)$.

8.7.1 Karhunen–Loeve Expansion Approach

The solution to the binary detection problem with white Gaussian noise was relatively simple since the coefficients of the Karhunen–Loeve expansion generated by any set of orthonormal basis function resulted in independent samples. The coefficients $Y_1, Y_2, \cdots, Y_K$ were statistically independent Gaussian random variables, and thus, the likelihood function was just the joint probability density function of these coefficients in the limit as $K \to \infty$. The goal is still to generate uncorrelated coefficients from the likelihood ratio and obtain the decison rule. That is, the corresponding orthonormal functions and the eigenfunctions satisfy the integral equation:

$$\int_0^T C_{NN}(t, u) f_k(u) du = \lambda_k f_k(t) \qquad , 0 \le t \le T \qquad (8.169)$$

where λ_k is the eigenvalue. This means that the coefficients:

$$
\begin{aligned}
Y_k &= \int_0^T Y(t) f_k(t) dt \\
&= \int_0^T \{s(t) + N(t)\} f_k(t) dt \\
&= s_k + N_k \qquad (8.170)
\end{aligned}
$$

where

$$s_k = \int_0^T s(t) f_k(t) dt \qquad (8.171)$$

and

$$N_k = \int_0^T N(t) f_k(t) dt \qquad (8.172)$$

are obtained by the correlation operation, and are uncorrelated. The noise components are uncorrelated Gaussian random variables with zero mean, such that

$$E[N_k] = \int_0^T f_k(t) E[N(t)] dt = 0 \qquad (8.173)$$

and

$$E[N_j N_k] = \lambda_k \delta_{jk} \tag{8.174}$$

as shown in Section 6.5. The series expansion of the noise is

$$N(t) = \lim_{K \to \infty} \sum_{k=1}^{K} N_k f_k(t) \tag{8.175}$$

The Karhunen–Loeve coefficients under hypotheses H_1 and H_0 are

$$H_1 : Y_k = s_{1k} + N_k \tag{8.176}$$

$$H_0 : Y_k = s_{0k} + N_k \tag{8.177}$$

with means:

$$E[Y_k|H_1] = s_{1k} \tag{8.178}$$

$$E[Y_k|H_0] = s_{0k} \tag{8.179}$$

and variances:

$$\text{var}(Y_k|H_1) = \text{var}(Y_k|H_0) = \lambda_k \tag{8.180}$$

since the coefficients Y_k, $k = 1, 2, \cdots, K$, are statistically independent under each hypothesis, the conditional density functions are given by

$$f_{\mathbf{Y}|H_1}(\mathbf{y}|H_1) = \prod_{k=1}^{K} \frac{1}{\sqrt{2\pi\lambda_k}} \, e^{-\frac{(y_k - s_{1k})^2}{2\lambda_k}} \tag{8.181}$$

$$f_{\mathbf{Y}|H_0}(\mathbf{y}|H_0) = \prod_{k=1}^{K} \frac{1}{\sqrt{2\pi\lambda_k}} \, e^{-\frac{(y_k - s_{0k})^2}{2\lambda_k}} \tag{8.182}$$

Consequently, the K-term approximation of the likelihood ratio is

$$\begin{aligned}
\Lambda[y_K(t)] &= \frac{\prod_{k=1}^{K} \frac{1}{\sqrt{2\pi\lambda_k}} \, e^{-\frac{(y_k - s_{1k})^2}{2\lambda_k}}}{\prod_{k=1}^{K} \frac{1}{\sqrt{2\pi\lambda_k}} \, e^{-\frac{(y_k - s_{0k})^2}{2\lambda_k}}} \\
&= e^{\sum_{k=1}^{K} \frac{1}{2\lambda_k} s_{1k}(2y_k - s_{1k}) - \sum_{k=1}^{K} \frac{1}{2\lambda_k} s_{0k}(2y_k - s_{0k})}
\end{aligned} \tag{8.183}$$

Taking the logarithm, we obtain

$$
\ln\Lambda[y_K(t)] = \sum_{k=1}^{K} \frac{1}{2\lambda_k} s_{1k}(2y_k - s_{1k}) - \sum_{k=1}^{K} \frac{1}{2\lambda_k} s_{0k}(2y_k - s_{0k}) \underset{H_0}{\overset{H_1}{\underset{<}{\overset{>}{}}}} \ln\eta
$$

$$(8.184)$$

or, letting $K \to \infty$, the log-likelihood ratio is

$$
\ln\Lambda[y_K(t)] = \sum_{k=1}^{\infty} \frac{1}{2\lambda_k} s_{1k}(2y_k - s_{1k}) - \sum_{k=1}^{\infty} \frac{1}{2\lambda_k} s_{0k}(2y_k - s_{0k}) \quad (8.185)
$$

Substituting Equations (8.170) and (8.171) into (8.185), we obtain the log-likelihood ratio in terms of the correlation to be

$$
\begin{aligned}
\ln\Lambda[y(t)] &= \frac{1}{2}\int_0^T \int_0^T s_1(t)\{2y(u) - s_1(u)\} \sum_{k=1}^{\infty} \frac{f_k(t)f_k(u)}{\lambda_k} dt\,du \\
&\quad - \frac{1}{2}\int_0^T \int_0^T s_0(t)\{2y(u) - s_0(u)\} \sum_{k=1}^{\infty} \frac{f_k(t)f_k(u)}{\lambda_k} dt\,du
\end{aligned}
$$

$$(8.186)$$

Define

$$
h_1(u) = \int_0^T s_1(t) \sum_{k=1}^{\infty} \frac{f_k(t)f_k(u)}{\lambda_k} dt \qquad (8.187)
$$

and

$$
h_0(u) = \int_0^T s_0(t) \sum_{k=1}^{\infty} \frac{f_k(t)f_k(u)}{\lambda_k} dt \qquad (8.188)
$$

Substituting Equations (8.187) and (8.188) into (8.186), the log-likelihood ratio test becomes

$$
\frac{1}{2}\int_0^T h_1(t)\{2y(t) - s_1(t)\}dt - \frac{1}{2}\int_0^T h_0(t)\{2y(t) - s_0(t)\}dt \underset{H_0}{\overset{H_1}{\underset{<}{\overset{>}{}}}} \ln\eta
$$

$$(8.189)$$

or

$$\int_0^T y(t)h_1(t)dt - \int_0^T y(t)h_0(t)dt \underset{H_0}{\overset{H_1}{\underset{<}{>}}} \gamma \qquad (8.190)$$

where

$$\gamma = \ln\eta + \frac{1}{2}\int_0^T s_1(t)h_1(t) - \frac{1}{2}\int_0^T s_0(t)h_0(t) \qquad (8.191)$$

Hence, from Equation (8.189), we see that the receiver for detection of signals in colored noise can be interpreted as a correlation detector as shown in Figure 8.16.

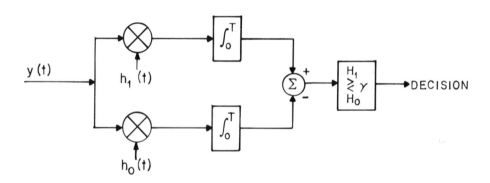

Figure 8.16 Correlation receiver for signal in colored noise.

To build such a receiver, we need to determine $h_1(t)$ and $h_0(t)$. Substituting Equations (8.187) and (8.188) into Equation (8.169), we obtain

$$\int_0^T C_{NN}(t,u)h_1(u)du = \int_0^T C_{NN}(t,u)\int_0^T \sum_{k=1}^{\infty} \frac{f_k(u)}{\lambda_k}s_1(t)f_k(t)dtdu$$

$$= \sum_{k=1}^{\infty} \frac{s_{k1}}{\lambda_k}\int_0^T C_{NN}(t,u)f_k(u)du$$

$$= \sum_{k=1}^{\infty} s_{k1}f_k(t)$$

$$= s_1(t) \tag{8.192}$$

and similarly,

$$\int_0^T C_{NN}(t,u)h_0(u)du = s_0(t) \tag{8.193}$$

That is, $h_1(t)$ and $h_0(t)$ are solutions to the integral Equations in (8.192) and (8.193), respectively.

8.7.2 Whitening Approach

Another approach to detect signals in colored Gaussian noise is to do a preliminary processing of the colored noise. The received signal is passed through a linear time-invariant filter, such that the noise at the output of the filter is white, as shown in Figure 8.17.

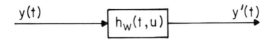

Figure 8.17 Whitening filter.

The process of converting colored noise to white noise is referred to as whitening. Once the noise is white, the problem becomes a detection of known signals in additive white Gaussian noise, which we covered in the previous sections.

We now solve the binary detection problem in colored noise. The output of the whitening filter under hypothesis H_1 is given by

$$Y'(t) = \int_0^T h_w(t, u)Y(u)du \qquad (8.194a)$$

$$= \int_0^T h_w(t, u)\{s_1(u) + N(u)\}du \qquad (8.194b)$$

$$= s_1'(t) + N'(t) \qquad (8.194c)$$

where

$$s_1'(t) = \int_0^T h_w(t, u)s_1(u)du \qquad ,0 \leq t \leq T \qquad (8.195)$$

and

$$N'(t) = \int_0^T h_w(t, u)N(u)du \qquad ,0 \leq t \leq T \qquad (8.196)$$

Under hypothesis H_0, $Y'(t)$ is

$$Y'(t) = \int_0^T h_w(t, u)\{s_0(u) + N(u)\}du$$

$$= s_0'(t) + N'(t) \qquad ,0 \leq t \leq T \qquad (8.197)$$

where

$$s_0'(t) = \int_0^T h_w(t, u)s_0(u)du \qquad ,0 \leq t \leq T \qquad (8.198)$$

Since $N'(t)$ is the white Gaussian noise, its covariance function is given by

$$C_{N'N'}(t, u) = E[N'(t)N'(u)] = \delta(t - u) \qquad ,0 \leq t, u \leq T \qquad (8.199)$$

where we have assumed $N_0 = 2$. Thus, we have reduced the problem to general binary detection in white Gaussian noise. The equivalent problem is summarized as

$$H_1 : Y'(t) = s_1'(t) + N'(t)$$

$$H_0 : Y'(t) = s_0'(t) + N'(t) \qquad (8.200)$$

This problem was solved in Section 8.2.2. Thus, by analogy to Equations (8.71) and (8.72), the decision rule can be written as

$$\int_0^T y'(t)\{s_1'(t) - s_0'(t)\}dt \underset{H_0}{\overset{H_1}{\underset{<}{>}}} \gamma \qquad (8.201)$$

where

$$\gamma = \frac{1}{2}\ln\eta - \frac{1}{2}\int_0^T \{[s_0'(t)]^2 - [s_1'(t)]^2\}dt \qquad (8.202)$$

Note that $y'(t), s_1'(t)$ and $s_0'(t)$ are given in Equations (8.194a), (8.195), and (8.198), respectively, in terms of the original signals $y(t), s_1(t)$ and $s_0(t)$. Rewriting Equations (8.201) and (8.202) in terms of the original signals, we obtain

$$\int_0^T \int_0^T h_w(t,u)Y(u)\int_0^T h_w(t,v)\{s_1(v) - s_0(v)\}dv\,du\,dt \underset{H_0}{\overset{H_1}{\underset{<}{>}}} \gamma \quad (8.203)$$

which can be implemented as shown in Figure 8.18.

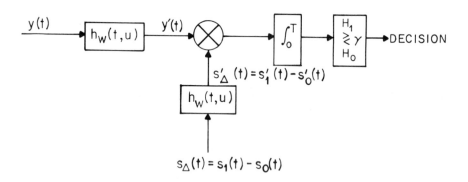

Figure 8.18 Receiver for colored noise using whitening.

Again, construction of the receiver requires knowledge of the impulse response $h_w(t, u)$ which can be obtained by substituting Equation (8.196) into Equation (8.199). We have

$$
\begin{aligned}
E[N'(t)N'(u)] &= E[\int_0^T \int_0^T h_w(t, \alpha)h_w(u, \beta)N(\alpha)N(\beta)d\alpha d\beta] \\
&= \int_0^T \int_0^T h_w(t, \alpha)h_w(u, \beta)C_{NN}(\alpha, \beta)d\alpha d\beta \\
&= \delta(t - u)
\end{aligned}
\tag{8.204}
$$

The solution to the integral equation in (8.204) yields $h_w(t, w)$.

Another way to define the integral equation is in terms of the function $Q(\alpha, \beta)$. In some applications, we may need to express the colored noise as the sum of two components, such as

$$
N(t) = N_C(t) + N'(t)
\tag{8.205}
$$

where $N_C(t)$ is not known. In this case, the function $Q(\alpha, \beta)$ is useful in obtaining the minimum mean-square error, $\hat{N}_C(t)$, of $N_C(t)$. We define

$$
Q_{NN}(\alpha, \beta) = \int_0^T h_w(t, \alpha)h_w(t, \beta)dt = Q_{NN}(\beta, \alpha)
\tag{8.206}
$$

In order to write the integral equation in terms of $Q(\alpha, \beta)$, we multiply both sides of Equation (8.204) by $h_w(t, v)$, and integrate with respect to t, to obtain

$$
\int_0^T h_w(t, u)\delta(t, u)dt
$$

$$
= h_w(u, v)
\tag{8.207a}
$$

$$
= \int_0^T \int_0^T \int_0^T h_w(t, v)h_w(t, \alpha)h_w(u, \beta)C_{NN}(\alpha, \beta)d\alpha d\beta dt
\tag{8.207b}
$$

$$
= \int_0^T h_w(u, \beta) \int_0^T C_{NN}(\alpha, \beta) \int_0^T h_w(t, v)h_w(t, \alpha)dt d\alpha d\beta
\tag{8.207c}
$$

Substituting Equation (8.206) into Equation (8.207c) results in

$$h_w(u,v) = \int_0^T h_w(u,\beta) \int_0^T C_{NN}(\alpha,\beta) Q_{NN}(v,\alpha) d\alpha d\beta \qquad (8.208)$$

From Equations (8.207a) and (8.208) we deduce that

$$\delta(\beta - v) = \int_0^T C_{NN}(\alpha,\beta) Q_{NN}(v,\alpha) d\alpha \qquad (8.209)$$

which means that given the covariance function $C_{NN}(\alpha,\beta)$, we solve the integral equation in (8.209) to yield $Q_{NN}(v,\alpha)$.

8.7.3 Detection Performance

In this section, we study how the colored noise affects the performance. Recall that for binary detection in white noise, the decision rule, from equations (8.71) and (8.72), was

$$T(y) = \int_0^T y(t)\{s_1(t) - s_0(t)\}dt - \frac{1}{2}\int_0^T \{s_1^2(t) - s_0^2(t)\}dt \underset{H_0}{\overset{H_1}{\underset{<}{>}}} \frac{N_0}{2}\ln\eta$$

$$(8.210)$$

Using the whitening approach, the nonwhite noise $N(t)$ is transformed into white noise $N'(t)$, with $N_0 = 2$. The received waveform $Y(t)$ is transformed into $Y'(t)$, and the transmitted signals $s_1(t)$, and $s_0(t)$ are transformed into $s_1'(t)$, and $s_0'(t)$, respectively. Assuming minimum probability of error criterion, and that the hypotheses are equally likely, the test may be expressed as

$$T(y') = \int_0^T y'(t)s_1'(t)dt - \int_0^T y'(t)s_0'(t)dt \underset{H_0}{\overset{H_1}{\underset{<}{>}}} \frac{1}{2}\int_0^T \{[s_1'(t)]^2 - [s_0'(t)]^2\}dt$$

$$(8.211)$$

The sufficient statistic $T(Y')$ is Gaussian with mean:

$$
\begin{aligned}
E[T|H_1] &= \int_0^T [s_1'(t)]^2 dt - \int_0^T s_1'(t)s_0'(t)dt \\
&= \int_0^T dt \int_0^T \int_0^T h_w(t,u)s_1(u)h_w(t,v)s_1(v)dudv \\
&\quad - \int_0^T dt \int_0^T \int_0^T h_w(t,u)s_1(u)h_w(t,v)s_0(v)dudv \quad (8.212)
\end{aligned}
$$

$$
\begin{aligned}
E[T|H_0] &= \int_0^T s_0'(t)s_1'(t)dt - \int_0^T [s_0'(t)]^2 dt \\
&= \int_0^T dt \int_0^T \int_0^T h_w(t,u)s_0(u)h_w(t,v)s_1(v)dudv \\
&\quad - \int_0^T dt \int_0^T \int_0^T h_w(t,u)h_w(t,v)s_0(u)s_0(v)dudv \quad (8.213)
\end{aligned}
$$

The variances under hypotheses H_1 and H_0 are the same. The expression is cumbersome. However, it can be shown [17] to have a value of twice the mean of T under H_1. Denote this variance by σ^2, then the probability of error is

$$
\begin{aligned}
P(\varepsilon) &= \int_0^\infty \frac{1}{\sqrt{2\pi}\sigma} e^{-\frac{(t+\frac{1}{2}\sigma^2)^2}{2\sigma^2}} dt \\
&= \int_{\frac{\sigma}{2}}^\infty \frac{1}{\sqrt{2\pi}} e^{-\frac{u^2}{2}} du \quad (8.214)
\end{aligned}
$$

where σ is given by Equation (8.212). The calculation of Equation (8.214) is involved. However, we observe that the probability of error is a function of the signals shape, unlike the case of detection in white noise, where the performance was a function of the signal-to-noise ratio only. Consequently, to minimize the probability of error we need to find the signals shape. We also see, from Equation (8.214), that the probability of error is minimized if σ is maximized, subject to the constraint that the energy is fixed to a value $\mathcal{E}$. Hence, we form the objective function J and solve the equation:

$$
J = \sigma^2 - \lambda\mathcal{E} \quad (8.215)
$$

where λ is the Lagrange multiplier and $\mathcal{E}$ is given by

$$\mathcal{E} = \frac{1}{2} \int_0^T \{s_1^2(t) - s_0^2(t)\} dt \tag{8.216}$$

The solution of Equation (8.215) results in the optimum signals shape which are obtained to be

$$s_1(t) = -s_0(t) \quad , 0 \le t \le T \tag{8.217}$$

that is, we have optimum performance when the correlation coefficient $\rho = -1$, which is the same result obtained for binary detection in white noise.

8.8 SUMMARY

In this chapter, we have discussed the problem of detection of signal waveforms, and parameter estimation of signals in the presence of additive noise. We first covered binary and M-ary detection. The approach adopted was to decompose the signal waveform into a set of K independent random variables, and write the signal in a Karhunen–Loeve expansion. The coefficients of the Karhunen–Loeve expansion are in a sense samples of the received signal. Since the additive noise was white and Gaussian, the coefficients of the Karhunen–Loeve expansion were uncorrelated and jointly Gaussian. Consequently, the problem was reduced to an equivalent decision problem as developed in Chapter 3.

In Sections 8.4 and 8.5, we assumed that the received signals may contain some unknown parameters which needed to be estimated. Linear and nonlinear estimation were considered. When the parameter to be estimated was nonrandom, we used the maximum likelihood estimation. The maximum *a posteriori* estimation was used for random parameter. The "goodness" of the estimation techniques was studied as well.

The general binary detection with unknown parameters was presented in Section 8.6. Again, using the Karhunen–Loeve coefficients, we obtained the approximate K-term likelihood ratio and then, we let $K \to \infty$ to obtain the likelihood ratio. This approach of obtaining a K-term approximation of the Karhunen–Loeve coefficients and letting $K \to \infty$ was also used in solving for the parameter-estimates discussed in Sections 8.4 and 8.5.

We concluded the chapter with a section on binary detection in colored noise. Since the noise was not white anymore, the generated Karhunen–Loeve coefficients were not uncorrelated anymore. In solving this problem we first used the K-term approximation from the Karhunen–Loeve coefficients. However, due to the nature of noise, some integral equations needed to be solved in order to design the optimum receiver. The second approach used to solve this problem was whitening. That is, we did a preliminary processing by passing the received signal through a linear time-invariant system, such that the noise at the output of the filter was white. Once the noise became white, the techniques developed earlier for binary detection were then used to obtain the optimum receiver. A brief study on the performance of detection of signals in colored noise was also presented.

PROBLEMS

8.1 A signal source generates signals as shown in Figure *P8.1*.

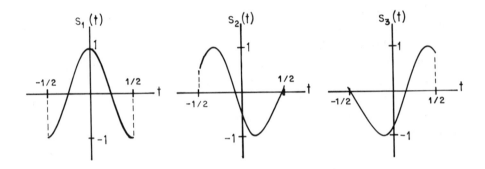

Figure P8.1 Signal set.

The signals are expressed as $s_1(t) = \cos(2\pi t)\,\text{rect}(t)$,
$s_2(t) = \cos(2\pi t + \frac{2}{3}\pi)\,\text{rect}(t)$ and $s_3(t) = \cos(2\pi t - \frac{2}{3}\pi)\,\text{rect}(t)$.
(a) Describe a correlation receiver for these signals.
(b) Draw the corresponding decision regions on a signal space.

8.2 A rectangular pulse of known amplitude A is transmitted starting at time instant t_0 with probability $1/2$. The duration T of the pulse is a random variable uniformly distributed over the interval $[T_1, T_2]$. The additive noise to the pulse is white, and Gaussian with mean zero, and variance $N_0/2$.
(a) Determine the likelihood ratio.
(b) Describe the likelihood ratio receiver.

8.3 Consider the general binary detection problem:

$$H_1 : Y(t) = s_1(t) + W(t) \quad , 0 \le t \le T$$

$$H_0 : Y(t) = s_0(t) + W(t) \quad , 0 \le t \le T$$

where $s_1(t)$ and $s_0(t)$ are as shown in Figure P8.3 and $W(t)$ is a white Gaussian noise with mean zero and power spectral density $N_0/2$.

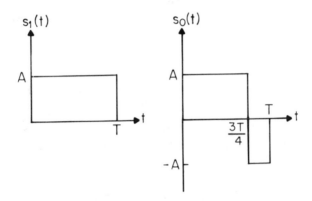

Figure P8.3 Signal set.

(a) Determine the probability of error. Assume minimum probability of error criterion, and $P(H_0) = P(H_1) = 1/2$.
(b) Draw a block diagram of the optimum receiver.

8.4 In a binary detection problem, the transmitted signal under hypothesis H_1 is either $s_1(t)$ or $s_2(t)$, with respective probabilities P_1 and P_2. Assume $P_1 = P_2 = 1/2$, and $s_1(t)$ and $s_2(t)$ orthogonal over the observation time $t\epsilon[0, T]$. No signal is transmitted under hypothesis H_0. The additive noise is white Gaussian with mean zero, and power spectral density $N_0/2$.

(a) Obtain the optimum decision rule assuming minimum probability of error criterion and $P(H_1) = P(H_0) = 1/2$..
(b) Draw a block diagram of the optimum receiver.

8.5 Consider the binary detection problem:

$$H_1 : Y(t) = s_1(t) + W(t) \quad ,0 \le t \le 2$$

$$H_0 : Y(t) = s_0(t) + W(t) \quad ,0 \le t \le 2$$

where $s_1(t) = -s_0(t) = e^{-t}$ and $W(t)$ is an additive white Gaussian noise with mean zero and covariance function $C_{NN}(t, u) = (N_0/2)\delta(t{-}u)$.
(a) Determine the probability of error assuming minimum probability of error criterion.
(b) Draw a block diagram of the optimum receiver.

8.6 During transmission of 16 quadrature amplitude modulated signals an additive white Gaussian noise, with mean zero and power spectral density $N_0/2$ is superimposed on the signals. The signal space is shown in Figure $P8.6$. The signal points are spaced d unit apart. They are given by

$$s_k(t) = a_k\phi_1(t) + b_k\phi_2(t) \quad , \quad \begin{matrix} -\frac{T}{2} \le t \le \frac{T}{2} \\ k = 1, 2, \cdots, 16 \end{matrix}$$

where,

$$\phi_1(t) = \sqrt{\frac{2}{T}} \cos 2\pi f_0 t \quad \text{and} \quad \phi_2(t) = \sqrt{\frac{2}{T}} \sin 2\pi f_0 t$$

Assume minimum probability of error criterion.
(a) Draw a block diagram of the optimum receiver.
(b) Show the decision regions in the signal space.
(c) Determine the probability of error.

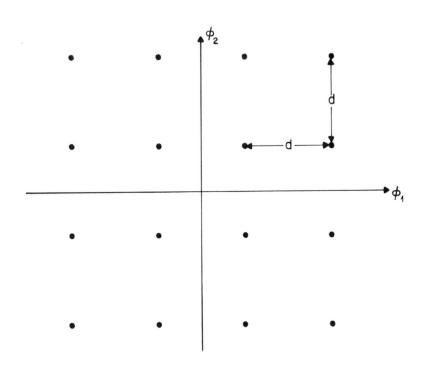

Figure P8.6 Signal space.

8.7 Starting from Equation (8.102), derive the expression in (8.104).

8.8 Consider the situation where the received signal is given by

$$H_1 : Y(t) = As(t) + W(t) \quad , 0 \le t \le T$$

$$H_0 : Y(t) = \qquad\quad W(t) \quad , 0 \le t \le T$$

Let A be an unknown constant and $W(t)$ a white Gaussian noise process with mean zero, and power spectral density $N_0/2$. Design the optimum receiver assuming minimum probability of error criterion.

8.9 Consider the estimation problem:

$$Y(t) = s(t, \theta) + W(t) \qquad , 0 \leq t \leq T$$

where $s(t, \theta) = [1/\theta]s(t)$. θ is an unknown constant whereas $s(t)$ is a known signal with energy $\mathcal{E}$. $W(t)$ is a white Gaussian noise with mean zero, and covariance function $C_{NN}(t, u) = (N_0/2)\delta(t - u)$. Determine $\hat{\theta}_{ml}$ the maximum likelihood estimate of θ.

8.10 Consider the problem given in 8.9 where θ is now a Gaussian random variable, with mean zero, and variance σ_θ^2. Determine the equation for a which a solution is $\hat{\theta}_{map}$, the maximum *a posteriori* estimate of θ, and show that this equation also gives $\hat{\theta}_{ml}$ as the variance $\sigma_\theta^2 \to \infty$.

8.11 Consider the estimation problem:

$$Y(t) = s(t, \theta) + N(t) \qquad , 0 \leq t \leq T$$

where $N(t)$ is a nonwhite Gaussian noise with mean zero and covariance function:

$$C_{NN}(t, u) = C_{N'N'}(t, u) + C_{N_C N_C}(t, u)$$

where $C_{N'N'}(t, u) = (N_0/2)\delta(t - u)$. The received waveform is passed through a correlation operation to yield

$$Y_k = s_k(\theta) + N_k \qquad , k = 1, 2, \cdots, K$$

such that

$$s_k(\theta) = \int_0^T s(t, \theta)\phi_k(t)dt$$

and N_k, $k = 1, 2, \cdots, K$, are random variables. $\phi_k(t)$, $k = 1, 2, \cdots, K$, are eigenfunctions corresponding to λ_k, $k = 1, 2, \cdots, K$, associated with the covariance function $C_{N_C N_C}(t, u)$. In the limit as $K \to \infty$, we have

$$Y(t) = \lim_{K \to \infty} \sum_{k=1}^K s_k(\theta)\phi_k(t) + \lim_{K \to \infty} \sum_{k=1}^K N_k \phi_k(t)$$

(a) Determine the mean and variance of the Karhunen–Loeve coefficients N_k's.

(b) Are the noise components N_k, $k = 1, 2, \cdots$, statistically independent?

(c) If $Y(t)$ is passed through a whitening filter to obtain

$$
\begin{aligned}
Y'(t) &= s'(t,\theta) + N'(t) \qquad , 0 \le t \le T \\
&= \lim_{K \to \infty} \sum_{k=1}^{K} s'_k(\theta)\phi_k(t) + \lim_{K \to \infty} \sum_{k=1}^{K} N'_k \phi_k(t)
\end{aligned}
$$

Determine the mean and variance of the white noise components N_k's.

8.12 Consider a noise process such that $N_1(t) = W$ in the interval $t\epsilon[0,T]$. W is Gaussian with mean zero and variance σ_w^2.

(a) Can $N_1(t)$ be whitened in the given interval? Explain.

(b) Repeat (a) assuming that another independent noise process $N_2(t)$ is superimposed on $N_1(t)$ such that the new noise process is

$$
N(t) = N_1(t) + N_2(t) \qquad , 0 \le t \le T
$$

and the covariance function of $N_2(t)$ is

$$
C_{N_2 N_2}(t, u) = \frac{N_0}{2}\delta(t - u) \qquad , 0 \le t, u \le T
$$

REFERENCES

1. Davenport, Jr., W.B., and W.L. Root, *An Introduction to the Theory of Random Signals and Noise*, McGraw-Hill, New York, 1958.

2. Franks, L.E., *Signal Theory*, Prentice-Hall, Englewood Cliffs, NJ, 1969.

3. Haykin, S., *Communication Systems*, John Wiley and Sons, New York, 1983.

4. Haykin, S., *Digital Communications*, John Wiley and Sons, New York, 1988.

5. Helstrom, C.W., *Statistical Theory of Signal Detection*, Pergamon, New York, 1960.

6. Lee, Y.W., *Statistical Theory of Communication*, John Wiley and Sons, New York, 1960.

7. Melsa, J.L., and D.L. Cohn, *Decision and Estimation Theory*, McGraw-Hill, New York, 1978.

8. Mohanty, N., *Signal Processing: Signals, Filtering, and Detection*, Van Nostrand Reinhold, New York, 1987.

9. Nahi, N.E., *Estimation Theory and Applications*, John Wiley and Sons, New York, 1969.

10. Papoulis, A., *Probability, Random Variables, and Stochastic Processes*, McGraw-Hill, New York, 1984.

11. Proakis, J.G., *Digital Communications*, McGraw-Hill, New York, 1989.

12. Schwartz, M., *Information Transmission, Modulation, and Noise*, McGraw-Hill, New York, 1980.

13. Srinath, M.D., and P.K. Rajasekaran, *An Introduction to Statistical Signal Processing with Applications*, John Wiley and Sons, New York, 1979.

14. Urkowitz, H., *Signal Theory and Random Processes*, Artech House, Norwood, MA, 1983.

15. Van Trees, H.L., *Detection, Estimation, and Modulation Theory: Part I*, John Wiley and Sons, New York, 1968.

16. Weber, C.L., *Elements of Detection and Signal Design*, McGraw-Hill, New York, 1968.

17. Whalen, A.D., *Detection of Signals in Noise*, Academic, New York, 1971.

18. Wozencraft, J.M., and I.M. Jacobs, *Principles of Communication Engineering*, John Wiley and Sons, New York, 1965.

Appendix A

Rayleigh and Rice Distributions

In this appendix, we list two density functions, which may arise in practice, because they are related to the Gaussian distribution.

A.1 RAYLEIGH DISTRIBUTION

The Rayleigh density and distribution functions are given by

$$f_X(x) = \begin{cases} \frac{1}{\sigma^2}(x - \alpha)\, e^{-\frac{(x-\alpha)^2}{2\sigma^2}} & , x \geq \alpha \\ 0 & , \text{ otherwise} \end{cases} \tag{A.1}$$

$$F_X(x) = \begin{cases} 1 - e^{-\frac{(x-\alpha)^2}{2\sigma^2}} & , x \geq \alpha \\ 0 & , \text{ otherwise} \end{cases} \tag{A.2}$$

and are shown in Figures A.1 and A.2

418

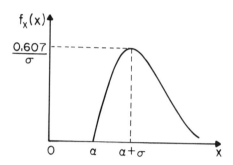

Figure A.1 Rayleigh density function.

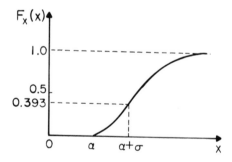

Figure A.2 Rayleigh distribution function.

The Rayleigh distribution can be obtained from the chi-square distribution with n degrees of freedom given in (1.81) by setting $n = 2$. When $n = 3$ in (8.181), the distribution obtained is known as Maxwell's distribution.

A.2 RICE DISTRIBUTION

Let the random variable $Z = \sqrt{X^2 + Y^2}$ where X and Y are independent Gaussian random variables with means m and zero, respectively, and both have the same variance σ^2. Then, the density function of Z, known as Rician, is

$$f_Z(z) = \begin{cases} \frac{z}{\sigma^2} e^{-\frac{z^2 + m^2}{2\sigma^2}} I_0(\frac{zm}{\sigma^2}) & , z \geq 0 \\ \\ 0 & , \text{otherwise} \end{cases} \qquad (A.3)$$

where

$$I_0(x) \triangleq \frac{1}{2\pi} \int_0^{2\pi} e^{x \cos\theta} d\theta \qquad (A.4)$$

is called the *zero-order modified Bessel function of the first kind*. The distribution function of Z is

$$F_Z(z) = \begin{cases} \frac{1}{\sigma^2} e^{-\frac{m^2}{2\sigma^2}} \int_0^z r I_0(\frac{rm}{\sigma^2}) e^{-\frac{1}{2}\frac{(z-m)^2}{\sigma^2}} dr & , z \geq 0 \\ \\ 0 & , \text{otherwise} \end{cases} \qquad (A.5)$$

where the following cartesian-to-polar transformation was used.

$$r = x^2 + y^2 \qquad (A.6)$$

$$x = r \cos\theta \qquad (A.7)$$

$$y = r \sin\theta \qquad (A.8)$$

and

$$\theta = \tan^{-1}\frac{y}{x} \qquad (A.9)$$

Note that when $m = 0$, we obtain the Rayleigh distribution, since $I_0(0) = 1$.

When $mz \gg \sigma^2$, the Rician distribution is almost Gaussian given by

$$f_Z(z) = \frac{1}{\sqrt{2\pi}\sigma}\sqrt{\frac{z}{m}} e^{-\frac{(z-m)^2}{2\sigma^2}} \qquad (A.10)$$

where $I_0(x)$ was approximated to be

$$I_0(x) = \frac{e^x}{\sqrt{2\pi x}} \qquad (A.11)$$

Appendix B

Error Function

In Chapter 1, we defined the error functions as

$$\text{erf}(x) = \frac{2}{\sqrt{\pi}} \int_{-\infty}^{x} e^{-u^2} du \qquad (B.1)$$

and the Q-function as

$$Q(x) = \frac{1}{\sqrt{2\pi}} \int_{x}^{\infty} e^{-\frac{u^2}{2}} du \qquad (B.2)$$

The above functions are related by the equation:

$$Q(x) = \frac{1}{2}[1 - \text{erf}(\frac{x}{\sqrt{2}})] \qquad (B.3)$$

We also defined the complements to be

$$\begin{aligned}
\text{erfc}(u) &= \frac{2}{\sqrt{\pi}} \int_{x}^{\infty} e^{-u^2} du \\
&= 1 - \text{erf}(x) \qquad (B.4)
\end{aligned}$$

and

$$\begin{aligned}
Q(-x) &= \frac{1}{\sqrt{2\pi}} \int_{-x}^{\infty} e^{-\frac{u^2}{2}} du \\
&= 1 - Q(x) \qquad \qquad , \text{ for } x > 0 \qquad (B.5)
\end{aligned}$$

Also

$$Q(x) \simeq \frac{1}{x\sqrt{2\pi}} \; e^{-\frac{x^2}{2}} \qquad \text{for } x > 4 \qquad (B.6)$$

For $x > 0$, we have the bounds:

$$(1 - \frac{1}{x^2})\frac{1}{\sqrt{2\pi}x} \; e^{-\frac{x^2}{2}} < Q(x) < \frac{1}{\sqrt{2\pi}x} \; e^{-\frac{x^2}{2}} \qquad (B.7)$$

which are shown in Figure B.1.

The function that is usually tabulated is not $Q(x)$, but erf (x) defined in $(B.1)$. Some values are given in Table[1] B.1.

---

[1] Abramowitz, M., and I.A. Stegun, *Handbook of Mathematical Functions with Formulas, Graphs, and Mathematical Labels,* United States Department of Commerce, Washington, DC, December 1972.

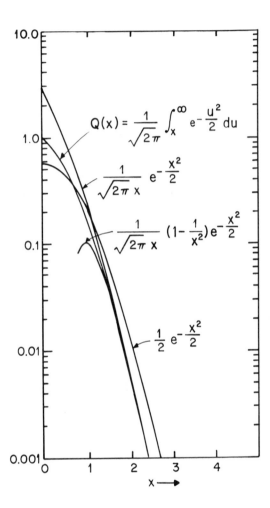

Figure B.1 Q-function and its bounds.

Table B.1 Error function

x	$erf(x)$	x	$erf(x)$	x	$erf(x)$	x	$erf(x)$
0.00	0.00000	0.25	0.27633	0.50	0.52050	0.75	0.71116
0.01	0.01128	0.26	0.28690	0.51	0.52924	0.76	0.71754
0.02	0.02256	0.27	0.29742	0.52	0.53790	0.77	0.72382
0.03	0.03384	0.28	0.30788	0.53	0.54646	0.78	0.73001
0.04	0.04511	0.29	0.31828	0.54	0.55494	0.79	0.73610
0.05	0.05637	0.30	0.32863	0.55	0.56332	0.80	0.74210
0.06	0.06762	0.31	0.33891	0.56	0.57162	0.81	0.74800
0.07	0.07885	0.32	0.34913	0.57	0.57982	0.82	0.75381
0.08	0.09007	0.33	0.35928	0.58	0.58792	0.83	0.75952
0.09	0.10128	0.34	0.36936	0.59	0.59594	0.84	0.76514
0.10	0.11246	0.35	0.37938	0.60	0.60386	0.85	0.77067
0.11	0.12362	0.36	0.38933	0.61	0.61168	0.86	0.77610
0.12	0.13476	0.37	0.39921	0.62	0.61941	0.87	0.78144
0.13	0.14587	0.38	0.40901	0.63	0.62705	0.88	0.78669
0.14	0.15695	0.39	0.41874	0.64	0.63459	0.89	0.79184
0.15	0.16800	0.40	0.42839	0.65	0.64203	0.90	0.79691
0.16	0.17901	0.41	0.43797	0.66	0.64938	0.91	0.80188
0.17	0.18999	0.42	0.44747	0.67	0.65663	0.92	0.80677
0.18	0.20094	0.43	0.45689	0.68	0.66378	0.93	0.81156
0.19	0.21184	0.44	0.46623	0.69	0.67084	0.94	0.81627
0.20	0.22270	0.45	0.47548	0.70	0.67780	0.95	0.82089
0.21	0.23352	0.46	0.48466	0.71	0.68467	0.96	0.82542
0.22	0.24430	0.47	0.49375	0.72	0.69143	0.97	0.82987
0.23	0.25502	0.48	0.50275	0.73	0.69810	0.98	0.83423
0.24	0.26570	0.49	0.51167	0.74	0.70468	0.99	0.83851
0.25	0.27633	0.50	0.52050	0.75	0.71116	1.00	0.84270

Table B.1 (continued)

x	$erf(x)$	x	$erf(x)$	x	$erf(x)$	x	$erf(x)$
1.00	0.84270	1.25	0.92290	1.50	0.96611	1.75	0.98667
1.01	0.84681	1.26	0.92524	1.51	0.96728	1.76	0.98719
1.02	0.85084	1.27	0.92751	1.52	0.96841	1.77	0.98769
1.03	0.85478	1.28	0.92973	1.53	0.96952	1.78	0.98817
1.04	0.85865	1.29	0.93190	1.54	0.97059	1.79	0.98864
1.50	0.86244	1.30	0.93401	1.55	0.97162	1.80	0.98909
1.06	0.86614	1.31	0.93606	1.56	0.97263	1.81	0.98952
1.07	0.86977	1.32	0.93807	1.57	0.97360	1.82	0.98994
1.08	0.87333	1.33	0.94002	1.58	0.97455	1.83	0.99035
1.09	0.87680	1.34	0.94191	1.59	0.97546	1.84	0.99074
1.10	0.88021	1.35	0.94376	1.60	0.97635	1.85	0.99111
1.11	0.88353	1.36	0.94556	1.61	0.97721	1.86	0.99147
1.12	0.88679	1.37	0.94731	1.62	0.97804	1.87	0.99182
1.13	0.88997	1.38	0.94902	1.63	0.97884	1.88	0.99216
1.14	0.89308	1.39	0.95067	1.64	0.97962	1.89	0.99247
1.15	0.89612	1.40	0.95229	1.65	0.98038	1.90	0.99279
1.16	0.89910	1.41	0.95385	1.66	0.98110	1.91	0.99308
1.17	0.90200	1.42	0.95530	1.67	0.98181	1.92	0.99338
1.18	0.90484	1.43	0.95686	1.68	0.98249	1.93	0.99366
1.19	0.90761	1.44	0.95830	1.69	0.98315	1.94	0.99392
1.20	0.91031	1.45	0.95970	1.70	0.98379	1.95	0.99418
1.21	0.91296	1.46	0.96105	1.71	0.98441	1.96	0.99442
1.22	0.91553	1.47	0.96237	1.72	0.98500	1.97	0.99466
1.23	0.91805	1.48	0.96365	1.73	0.98558	1.98	0.99489
1.24	0.92051	1.49	0.96490	1.74	0.98613	1.99	0.99511
1.25	0.92290	1.50	0.96611	1.75	0.98667	2.00	0.99532

Index

The Artech House Radar Library

David K. Barton, *Series Editor*

Modern Radar System Analysis Software and User's Manual by David K. Barton and William F. Barton

Monopulse Principles and Techniques by Samuel M. Sherman

Monopulse Radar by A.I. Leonov and K.I. Fomichev

MTI and Pulsed Doppler Radar by D. Curtis Schleher

Multifunction Array Radar Design by Dale R. Billetter

Multisensor Data Fusion by Edward L. Waltz and James Llinas

Multiple-Target Tracking with Radar Applications by Samuel S. Blackman

Multitarget-Multisensor Tracking: Advanced Applications, Yaakov Bar-Shalom, ed.

Over-The-Horizon Radar by A.A. Kolosov, et al.

Principles and Applications of Millimeter-Wave Radar, Charles E. Brown and Nicholas C. Currie, eds.

Principles of Modern Radar Systems by Michel H. Carpentier

Pulse Train Analysis Using Personal Computers by Richard G. Wiley and Michael B. Szymanski

Radar and the Atmosphere by Alfred J. Bogush, Jr.

Radar Anti-Jamming Techniques by M.V. Maksimov, *et al.*

Radar Cross Section by Eugene F. Knott, *et al.*

Radar Detection by J.V. DiFranco and W.L. Rubin

Radar Electronic Countermeasures System Design by Richard J. Wiegand

Radar Evaluation Handbook by David K. Barton, *et al.*

Radar Evaluation Software by David K. Barton and William F. Barton

Radar Propagation at Low Altitudes by M.L. Meeks

Radar Range-Performance Analysis by Lamont V. Blake

Radar Reflectivity Measurement: Techniques and Applications, Nicholas C. Currie, ed.

Radar Reflectivity of Land and Sea by Maurice W. Long

Radar System Design and Analysis by S.A. Hovanessian

Radar Technology, Eli Brookner, ed.

Receiving Systems Design by Stephen J. Erst

Radar Vulnerability to Jamming by Robert N. Lothes, Michael B. Szymanski, and Richard G. Wiley

RGCALC: Radar Range Detection Software and User's Manual by John E. Fielding and Gary D. Reynolds

SACALC: Signal Analysis Software and User's Guide by William T. Hardy

Secondary Surveillance Radar by Michael C. Stevens

SIGCLUT: Surface and Volumetric Clutter-to-Noise, Jammer and Target Signal-to-Noise Radar Calculation Software and User's Manual by William A. Skillman

Signal Theory and Random Processes by Harry Urkowitz

Solid-State Radar Transmitters by Edward D. Ostroff, *et al.*

Space-Based Radar Handbook, Leopold J. Cantafio, ed.

Spaceborne Weather Radar by Robert M. Meneghini and Toshiaki Kozu

Statistical Theory of Extended Radar Targets by R.V. Ostrovityanov and F.A. Basalov

The Scattering of Electromagnetic Waves from Rough Surfaces by Peter Beckmann and Andre Spizzichino

VCCALC: Vertical Coverage Plotting Software and User's Manual by John E. Fielding and Gary D. Reynolds